The New Communications Technologies:
Applications, Policy, and Impact

Fourth Edition

Michael M. A. Mirabito
Barbara L. Morgenstern

Focal Press

Boston   Oxford   Auckland   Johannesburg   Melbourne   New Delhi

Focal Press is an imprint of Butterworth–Heinemann.

 A member of the Reed Elsevier group

 Recognizing the importance of preserving what has been written, Butterworth–Heinemann prints its books on acid-free paper whenever possible.

 Butterworth–Heinemann supports the efforts of American Forests and the Global ReLeaf program in its campaign for the betterment of trees, forests, and our environment.

Library of Congress Cataloging-in-Publication Data

Mirabito, Michael M., 1956–
     The new communications technologies : applications, policy, and
impact / Michael M. A. Mirabito, Barbara L. Morgenstern.—4th ed.
        p.   cm.
     Includes bibliographical references and index.
     ISBN 0-240-80429-5 (pbk. : alk. paper)
     1. Telecommunication systems—Technological innovations.
I. Morgenstern, Barbara L.   II. Title
TK5101 .M545 2000
621.382—dc21

                                                              00-042699

British Library Cataloguing-in-Publication Data
A catalogue record for this book is available from the British Library.

The publisher offers special discounts on bulk orders of this book.
For information, please contact:
Manager of Special Sales
Butterworth–Heinemann
225 Wildwood Avenue
Woburn, MA 01801-2041
Tel: 781-904-2500
Fax: 781-904-2620

For information on all Focal Press publications available, contact our World Wide Web
home page at: http://www.focalpress.com

10 9 8 7 6 5 4 3 2 1

Printed in the United States of America

To our families

To the men and women of the Space Shuttle Challenger
and all their fellow travelers.
May their dream live forever in the minds, hearts, and works
of the present and all future generations.

Table of Contents

Foreword

In today's fast-paced media environment, every journalist—and every communications professional—needs to be familiar with the continuing impact of technological change on their work. With *The New Communications Technologies,* Professors Mirabito and Morgenstern bring together in one concise and well-organized review many of the most important developments to date.

Professors Mirabito and Morgenstern cover a wide array of communications technologies in this text, touching on many of the important issues that today's news executives and journalists grapple with daily. At the Radio-Television News Directors Association, the world's largest professional organization devoted exclusively to electronic journalism, we are keenly interested in the impact these rapidly changing communication technologies have on the gathering, editing, and dissemination of news. From the rapid growth of the Internet to the development of satellite communications, the whirlwind of technological change is affecting the lives of both journalists and the citizens they serve.

The New Communications Technologies provides the reader with a snapshot of how everything from computers to fiber optics to communication satellites work, and why these technologies are important. Individual chapters also include discussions about the important philosophical implications of these technologies and extensive glossaries that explain the jargon. There is also an insightful chapter examining what impact these technologies may have on our personal privacy and on the First Amendment rights of free speech and freedom of the press as we move forward in the new century.

With this book, Professors Mirabito and Morgenstern have created a text that journalists and students will value as a technological resource and philosophical guide.

Barbara Cochran
President
Radio-Television News Directors Association
1000 Connecticut Avenue, NW, Suite 615
Washington, DC 20036
March 31, 2000

Preface

WHAT'S NEW

Readers who are familiar with previous editions of *The New Communications Technologies* will see a number of changes in the fourth edition, including these:

- The legal discussions have been expanded. The topics range from First Amendment issues to copyright to encryption to Internet filtering software.
- The chapters have been updated to explore new applications and technological trends.
- New topics have been added. For example, the Y2K scenario and e-business are covered in Chapters 4 and 18, respectively.
- New illustrations are presented.
- As with the third edition, personal viewpoints about the new technologies are presented. In one case, an article by Barbara Cochran, president of the RTNDA and the author of the book's Foreword, highlights some of the problems faced by the media in the satellite imaging arena. Besides being insightful, these statements present us with a slice of time—how a technology and its applications were viewed during the late twentieth/early twenty-first century.
- For PowerPoint presentations, legal citations, and review questions, visit the newly updated companion site at www.focalpress .com/companions.

They can also point the way to future developments. For instance, Don West's piece about the V-chip, originally printed in *Broadcasting & Cable*, presents us with a timeline that looks at how a First Amendment question was handled in the 1960s and some years later in an analogous situation. More ominously, it also points the way to a potential, future trend.

What's Unchanged

As in previous editions, the book still explores the new technologies from a broad perspective. This includes the convergence factor—the relationships between different fields and how developments in one area can affect another.

Applications and underlying concepts also remain a focal point. The latter is particularly salient. Although new products may be introduced, fundamental principles usually remain unaltered. Thus, by learning the concepts now, you may be able to work with new applications well into the future.

THE POTENTIAL READERSHIP

The book is appropriate for communications technology courses in TV/radio, communication, journalism, and corporate communication departments. It can also serve as a primer for graduate courses in the same departments and as a supplementary text for public relations/advertising, management, and instructional technology courses.

The book may also prove useful to communications professionals, including journalists, who want to gain a broad overview of the field. I also hope it will appeal to those who are interested in the communication revolution and its impact.

ACKNOWLEDGMENTS

As with the first edition, I want to acknowledge those pioneers, those individuals, who helped launch the communication revolution itself.

My students deserve thanks for their willingness to serve as sounding boards when new material was introduced and for providing me with valuable feedback. I also want to

Figure 1

The new communications technologies have altered the way we can gain access to information. In this example, you can explore the world—Stromboli and its active volcano—via the Internet. (Courtesy of Juerg Alean and Stromboli.net.)

thank the readers whose comments about the third edition, and this book's proposal, proved insightful. I also appreciate the ongoing efforts and support of Ayn Miralano, my colleagues at Marywood, and the numerous individuals at different companies who responded to photo requests. Edie Emery and Barbara Cochran of the RTDNA were also gracious in their support of this project. Thank you all.

The editorial and production staffs at Focal Press helped make this book a reality. I espe-cially want to thank Marie Lee, Lilly Roberts, Lauren Lavery, Maura Kelly, and Lorretta Palagi.

DISCLAIMER

Some of the companies and products men-tioned herein are trademarks or service marks. Such usage of these terms does not imply endorsements or affiliations.

1 Communication in the Modern Age

Much has been written about the communication revolution—some of it realistic, some of it not. Authors predicted we would be using videophones, the country would be crisscrossed by fiber-optic lines, satellites would create worldwide communications nets, and we would be watching 3-D television and using personal computers (PCs).

Even though all of these predictions have not fully materialized, many have. Sophisticated satellites ring the globe, information is relayed in the form of light, and PCs have forever altered the way we work.

We are clearly living during a communication revolution. New and existing technologies and applications are shaping the communications industry and society. For example:

- The latest generation of PCs can produce sophisticated multimedia presentations.
- Optical disks offer increased storage capabilities.
- Telephone lines are channels for vast information pools.
- Satellites can serve as personal communications platforms.

Despite this revolution, some changes are actually evolutionary. As covered in different chapters, this may include the upgrading or modernization of an industry's infrastructure.

Consequently, the communication revolution can actually be viewed as a mix of revolution and evolution—the influx of new and the enhancement of older products. Together, they have brought about massive changes in the world around us.

Chapters 1 and 2 provide a foundation for the exploration of these topics and changes. Key concepts are introduced, and digital technology, the driving force behind many of the changes, is covered.

BASIC CONCEPTS

Communications System— An Extended Definition

A communications system is one such key concept. It provides the means by which information, coded in signal form, can be exchanged. If we decide to telephone a friend, the communications system would include the telephone receivers, the telephone line, and other components. We use a variety of communications systems to exchange information. As covered throughout this book, they range from satellites to underwater fiber-optic lines.

Figure 1.1
Satellite applications enable us to see our world in new ways. This 1-meter resolution image of Taipei, Taiwan, collected October 21, 1999, by Space Imaging's IKONOS satellite, features the Grand Hotel and Sun Yat-Sen Freeway. The satellite is discussed in Chapter 7. (Courtesy of Space Imaging.)

In the context of this book, the term *communications system* also has a broad definition. It isn't limited to a description of information exchange systems. It also includes the communications tools we use, their applications, and the implications that arise from the production, manipulation, and potential exchange of information.

Information

Information can be defined as a collection of symbols that, when combined, communicates a message or intelligence. When you write a note to a friend about the Yankees, you combine letters and numbers to convey your thoughts or ideas, a message that has meaning to both of you. The combination of the letters W-o-r-l-d S-e-r-i-e-s is not just a collection of letters. It represents a concept.[1]

In our communications system, this information may be coded in a standardized form: in an electronic or electrical signal analogous to the coding of information by the printed letters and, ultimately, words in the note.

Next, the information can be relayed via a telephone line, satellite, or other communications channel. After it is received, it can be decoded or converted back into its original form.

This series of steps follows the traditional model of a communications system devised by Claude Shannon and Warren Weaver. The system consists of an information source, a transmitter, a channel to deliver the information, a receiver, and a destination. Noise, another element, is described in the next chapter.

Information As a Signal.   When coded in signal form, the information—a person's voice or a video camera's view of the Grand Canyon—is also compatible with communications equipment. The camera's picture can be transmitted, stored on videotape, or altered by computer.

The communications system is also quite flexible in the representation of information. In one application, light can convey information in a fiber-optic system. A laser's light is also used to retrieve information in optical storage media. These topics are covered in Chapters 5 and 10, respectively.

If you're interested in reading a more detailed explanation of what constitutes information, please see the books by Shannon and Weaver and by Rogers listed at the end of this chapter. They also cover the historical and technical elements of information theory and the mathematical theory of communication and information.

Information—An Extended Definition. Information also extends beyond the definition used in the general communications industry. Information is not only television programs, telephone conversations, and music stored on CDs. Information is also a picture manipulated by a computer that highlights details of the human body. Information is a library of books stored on optical disks. Information is stock market facts that can be accessed at home with a computer.

Information can also be viewed as a commodity. Traditionally, a television program has a financial value that varies with its ratings. The information generated by nontraditional tools is also valuable.

Companies use the telephone system to sell financial data to computer owners. The television networks purchase images created by remote sensing satellites to shed light on national and international news events.

Finally, information can be equated with power. If you know how to use a computer, you will potentially be able to tap a greater wealth of information than a nonuser. Such access might give you a political or economic edge.

Information Society.   Some technological developments, including the notion that information can be equated with power, have contributed to the creation of an information society. Even though this term has become a cliché, it is nevertheless accurate. Our society is driven by information, be it the latest international and financial news needed to keep abreast of a volatile world market or the creation of databases that can be accessed by computer owners.

For example, as described in Chapter 4, computers have been programmed to solve various problems. Much like human experts, they can be used to help pinpoint a malfunctioning component in complex machinery. In this situation, computers and expert system software have influenced the way we work.

Our information society has also ushered in new job categories. Knowledge engineers help create expert systems while specialists are hired to retrieve information from computer databases.

The U.S. economy has also been influenced. We are becoming a nation based on service rather than manufacturing. Tubing and steel plants have given way to the service industry. Hospitals, banks, and computer information companies fall into the latter category. Most, if not all, of these companies use and process information as integral elements of the services they offer.

Other companies produce the tools that sustain this society as new highways have been constructed. But instead of carrying cars and trucks, these new "highways" relay information to our work sites and homes.

IMPLICATIONS OF THE COMMUNICATION REVOLUTION

The new information and communications technologies have profoundly affected our social structure. There's also a growing interdependence among technology, information, and society.

For example, the new technologies have raised a series of ethical questions. As detailed in Chapter 12, you can use a scanner, a device that inputs graphic information to a computer, to copy someone else's work.

As presented in Chapter 4, some individuals are afraid that automation will accelerate the loss of jobs. This idea is typically linked with the belief that society is becoming increasingly dehumanized as the market is flooded with computers and computer-controlled systems.

A new social class is also forming in the United States. In the past, distinctions be-

tween social groups were influenced by economic and educational factors. These same forces are present, but our information society makes these distinctions even more pronounced. This is especially true if you do not have information and computer management skills.

Look at the world around you. Banks are shifting to computer-based teller systems, and telephone lines are the keys to information pools. But unless you know how to tap these resources, you may join the ranks of the information poor. You may not be able to compete for jobs that require computer proficiency.

This scenario also reaches beyond the United States. It influences nations that may not have the economic, political, or educational resources to participate in the information age. This situation presents us with a paradox. Although more people have access to information, whole segments of society may not be able to partake of this information bonanza as we accelerate toward a world where information is a lifeline.

But there's also a flip side. The communication revolution has numerous positive implications.

The same computer systems that some people fear have improved the treatment of cancer patients. Doctors have drawn on

Figure 1.2
Computers have played a major role in the broadcast industry. This shot illustrates a virtual set— a computer-generated set that could be used for a news program. (Courtesy of Evans & Sutherland. Copyright 2000, Evans and Sutherland Computer Corporation.)

expert advice by using computer software. Other computer systems have helped individuals with physical handicaps to better communicate with the world around them. It can also be argued that despite the unequal distribution of information, more people have greater access than at any other time in human history.

As stated, the new technologies have societal implications. While one of the book's focus points is applications, the other is the communication revolution's fallout. Our discussions range from ethical to legal to political considerations.

The implications also play key roles in defining our information society. This section serves as an introduction to some of these topics, which are discussed at greater length in subsequent chapters.

Systems Approach

This book uses a systems approach.[2] Although you can examine individual applications and implications, you should also explore related areas to gauge interrelationships and their impact on the overall communications system. Without this broad perspective, we may be looking at an incomplete picture.

In one example, our society's vulnerability to world events vividly highlights the importance of this approach. It also points out the interdependence between technology and our information society.

In brief, our information and communications infrastructure could be crippled and silenced by a single high-altitude nuclear explosion.[3] This paralysis would be caused by a powerful electromagnetic pulse (EMP), a by-product of a nuclear detonation.

Scientists have observed this phenomenon during weapons tests. They have subsequently used simulations to gauge the vulnerability of electronic components, equipment, and systems that could be affected by this pulse, with the end goal of developing protection mechanisms.

A step in this process was the Federal Emergency Management Agency's program to protect radio stations in the Emergency Broadcast System (EBS) and other communications centers. Theoretically, if a nuclear burst ever paralyzed our communication capabilities, the protected EBS stations would be activated to supplement the damaged channels.

A potential problem, though, is the receiving device. To pick up a broadcast, you may have to rely on a portable, battery-operated radio, which is one of the few modern electronic devices that may not be "readily damaged" by the EMP.[4] But many radios are part of a stereo system, which may be damaged. If not, there is a good chance that the power plant that supplies the electricity to run the stereo would not function.

Thus, as a society that depends on and is driven by information, the very tools we use to create, manipulate, and deliver this information would be disrupted, if not rendered inoperable. The issue is multifaceted, and a broad systems perspective enables us to examine it from different angles. We can also cover the key implications both inside and outside the communications field.

In keeping with this approach, it is also ironic to note that vacuum tubes are relatively impervious to EMP effects. But in our contemporary communications system, they

Figure 1.3

Software and hardware developments provide us with new tools to manipulate audio and video information. In this case, an audio editing program can be used to create multitrack audio pieces. (Software courtesy of Syntrillium Software Corp.; Cool Edit Pro.)

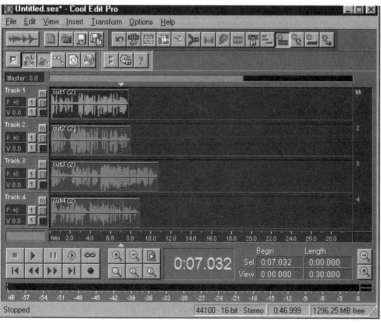

have been universally replaced by solid-state chips, one of an EMP's favorite snacks. As our communications system was modernized, it was also made more vulnerable to world events and, potentially, to random acts of terrorism.

Finally, in another example, the Y2K bug promised to wreak global havoc. Companies and countries launched operations to correct, or to at least contain, the fallout created by the possible disruption of computer services. Like the EMP scenario, it is a multifaceted issue that would benefit from a systems approach. This topic is covered in Chapter 4.

Convergence of Technologies

The communication revolution has accelerated the convergence of technologies and applications. As described in Chapter 13, powerful PCs, when mated with the appropriate hardware and software, have created integrated desktop video systems. These tools allow us to create video productions. When this capability is combined with desktop publishing and other applications, we are no longer just media consumers. We also become producers and editors.

This newfound capability also has aesthetic implications, as discussed in Chapters 3, 12, and 13. Simply stated, the technical capacity to create a project does not supplant underlying aesthetic principles.

Intellectual Property

The new communications technologies raise intellectual property questions, essentially the "rights of artists, authors, composers, and designers of creative works."[5] These questions include the electronic copying of someone else's work as well as copyright and patent issues.

A related factor is information's malleability. In our current system, it can be readily manipulated. While this capability can be used for creative purposes, it also highlights a problem: The same family of tools that can create a graphic or other product can be used to copy it illegally. Because the tools, such as computers, are ubiquitous, it is almost im-

Diane L. Morgenstein in *Photonics Spectra*

Photonics — the technology of generating and harnessing light and other forms of radiant energy whose quantum unit is the photon. The range of applications of photonics extends from energy generation to communications and information processing.

PHOTONICS SPECTRA
VOICE OF THE PHOTONICS TECHNOLOGY

© Laurin Publishing Co. Inc., 1982

Figure 1.4
This drawing highlights the interrelationships between photonics technology and its various fields. As stated in the caption, "photonics is the technology of generating and harnessing light and other forms of radiant energy whose quantum unit is the photon. The range of applications of photonics extends from energy generation to communications and information processing," some applications of which will be covered in the book. (Courtesy of Photonics Spectra; *Diane L. Morgenstein in* Photonics Spectra.*)*

possible, as well as undesirable in most cases, to monitor their use.[6] Copyright laws address these issues, and violators can be prosecuted.

The difficulty, though, may lie in enforcing the laws.

Problems also arise over attempts to recognize intellectual property as actual property. To many people, it may be intangible. You could spend a year writing a computer program that fits on a floppy disk. On the surface, the program may not look like much, perhaps just a single disk you can hold in your hand. But in reality, it may represent the "product of the creative intellect," that is, intellectual property, another form or type of property.[7]

The Democratization of Information

The communication revolution has promoted the free flow of information. You can use a desktop publishing system to print a newsletter, and when combined with other equipment, to keep the world community apprised of fast-breaking political events.

This concept has been extended to other media. As outlined in Chapters 17 and 18, an electronic democracy can be supported by holding electronic meetings between groups of people thousands of miles apart. A key element in this process is providing broad, affordable access to the system.

First Amendment Issues

The new communications technologies have raised First Amendment questions that are covered in different chapters:

1.   Should an electronic information service be treated as a publisher or a distributor?

2.   What legal measures can help ensure that First Amendment rights are supported and nurtured?

3.   How do you reconcile First Amendment rights and privileges with the V-chip and potential Internet censorship?

4.   The broadcast industry has never enjoyed the print media's level of protection, and this philosophy could be extended to other electronic venues. What are the implications?

5.   Has global censorship become more prevalent, as reported in an article in *Inter-@ctive Week*.[8]

Privacy

As our communications systems become more sophisticated, so too does the potential to invade our privacy. Chapter 20 examines privacy and electronic mail, messages or mail relayed in electronic rather than print (hard-copy) form.

Another issue is electronic eavesdropping. Although new communications systems may be harder to tap and special encryption schemes can be introduced, legislative forces can potentially make a system more vulnerable to surveillance.

Telecommunications Act of 1996

The Telecommunications Act of 1996 has had an impact on the information and communications industries. This impact cuts across economic, political, and social lines. The act is discussed in appropriate chapters.

Economic Implications

The communication revolution has economic implications that can affect entire industries. In the United States, one issue has to do with spectrum allocations. We use the spectrum as a means to relay information. In our society, this capability also has an economic fallout.

Other Issues

Our information society will also be influenced by other factors. Will teleconferencing and telecommuting reduce human communication to a machine-dominated format? Do computers dehumanize the creative process? Or do they extend our creative capabilities by helping us to transform an idea or vision into an actual product?

CONCLUSION

Chapter 1 has served as an introduction to the new technologies' universe and some of the implications raised by the communication revolution. The other chapters focus on more specific topics and will follow a preset pattern.

If relevant, we discuss the technical underpinnings of a given technology, which will help us to explore and manage its applications. This knowledge may also prove valuable when deciding if a technology could be used for a specific application and, possibly, for developing applications that are as yet undiscovered.

Next, the implications raised by the technology and its applications will be examined.

REFERENCES/NOTES

1. Frederick Williams, *The New Communications* (Belmont, Calif.: Wadsworth Publishing Co., 1984), 9.
2. For more information, see Ervin Laszlo, *The Systems View of the World* (New York: George Braziller, Inc., 1972).
3. A single nuclear explosion could theoretically blanket the United States and parts of Canada and Mexico. It could disrupt the countries' communications infrastructure even if the weapon did not cause any direct damage.
4. Samuel Glasstone and Philip J. Dolan, *The Effects of Nuclear Weapons* (Washington, D.C.: U.S. Government Printing Office, 1977), 521.
5. Westlaw, Intellectual Property Database, downloaded December 1, 1992.
6. Monitoring, for example, could lead to censorship.
7. "Patents: Protecting Intellectual Property," *OE Reports 95* (November 1991): 1.
8. Steven Vonder Haar, "Censorship Wave Spreading Globally," *Inter@ctive Week* 3 (February 12, 1996): 6.

SUGGESTED READINGS

Carter, A. H., and members of the Electrical Protection Department. *EMP Engineering and Design Principles.* Whippany, N.J.: Bell Telephone Laboratories, Inc., Technical Publication Department, 1984; Pierce, John R., Chairman, Committee on Electromagnetic Pulse Environment; Energy Engineering Board; Committee on Engineering and Technical Systems; National Research Council. *Evaluation of Methodologies for Estimating Vulnerability to Electromagnetic Pulse Effects.* Washington, D.C.: National Academy Press, 1984. These publications examine different elements of the EMP issue. The first provides a good overview of the creation and technical implications of an EMP as well as different equipment and protection schemes. The second covers a wide range of subjects, including the role of statistics in trying to predict potential equipment and system failures.

Rogers, Everett M. *Communications Technology.* New York: The Free Press, 1986. An excellent reference and resource book, it examines the history of communications science, the social impacts of communications technologies, and research methods for studying the technologies.

Shannon, Claude, and Warren Weaver. *The Mathematical Theory of Communication.* Urbana, Ill.: University of Illinois Press, 1949. The book about information theory. Also see Bharath, Ramachandran, "Information Theory," *Byte* 12 (December 1987): 291–298, for a discussion of the information theory; and Tufte, Edward, *Envisioning Information.* Cheshire, Conn.: Graphics Press, for an in-depth examination of the visual representation of information.

The following publications can be used to provide an overview of our information society.

Frezza, Bill. "China Begins Building the Wall of Cyber." *Communications Week* (February 26, 1996): 43.

Gross, Lynne Schafer. *The New Television Technologies.* Dubuque, Iowa: William C. Brown Publishers, 1990.

Figure 1.5
Mount St. Helens. A computer-generated image. You can use PCs to create realistic and surrealistic views of the earth and other worlds. (Software courtesy of Virtual Reality Laboratories; Vista Pro.)

Hanson, Jarice. *Connections*. New York: Harper-Collins College Publishers, 1994.

Inglis, Andrew F. *Behind the Tube*. Boston: Focal Press, 1990.

National Telecommunications and Information Administration. *Telecommunications in the Age of Information*. NTIA Special Publication 91–26, October 1991.

Pavlik, John V. *New Media Technology*. Boston: Allyn and Bacon. 1996.

Telecommunications Act of 1996, 104th Congress, 2nd session, January 3, 1996.

Tracey, Lenore. "A Brief History of the Communications Industry." *Telecommunications* 31 (June 1997): 25–36.

Truxal, John G. *The Age of Electronic Messages*. New York: McGraw-Hill Publishing Company, 1990.

U.S. Congress, Office of Technology Assessment. *Critical Connections: Communication for the Future*. Washington, D.C.: U.S. Government Printing Office, 1990.

GLOSSARY

Communications System: The means by which information, coded in signal form, can be exchanged. In the context of this book, the communications system also encompasses the communications tools we use, their applications, and the various implications that arise from the production, manipulation, and potential exchange of information.

Electromagnetic Pulse (EMP): A by-product of a nuclear explosion; a brief but intense burst of electromagnetic energy that can disrupt and destroy integrated circuits and related components.

Information: A collection of symbols that, when combined, communicates a message or intelligence.

Information Society: A society driven by the production, manipulation, and exchange of information. Information can be viewed as a social, economic, and political force.

Intellectual Property: The rights of artists, authors, and designers of creative works; the products of the creative intellect.

2 Technical Foundations of Modern Communication

This chapter introduces the technical elements that are the foundation of our modern communications system. It is also a continuation of Chapter 1, but concentrates more on technical information and digital communication.

The chapter concludes with a discussion of the importance of technical standards. Standards can promote the growth of a communications technology and industry. If standards are not adopted, the industry's growth could be hampered.

BASIC CONCEPTS

The Transducer

A transducer is a device that converts one form of energy into another form of energy. When you talk into a microphone, it converts your voice—sound or acoustical energy—into electrical energy, or in more familiar terms, an electrical signal. A speaker, also a transducer, can convert this signal back into your voice.

The "spoken words," the sound waves, can also be considered a natural form of information, information that our senses can "perceive."[1]

A transducer can convert this category of natural information, among others, into an electrical representation. The transducer acts as a link between our communications system and the natural world.

Certain transducers can also be considered extensions of our physical senses. They can convert what we say, hear, or see (for example, a camera) into signals that can be processed, stored, and transmitted.[2]

The Characteristics of a Signal

If a microphone's operation were visible, we would see what appears to be a series of waves traveling through the connecting line. The waves, the electrical representation of your voice, have distinct characteristics. Two that are pertinent to our discussion are amplitude and frequency.

The amplitude is a wave's height, and in our example corresponds to the signal's strength or the volume of your voice. The frequency, the pitch of the voice, can be defined as the number of waves that pass a point in 1 second. If a single wave passes the point, the signal is said to have a frequency of 1 cycle per second (cps). If a thousand waves pass the same point, the signal's frequency is 1000 cps.

The frequency, the cps, is usually expressed in hertz (Hz), after Heinrich Hertz, one of the pioneers whose work made it possible for us to use the electromagnetic spectrum, a keystone of our communications system. Hence, a frequency of 10 cps is written as 10 Hz. Higher frequencies are expressed in kilohertz (kHz) for every thousand cycles per second, megahertz (MHz) for every million cycles per second, and gigahertz (GHz) for every billion cycles per second.

Modulation, Bandwidth, and Noise

Modulation is associated with, but not limited to, the communications systems that are the most familiar to us, including AM and FM radio stations. It can be defined as the process by which information, such as a DJ's program, is superimposed or impressed on a carrier wave for transmission.[3] A physical characteristic of the carrier is altered to convey, or to act as a vehicle to carry, the information. Prior to this time, it did not convey any intelligence.

After the signal is received, the original information can be stripped, in a sense, from the carrier. We can hear the DJ over the radio.

For the purpose of our discussion, a communications channel's bandwidth, its capacity, dictates the range of frequencies and, to all intents, the categories and volume of information the channel can accommodate in a given time period.[4]

A relationship exists between a signal's frequency and its information carrying capability. As the frequency increases, so too does the capacity to carry information. The signal must then be relayed on a channel wide enough to accommodate the greater volume of information. A television broadcast signal, for example, has a higher bandwidth requirement than either a radio or telephone signal.[5] Consequently, under normal operating conditions, a standard telephone line cannot carry a television signal.

During the information exchange, noise may also be introduced, potentially affecting the transmission's quality. If the noise is too severe, it may distort the relayed signal or render it unintelligible, and the information may not be successfully exchanged. For example, if the snow on a television screen—the noise distorting the relayed signal—is very severe, it may become impossible to view the picture.

Noise can be internal, introduced by the communications equipment itself, or external, originating from outside sources. The noise may be machine generated or natural.

Lightning is one such natural source. Lightning is fairly common and may be manifested as static that disrupts a radio transmission. The sun is another source, creating interference through solar disturbances and other phenomena. Machine-generated noise, on the other hand, can derive from the electric motors in vacuum cleaners or large appliances.

When discussing noise relative to a communications system, the term *signal-to-noise ratio* is frequently encountered. It is a power ratio, that of the power or strength of the signal versus the noise. For information to be successfully relayed, the noise must not exceed a certain level. If it does, the noise will disrupt the exchange to a degree dependent on its strength. It will have an impact on a communications channel's quality and transmission capabilities.

The Electromagnetic Spectrum

The electromagnetic spectrum is the entire collection of frequencies of electromagnetic radiation, ranging from radio waves to X rays to cosmic waves. Infrared and visible light, radio waves, and microwaves are all well-known elements and forms of electromagnetic energy that compose the spectrum.

We tap into the spectrum with our communications devices and use the electromagnetic energy as a communications tool—a means to relay information. For example, radio stations employ the radio-frequency range of the spectrum to broadcast their programming.

Figure 2.1

Electromagnetic spectrum. (Courtesy of the Earth Observation Satellite Company, Lanham, Maryland, USA.)

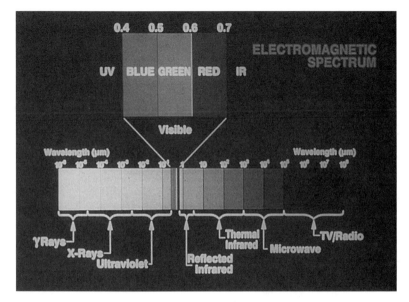

Spectrum space, allocated to television stations and other services, can also be viewed as a commodity in our information society. It can generate income, and like oil, gas, and other valuable natural resources, is scarce. In this light, the spectrum has an intrinsic worth and monetary value.

The issue of spectrum scarcity, which has also been used as a basis for broadcast regulation, is a reflection of our communications system. We cannot use all the spectrum for communications purposes, and the available portions are divided nationally and internationally. In the United States, allocations have been controlled by the National Telecommunications and Information Administration (NTIA) and the Federal Communications Commission (FCC).

New and emerging technologies are exacerbating this situation. The demand is so great that when a small portion of the spectrum opened up for mobile communications services in 1991, the FCC received approximately 100,000 applications for this space in 3 weeks.[6]

Because the spectrum is such a valuable resource, government agencies called for new spectrum management policies. These included imposing fees for spectrum usage and auctioning allocations.

Proponents initially claimed that auctioning would generate additional revenues, and mechanisms would ensure that the public interest mandate, an important regulatory ideal, would still be met.[7] Opponents countered that auctioning would conflict with the notion that the spectrum was a public resource. It was not private property that could be bought and sold.

Regardless of the viewpoint, Congress and the Federal Communications Commission (FCC) embraced auctioning in the 1990s. Although the debate continued to heat up in this decade, it certainly was not a new one. In 1964, Ayn Rand wrote about the broadcast industry that "the airwaves should be turned over to private ownership. The only way to do it now is to sell radio and television frequencies to the highest bidders (by an objectively defined, open, impartial process)."[8] Rand, an author, philosopher, and advocate of capitalism, indicated that private rather than public ownership of the airwaves would have protected the electronic media from government regulation and would have ensured an open, competitive, and free marketplace.

In another twist, some individuals called for switching services: Television should be delivered by cable, thus freeing the spectrum for wireless communication. The cable, owned by cable and potentially telephone companies, would most likely be part of a fiber-optic system. This proposal, also called the *Negroponte switch,* after Nicholas Negroponte, director of the Massachusetts Institute of Technology's Media Lab, was offered as a potential spectrum scarcity and management solution.[9]

This type of switch, though, could not take place overnight. Over-the-air television is also "free." Will we still receive free programming if it is relegated to a cable-based delivery system?

DIGITAL TECHNOLOGY

Digital technology has fueled the development of new communications lines, information manipulation techniques, and equipment. Preexisting communications channels and devices have also been affected. It is one of the communication revolution's driving forces.

Analog and Digital Signals: An Introduction

Many communications devices, such as telephones, video cameras, and microphones, are analog devices that create and process analog signals. As stated by Simon Haykin in his book *Communication Systems,* "Analog signals arise when a physical waveform such as an acoustic or light wave is converted into an electrical signal."[10]

For a microphone, this signal, an electrical representation of your voice, is said to be continuous in amplitude and time. The amplitude, for example, can assume an enormous range of variations within the communications system's operational bounds. The signal is an "analogue" (analog); that is, it is representative of the original sound waves. As the sound waves change, so too do the signal characteristics in a corresponding fashion.

A digital signal, in contrast, is "a non-continuous stream of on/off pulses. A digital signal represents intelligence by a code consisting of the sequence of discrete on or off states. . . ."[11] A digital system uses a sequence of numbers to represent information and, unlike an analog signal, a digital signal is noncontinuous.

Digital and analog signals, and ultimately equipment and systems, are generally not mutually compatible. This mandates the use of analog-to-digital and digital-to-analog conversion processes. They help us to use a mixed bag of analog and digital equipment in the overall communications system. This capability is crucial. Digital information has certain advantages, and the conversion processes allow us to tap these advantages even when using a microphone or other analog device.[12]

Our communications structure is also somewhat based on an analog standard. The conversions allow us to integrate digital technology in this system.

Finally, different categories of information can be represented and consequently transmitted over the appropriate channels, in both analog and digital forms. An analog audio signal, for instance, can be converted into a digital representation and subsequently relayed.

Analog-to-Digital Conversion

In an analog-to-digital conversion, the analog signal is converted into a digital signal. Binary language is the heart of digital communication. It uses two numbers, 1 and 0, arranged in different codes, to exchange information.

The 1s and 0s are also called *bits,* from the words *binary digits,* and they represent the smallest pieces of information in a digital system. They are also the basic building blocks for a widely used digital information system, pulse code modulation (PCM). A simplified explanation of the general principles that govern this process follows.

PCM is a coding method by which an analog signal can be converted into a digital representation, a digital signal.[13] PCM information consists of two states, either the presence or the absence of a pulse, which can also be expressed as "on" or 1 and "off" or 0.

When the analog signal is actually digitized, it is sampled at specific time intervals. Rather than converting the entire analog signal into a digital format, it is sampled or segmented, and only specific parts of the signal are examined and converted. Enough samples are taken, however, to obtain a sufficiently accurate representation of the original signal.

The samples are then compared to, for illustration purposes, a preset scale composed of a finite number of steps. The steps represent the different values or amplitudes the original analog signal could assume.[14] A sample is assigned, in a sense, to the step that matches or is closest to its amplitude. Each step, in turn, corresponds to a unique word composed of binary digits (for example, 11 or 01).

A sample is then coded and represented by the appropriate word. The word can be relayed as on and off pulses, and when the information reaches the end of the line, it is detected by the receiver. Ultimately, a value corresponding to the original analog signal at the sampling point has been transmitted since the word represents a known quantity. (*Note:* For more specific details, see the books and articles at the end of this chapter.)[15]

Morse code functions in a similar fashion. Information, a message, is coded as a series of dots and dashes. Following its relay, the original message can be reconstructed by an operator since the code represents a known value.

Sampling and Frequency. The sampling rate, which is the number of times the analog signal is sampled per second, is central to the reproduction process. The primary goal of sampling is to reflect accurately the original signal through a finite number of individual samples.

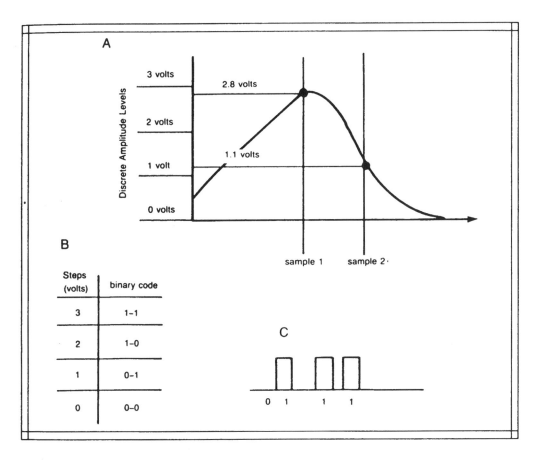

Figure 2.2
A PCM operation. The diagram has been simplified for illustration purposes. The two sampling points have been indicated by samples 1 and 2 (A). Sample 1 is 2.8 volts and is assigned to the nearest step, 3 volts. Sample 2, 1.1 volts, is assigned to the nearest step, 1 volt. Based on the chart (B), the samples are coded as 1-1 and 0-1, respectively. The result is the PCM signal (C).

A survey conducted across the United States to predict the outcome of a presidential election can serve as an analogy. A sample of individual voters represents the voting population just like the samples of the analog signal represent the original signal.

The sampling rate is based on a signal's highest frequency in a given communications system. If a signal is sampled at a rate that is at least twice its highest frequency, the analog signal could be accurately represented. This is called the *Nyquist rate*.

For a standard telephone line, the communications channel carries only frequencies below 4 kHz, and the sampling rate is 8000 samples per second. Enough samples are generated to reproduce the analog signal and a person's voice. Note that higher sampling rates can also be employed.

Steps. In many communications systems, when the analog signal is digitized, it is generally coded as either 7- or 8-bit words. A re-

lationship exists between the number of bits in a word and the number of steps: As bits are added, the number of steps that can represent the analog signal increases in kind.

With an 8-bit-per-word code, essentially 256 levels of the signal's strength can be represented. Derived from the binary system, an 8-bit code is the equivalent of two to the eighth power ($2 \times 2 \times 2 \times 2 \times 2 \times 2 \times 2 \times 2$). A different combination of eight 1s and/or 0s represents each step.

An increase in the number of steps can lead to a more precise representation of the original signal. Certain communications systems and data also require a large number of steps, and consequently levels, for an accurate representation. Nevertheless, they're limited in comparison with the range an analog signal could assume.

Analog-to-Digital and Digital-to-Analog Converters. For the purpose of our discussion, the analog signal is converted into a dig-

Figure 2.3

An ADC and DAC operation. An analog signal is converted into a digital domain via the ADC, and back into analog, with the DAC.

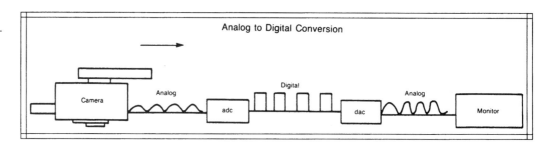

ital format by an analog-to-digital converter (ADC). Once the coded information is relayed, it can be reconverted into the original analog signal by a digital-to-analog converter (DAC) to make the signal compatible, once again, with analog equipment and systems. ADCs and DACs act as bridges between the analog and digital worlds.

The Transmission.   After the analog signal is sampled and coded, the final transmission can be composed of millions of bits. In one format, the telephone industry has employed an 8 bit-per-word code when the analog signal is digitized, and a telephone conversation can be transmitted at the rate of 64,000 bits per second (64 kilobits/second). Other digitized analog signals similarly generate high bit rates.

This high volume of information poses a problem for some communications systems. They may not have the capacity to transmit the information, and special lines are used. One such line is AT&T's T1 carrier. It can transmit 24 digitally coded telephone channels at a rate of 1.544 million bits per second (1.544 megabits/second). This data stream is also composed of bits that ensure the data's integrity and satisfy other technical and operational parameters.

The T1 line is one of the workhorse channels of the communications industry, and transmissions take place both inside and outside of the telephone system. It is also a flexible standard and can integrate voice and data so one communications channel can carry different types of information. It can also accommodate video.[16]

ADVANTAGES OF DIGITAL COMMUNICATIONS

Digital communications equipment and systems have numerous advantages. These features have and will continue to provide the impetus for their continued use and growth.

Computer Compatibility

Once digitized, a signal can be processed by a computer. The capability to manipulate digitally coded information, such as a video camera's pictures, is central to the video production industry and other industries.

For a camera, the video signal, an electrical representation of the light and dark variations, the brightness levels of the scene the camera is shooting, is converted into a digital form. The digital information represents picture elements (pixels), a number of small points or dots that actually make up the picture.[17] In a black-and-white system, a pixel assumes a specific level or shade of the gray scale. The gray scale refers to a series or range of gray shades and to the colors black and white, all of which compose and reproduce the scene. A pixel is represented by a binary word that is equivalent to a level of this scale.[18]

The actual number of gray levels is determined by the number of bits assigned to a word. If too few bits are used, only a limited number of gray shades will be supported, and the image may not be accurately reproduced.

Once the picture has been digitized, it can be computer manipulated. A special effect can be created, or in a science discipline, picture qualities and defects can be enhanced

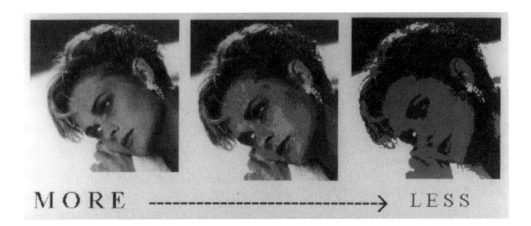

MORE ----------------------------→ LESS

Figure 2.4
The visual effect of reducing the number of gray levels in an image. (Software courtesy of Image-In Inc.; Image-In Color.)

and corrected—an unfocused picture can be sharpened.

Information in digital form also lends itself to mass storage, and the information can be duplicated, as is the case with video copies, without generation loss. Both topics are covered in later chapters.

The digitization of a video camera's signal is also not limited to black-and-white systems. Color is popular, but this information's more complex nature in turn makes a color system more complex.

Multiplexing

Digital and analog signals can be multiplexed; that is, multiple signals can be relayed on a single communications line. The signals share the line and, thus, fewer lines have to be constructed and maintained.

Two advantages offered by multiplexing are its cost and labor-saving properties. In frequency-division multiplexing (FDM), a communications line is divided into separate and smaller channels, each with its own unique frequency. For a telephone system, the various signals are assigned to these separate channels, processed, and relayed.

In time-division multiplexing (TDM), time, not frequency assignments, separates the different signals. A TDM scheme can make for a very effective relay when conducted in the digital domain. An example is the T1 line as employed by the telephone industry.

The T1 system is digital in form, but the typical telephone receiver and local telephone system work with analog information. This requires the analog signal to be digitized. Briefly, groups of signals, the different conversations, are divided into smaller pieces for transmission. Each signal is relayed a piece at a time, in specific time slots.[19] This digital multiplexing system is fast, efficient, and clean with regard to the signal's quality. It can also be controlled and monitored by computers with all the attending advantages—computer accuracy and speed.

Integrity of the Data When Transmitted

A relay may necessitate the use of a repeater(s), which with an analog signal, strengthens or amplifies the signal as it travels along the transmission path. But the signal may be ". . . degraded in this process because of the ever-present noise. . . ."[20] Noise can also accumulate, and the signal's quality can progressively deteriorate.

Digital systems are not similarly affected. Instead of boosting a signal, the pulses are regenerated. New pulses are created and relayed at each repeater site. This process makes a digital transmission much hardier. A digital transmission is also less susceptible to noise and interference in general, further contributing to its superior transmission capabilities.

This factor led to the adoption of a digital relay by the National Aeronautics and Space Administration (NASA) during the 1964 *Mariner 4* mission to Mars. *Mariner 4* provided us with the first close-up views of the

Figure 2.5
The Mariner 4 *space probe. As described in the text,* Mariner 4 *helped pioneer the exploration of Mars. (Courtesy of the National Space Science Data Center-NSSDC.)*

planet, and the digital relay helped preserve the information's integrity as it traveled through millions of miles of space.[21]

Flexibility of Digital Communications Systems

Digital systems are flexible communications channels that can carry information ranging from computer data to digitized voice and video. In an all-digital environment, analog signals, such as those produced by telephones, would be digitized, whereas computer data, already in digital form, would be accommodated without this type of processing. A digital configuration would also eliminate the use of a conventional modem, even though a special adapter may be required to connect a computer, for example, to the line.

The modem, an acronym of the words *modulator* and *demodulator*, is used to relay computer data over a standard telephone line. It converts a computer's digital information into a form that complies with the technical characteristics of the line. At the other end, the information is reconverted by a second modem.

In the early 1980s, typical consumer and business data relays were conducted at 300- to 1200-bit-per-second speeds. Contemporary relays greatly exceed this figure and equate to more efficient and cost-effective operations.[22] Nevertheless, digital communications systems can support much faster relays and benefit from a digital transmission's

superior performance. A digital system can also integrate a wide range of information on a single line, instead of using separate communications channels.

A popular digital platform or network, which can tap the current telephone infrastructure, is the Integrated Services Digital Network (ISDN). The ISDN "can be thought of as a huge information pipe, capable of providing all forms of communications and information (voice, data, image, signaling). . . . It is an information utility accessible from a wall outlet . . . with a variety of devices able to be simply plugged and unplugged."[23]

Basic and Primary ISDN services have supported the following transmission schemes:

- *Basic Rate Interface (2B+D):* two 64 kilobit/second B channels for information (for example, voice) and one 16 kilobit/second D channel for signaling and control information
- *Primary Rate Interface (23B+D):* twenty-three 64 kilobit/second B channels and one 64 kilobit/second D channel. For Europe, the Primary Rate Interface is 30B+D.[24]

This type of service can help revolutionize the way we communicate. We could exchange pictures, computer graphics, and data with the same regularity and ease with which we use a telephone for conversations. Developments of this nature will continue to reshape the communications industry. For ISDN and similar standards, we will be able to select the proper medium for the form of communication we are undertaking rather than being bound to a voice-only system.

The technology will also be more transparent. Just like we drive a car or use a telephone without thinking about the underlying technologies, the ISDN may make it possible to relay information without thinking about the actual process. We do not have to be concerned with how the information is relayed since it is automatically handled. In sum, the ISDN could provide us with a dynamic communications tool that works with preexisting and new lines.

There were and still are some problems, however, associated with the platform. The first was the uncertainty of standards, the technical parameters that govern a system's facilities and operation. For instance, before the early 1990s and the adoption of the national ISDN-1 standard, equipment from different manufacturers was not necessarily compatible.

A second problem has been availability. ISDN and other digital systems are not universal. Certain geographical restrictions may exist, and it may not be possible to establish a link between a customer and a telephone company's central office. The office must also be suitably equipped.

Consequently, it may be a number of years before certain customers are able to take advantage of an ISDN relay or another configuration, as the telephone industry upgrades its physical plant. When completed, even the local loop, essentially the line between your home or business and a central office, should be able to support digital services.

Two other contenders in providing us with high-speed communication links are Digital Subscriber Line (DSL) systems and cable modems. For the former, DSL technology can use "existing telephone lines to deliver Net [Internet] data at speeds up to 50 times faster than a 28.8 Kbps modem."[25] Different flavors of this standard exist, including ADSL or the Asymmetric Digital Subscriber Line. In one trial operation, Bell Atlantic experimented with an ADSL-based video delivery service.[26] The technology may also serve as a *broadband* communications highway that will support the high-speed delivery of information and entertainment services to consumers and businesses. This includes a high-speed link with the Internet, which, in fact, helped the service to gain broader consumer acceptance in the late 1990s. As described in the Internet chapter, the same scenario holds true for cable modems.

The ultimate goal is to break through bottlenecks caused by slower communication channels, an ever increasing data load, and data-intensive media types, such as video and audio clips. The adoption of DSL, cable modems, or other such technology is one step in this process.

Finally, the modernization of a nation's telecommunication infrastructure is crucial. As stated, information is a commodity. The country that can create an efficient national and international information platform will have a decided advantage when competing in the world market.

Cost Effectiveness

As digital equipment is mass produced and manufacturing costs are reduced, digital systems become increasingly cost efficient to build and maintain. They're also generally more stable and require less maintenance than comparable analog configurations.

This superior stability and durability is partly due to the integrated circuit (IC). An IC is a semiconductor, a solid-state device that is one of the driving forces behind our information and communications systems. In essence, the IC is one of the much talked about chips that have helped launch the communication revolution.

The creation of the IC chip had an enormous impact on technology. For example, instead of wiring thousands of individual components to build a piece of equipment, a single chip is used. If the equipment malfunctions, only one chip may have to be replaced, and it is easier to isolate component failures.

But ICs are not without their own set of problems. Initial development costs may be high, and until the chips are mass produced, the early units can be expensive. The chips also tend to be vulnerable to static charges and power surges. If exposed to such an environment, a chip's operational capabilities may be temporarily or permanently disrupted.

Political issues may also have an impact. In one example, the price of computer memory chips soared during the 1980s. The Reagan administration tried to protect American chip manufacturers from Japanese competitors, and a short-term outcome of this political maneuvering was a jump in chip prices.

At one point, a single, common memory chip was more expensive than an ounce of silver.

The 1990s witnessed a similar situation. Small, portable computers fitted with liquid crystal display (LCD) panels became popular. A group of American manufacturers subsequently charged the Japanese with dumping inexpensive panels on the market, and a tariff was levied on such imports to help support American manufacturers.

Although the tariff could have helped manufacturers in the long term, there was a consequence. Imported computers already fitted with the displays were exempt from the tariff, and American computer companies threatened to move their manufacturing facilities offshore to avoid the additional fee.[27] Thus, government intervention led to unforeseen circumstances.

DISADVANTAGES OF DIGITAL COMMUNICATIONS

With any technology, there are disadvantages as well as advantages. Digital technology is no exception.

Quantization Error
The digitization process may introduce a quantization error if not enough levels are used to represent the analog signal. If, for instance, a video system is governed by a 2-bit word code, only four distinct colors would be reproduced. The original analog signal and scene would not be accurately represented. To correct the problem, the number of levels can be raised. But since this may increase the relay and/or storage requirements, a compromise is usually made between these factors and the accuracy of the digitization process.

Dominance of the Analog World and Standards
We live, to a certain extent, in an analog world. Many forms of information, besides the devices and systems that produce and relay information, are analog. These include telephones, televisions, and radios. This necessitates the use of ADCs and DACs.

Public Investment
The public investment issue must also be weighed. An overnight switch to an all-digital standard would necessitate the immediate replacement of our telephones, television sets, and radios or the use of special converters. The same principle applies to communications organizations. The industry's retooling would cost billions of dollars and would disrupt both industry and society.

These are two of the reasons the change to a digital standard has generally been evolutionary rather than revolutionary. As described in different chapters, many of the new technologies and their products will be integrated in the current communications structure, which may ease the impact of their introduction.

A case in point is the recording industry and the compact disk (CD) market. If an individual owned a turntable and large record collection, an overnight switch to a CD format would have made the original system obsolete. The old records, the LPs, were not compatible with a CD player, and the owner may not have been able to purchase additional records.

Accordingly, CD players and disks were integrated in the industry. Companies began manufacturing CD players while maintaining their conventional equipment lines. The record industry likewise supported the new medium but continued to produce vinyl LPs.

As more CD players were sold, the CD market expanded. Record companies increased their CD production and curtailed their vinyl lines. This trend accelerated, and the digital format, along with audiocassettes, dominated the industry. But because the process was somewhat gradual, the industry was not disrupted. Owners did not suffer immediate losses, and equipment manufacturers continued their support of LP systems, albeit at a reduced level.

Other factors do play a role in the acceptance of a new technology. This may include whether a product is cost effective when compared with the product it is replacing. For the CD, a player's cost fell over a relatively short time.

Another potential factor is whether a product has certain advantages over preexisting ones. For the average user, the CD satisfied this criterion, which contributed to its popularity.

Finally, while the gradual integration of a new technology and its products may have benefits, it may not always be the best technological solution. The issue of backward compatibility, for instance, chained the color television standard to the past, to the detriment of a picture's perceived quality. Similar concerns were raised that this scenario could have been played out in the development of a new television standard, as discussed in Chapter 15.

On the one hand, a bold, technological step forward, free of past constraints, could revolutionize elements of the communications industry. The downside, though, could be industry disruption and the public investment factor.

munications industry, has also become more prevalent. In the view of the NTIA in a 1991 document,

[t]he task of standards-setting is best left to the private sector. . . . We recognize that there may be rare cases where FCC or NTIA action to expedite the standards process could be justified. . . . Government intervention could include mediation among conflicting interests or a mandate to industry to develop standards by a time certain [sic], leaving the actual development of standards to the private sector. . . . Any intervention, however, should be limited to cases where there is a specific and clearly identified market failure, and where the consequences of that failure outweigh the risk of regulatory failure (i.e., forcing the resolution of a standard too early in the development of a technology).[28]

The private sector should have the freedom to develop appropriate standards in an open market. But according to some individuals and agencies, if standards do not emerge—to the detriment of a technology

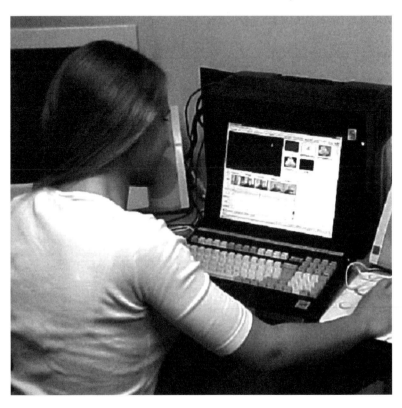

Figure 2.6
A fallout of the digital revolution is the introduction of cost-effective digital tools. These include portable video editing systems that can support high-quality output.

STANDARDS

This final section is devoted to the idea of standards, another central concept that will resurface in future chapters. Simply stated, standards are a series of technical parameters that govern communications equipment and systems. The standards dictate how information is generated, stored, and exchanged. The Society of Motion Picture and Television Engineers (SMPTE) and other national and international organizations have been engaged in this task.

A standard's influence can vary. Some are mandatory and legally enforced. Other standards are voluntarily supported or may be *de facto* in form. Various technical and economic forces may generate industry-wide support for the latter, thus helping to avoid the chaos that would arise if multiple standards were adopted.

The idea of allowing an industry to adopt its own standards, especially in the telecom-

and an industry—then government intervention may be appropriate.

The adoption of this regulatory stance could be important for new products that have to compete in a marketplace with established competitors. As briefly described in Chapter 17, this model of government regulation could have helped the developing U.S. teletext industry.

The standards issue is vital for several reasons. First, if standards did not exist, it would be almost impossible to develop an electronic communications system and to foster the idea of equipment compatibility. Without standards, the telephone and television you buy may not be compatible with the local telephone system and television station.

Second, standards can promote the growth of a communications system. If a standard is adopted, both manufacturers and consumers can benefit. A manufacturer would be ensured that its television set would work equally well in New York City and Alaska. The consumer would be willing to purchase the set for the same reason.

There can also be multiple standards in the same industry and, depending on the circumstances, they may not adversely affect its growth. The parallel development of the Beta and VHS videotape formats serves as an example. While incompatible, they created their own market niches, despite VHS's emergence as the *de facto* consumer standard.

The presence of two strong standards in a market may also accelerate an industry's growth. The competition may spur a manufacturer to introduce improved equipment to gain a larger market share.

Although multiple standards could be beneficial, this is not always true. This factor could have a devastating effect on an organization that is introducing a new application in a competitive field. In one case, competing standards may splinter a small, initial market.

Third, an accepted standard helps guarantee that a piece of equipment you buy today will generally not be replaced by a new and incompatible device tomorrow. While developments do take place and equipment may become obsolete, this process may be gradual. This protects industry and the consumer.

It is also appropriate to point out that obsolescence does not necessarily mean incompatibility. An older video camera may not have all the latest technological "bells and whistles," but it may still be compatible with your system.

There's a point, though, at which a standard may be replaced by another, as was the case with the $\frac{1}{2}$-inch black-and-white videotape recorder (VTR). In the 1970s, this portable VTR and companion camera were popular with video artists and schools. The system was easy to use and affordable. But it became obsolete. The equipment and governing standards were superseded.

Fourth, standards have an impact in the international video markets. For example, the industry has been dominated by three incompatible standards: NTSC, SECAM, and PAL. A program produced under one system must undergo a conversion process to make it compatible with another system.[29] This arrangement has posed some problems for the international market. It creates a roadblock in the exchange of programs and adds to a program's overall cost.

Thus, even though the adoption of a single standard might have fueled a country's domestic video industry, the variety of standards became a hindrance. This problem has assumed a greater degree of importance as the communications industry becomes increasingly international in scope, in keeping with the automotive, computer, and other major industries.

REFERENCES/NOTES

1. Frederick Williams, *The New Communications* (Belmont, Calif.: Wadsworth Publishing Co., 1984), 133.
2. Ken Marsh, *The Way the New Technology Works* (New York: Simon and Schuster, 1982), 26.
3. Herbert S. Dordick, *Understanding Modern Telecommunications* (New York: McGraw-Hill Book Publishing Co., 1986), 32.
4. Technically speaking, the bandwidth is the difference between the highest and lowest

frequencies (that is, the range of frequencies) a communications channel can accommodate. The term *bandwidth* is likewise applied to the signals that are relayed over these channels. Note also that a time factor plays a role. In essence, a communications channel can accommodate or relay a specific volume of information in a given time period based on the channel's capacity and the noise present on the line.

5. For example, it has a greater information content.

6. "Al Sikes's Grand Agenda," *The New York Times* (June 2, 1991), Sec. 3, 6. *Note:* All over-the-air communications systems do not require FCC allocations. Commercial over-the-air lasers discussed in a later chapter fall in this category.

7. Basically, the public would still be served by various types of programming, such as news and public information shows.

8. Ayn Rand, "The Property Status of Airwaves," *The Objectivist Newsletter* 3 (April 1964): 13; from a reprint of *The Objectivist Newsletter,* Vols. 1–4 (New York: The Objectivist, Inc., 1971).

9. Gary M. Kaye, "Fiber Could Be Winner in the Battle for the Spectrum," *Photonics Spectra* 25 (November 1991): 79.

10. Simon Haykin, *Communication Systems* (New York: John Wiley & Sons, 1983), 6.

11. Tom Smith, *Telecabulary 2* (Geneva, Ill.: abc TeleTraining, 1987), 28.

12. The analog information is converted into digital information, as described in the next section.

13. Bernhard E. Keiser and Eugene Strange, *Digital Telephony and Network Integration* (New York: Van Nostrand Reinhold Company, 1985), 19.

14. William Flanagan, "Digital Voice and Multiplexing," *Communications News* (March 1984): 38E.

15. Thanks to Jim Loomis, director of Telecommunications Facilities, Ithaca College, for his suggestions for this section.

In a more detailed look at this process, three phases are completed. In the sampling phase, the analog signal (continuous in time and amplitude) is sampled. This creates a pulse amplitude modulation (PAM) signal. The amplitudes of the individual and discrete PAM pulses correspond to the variable amplitude of the original analog signal at the sampling points. Thus, the amplitudes of the PAM pulses are continuously variable, like the original signal. In the next phase, the quantization stage, the PAM pulses are assigned to the nearest steps or levels in reflection of their amplitudes. This phase converts the wide range of amplitudes to a finite and limited number of amplitudes or values. The final phase is the coding process. The samples are coded in binary form. Consequently, the original analog signal, which is continuous in time and amplitude, is made noncontinuous by the sampling and quantization phases.

In this type of system, the information is also composed of pulses with identical amplitudes. This is a reflection of PCM information where the amplitudes do not convey the information.

16. The T1 line is only one of a digital family of lines.

17. Arch C. Luther, *Digital Video in the PC Environment* (New York: McGraw-Hill Book Publishing Co., 1989), 51. The picture is organized in memory as a series of pixels. *Note:* For specific technical details, see Frederick J. Kolb, Jr., *et al.,* "Annotated Glossary of Essential Terms," *SMPTE Journal* 100 (February 1991): 122.

18. The level, in turn, is delineated by the brightness value of the corresponding section of the original scene now represented by the pixel.

19. In this particular configuration, digitized information from 24 channels is transmitted one after the other on the line. The information is organized in a grouping (frame). Following the transmission, the 24 channels are separated. For details, see Tom Smith, *Anatomy of Telecommunications* (Geneva, Ill.: abc TeleTraining, 1987), 107.

20. Henry Stark and Franz Tuteur, *Modern Electrical Communications* (Englewood Cliffs, N.J.: Prentice-Hall, 1979), 162.

21. Michael Mirabito, *The Exploration of Outer Space with Cameras* (Jefferson, N.C.: McFarland and Co., 1983), 30. *Note:* Digital communications systems similarly played an important role in later missions. Digital tools were also used to enhance and manipulate the pictures produced by a spacecraft's camera system after the data were received on Earth.

22. The terms *bits per second* and *baud* are associated with a modem's relay speed. There are differences between the two, however. For a detailed discussion, see Brett Glass, "Buyer's Advisory," *InfoWorld* 13 (October 21, 1991): 147.

23. Don Wiley, "The Wonders of ISDN Begin to Turn into Some Real-World Benefits as Users Come On Line," *Communications News* (January 1987): 29.
24. Bill Baldwin, "Integrating ISDN Lines for Financial Users," *Telecommunications* 25 (June 1991): 34.
25. Michelle V. Rafter, "Users, Start Your Modems," *The Industry Standard* 2 (October 11,1999): 16.
26. Salvatore Salamone, "Higher Data Speeds Coming in Plain Phone Lines," *Byte* 21 (January 1996): 37.
27. "Dropping the Color LCD Tariff Will Save Jobs," *PC Week* 9 (February 10, 1992): 76. *Note:* American chip manufacturers experienced a recovery during the early 1990s (for example, a consortium of companies formed to promote research and development).
28. NTIA, Telecommunications in the Age of Information, NTIA Special Publication 91–26, October 1991, xvi.
29. The standards are national in origin and have been used by different countries. They are United States, NTSC; England, PAL; and France, SECAM.

SUGGESTED READINGS

Berst, Jessie. "The Broadband Contenders." *ZDNET* (October 6, 1999) (online service); and Rafter, Michelle V. "Users, Start your Modems." *Telecom* 2 (October 11, 1999): 116–121. Coverage of DSL systems.

Bigelow, Stephen J. *Understanding Telephone Electronics.* Carmel, Ind.: SAMS, 1991. A comprehensive overview of the telephone, from basic concepts to digital and network operations.

Clement, Fran. "Digital Made Simple." *Instructional Innovator* (March 1982): 18–20. A good primer about digital information and signals.

Edwards, Morris. "What Telecomm Managers Need to Know About Europe '92." *Communications News* (June 1991): 64–68. An early look at European communications systems.

Epstein, Steve. "Digital Basics, Part 1." *Broadcast Engineering* (August 1995): 12–14. A digital series. This article covers basic, digital principles.

Haykin, Simon. *Communication Systems.* New York: John Wiley & Sons, 1983, 408–428. Technical descriptions of PCM and multiplexing operations, among other communications topics.

"Radio Wave Propagation." *Hands-On Electronics* (November/December 1985): 32–38, 104. A primer on how a radio wave can travel from its source to the receiver.

Rogers, Tom. *Understanding PCM.* Geneva, Ill.: abc TeleTraining, 1982. One of a series of abc TeleTraining publications. This particular publication provides an excellent overview and explanation of PCM.

Rubacky, Sean. "Bell Atlantic's ISDN Prices Plummet." *Computer Telephony* (November 1995): 104–112. See also Tropiano, Lenny, and Dinah McNutt. "How to Implement ISDN." *Byte* 20 (April 1995): 67–74. ISDN growth and setting-up a connection.

Streeter, Richard. "Is Standardization Obsolete?" *Broadcasting* (February 9, 1987): 30. A brief about the importance of standards.

Winch, Robert G. *Telecommunication Transmission Systems.* New York: McGraw-Hill, 1998. An excellent examination of telecommunications systems and their operation.

GLOSSARY

Analog Signal: A continuously variable and varying signal. The communications devices and systems we are most familiar with, such as video cameras and radio stations, produce and process analog signals.

Analog-to-Digital Converter (ADC): An ADC converts analog information into a digital form. ADCs work with DACs to bridge the gap between analog and digital equipment and systems.

Binary Digit (Bit): A bit is the smallest piece of information in a digital system and has a value of either 0 or 1. Bits are also combined in our communications systems to create codes to represent specific information values.

Channel: A communications line. The path or route by which information is relayed.

Communications Channel Bandwidth: A communications channel's bandwidth, its capacity, dictates the range of frequencies and, to all intents and purposes, the categories and volume of information the channel can accommodate in a given time period.

Digital Signal: A digital signal is noncontinuous and assumes a finite number of discrete values. Digital information is represented by bits, 1s and 0s.

Digital-to-Analog Converter (DAC): A DAC converts digital information into an analog form.

Electromagnetic Spectrum: The entire collection of frequencies of electromagnetic radiation.

Frequency: The number of waves that pass a point in a second. The frequency of a signal is expressed in cycles per second or, more commonly, in hertz.

Integrated Services Digital Network (ISDN): A digital communications platform that could seamlessly handle different types of information (for example, computer data and voice).

Modem: The device used to relay computer information over a voice-grade telephone line. A modem is used at each end of the relay.

Modulation: The process by which information is impressed on a carrier signal for relay purposes.

Multiplexing: The process whereby multiple signals are accommodated on a single communications channel.

Picture Element (Pixel): A pixel is a segment of a scan line.

Pulse Code Modulation (PCM): A digital coding system.

Standards: The technical parameters that govern the operation of a piece of equipment or an entire industry. A standard may be mandated by law, voluntarily supported, or *de facto* in nature.

T1 Line: One of the workhorse and important digital communications channels. Information is relayed at a rate of 1.544 megabits/second.

Transducer: A transducer changes one form of energy into another form of energy. Transducers are the core of our communications system and include microphones and video cameras.

3 Computer Technology Primer

The computer has played a pivotal role in launching the communication revolution. This is particularly true of the microcomputer, also known as the personal computer. The PC has made it possible for us to complete jobs ranging from processing letters to designing spreadsheets to creating graphics for a news show.

PCs have also influenced a new generation of audio and video equipment. As described in later chapters, PCs, other computer systems, and microprocessors have touched almost every phase of the production process.[1] Consequently, this chapter serves as an introduction to computer technology and its complementary communications applications. It also provides a foundation for exploring the computer's roles at work, at home, and at play. But before we begin, three points about terms and other matters, as used by the book, must be made:

1. The generic term *personal computer* (PC) is used for IBM and Apple's Macintosh series (Macs) as well as clones (e.g., Compaq and Dell). The Amiga also falls under this generic term.[2]

2. The terms *data* and *information* are used as has been defined by Alan Freedman in his valuable reference work *The Computer Glossary*. Information is "the summarization of data. Technically, data are raw facts and figures that are processed into information. . . . But since information can also be raw data for the next job or person, the two terms cannot be precisely defined. Both terms are used synonymously and interchangeably."[3] This book follows suit.

3. When appropriate, specific computer products may be used to describe applications. They simply serve as springboards to explore these topics.

HARDWARE

PCs are equipped with a number of components. Some of the important ones, forms of which may be used by other computer systems, are discussed in the following subsections.

Memory

A computer's memory is its internal storage system and work space. The most important memory is random access memory (RAM). A program and the data generated during a work session are stored in RAM. These may include a word processing program and a letter you are writing to a friend. The RAM work space is also considered a temporary storage area. Once the computer is turned off, the program and letter are cleared from memory and cannot be recalled unless they were previously saved on a permanent storage system.

Contemporary PCs are generally furnished with 128 or more megabytes (MBs) of RAM. A computer must also be equipped with a specific quantity of RAM to run a given program, and many operations are enhanced if additional memory is available.[4]

Central Processing Unit

The central processing unit (CPU) is the computer's brain. It also dictates the type of software the computer can run and affects the computer's overall data processing speed. Additional performance factors include the computer's storage system, bus speed, and the use of graphics accelerators and other auxiliary systems. The former can speed up the graphics creation process.

Expansion Boards

Some computers are designed with internal slots that can be fitted with expansion boards or cards. The cards either support basic functions, such as generating a monitor's display, or enhance the computer's operation. They may be required or they may supplement built-in capabilities.

A video display or graphics card for an IBM PC, for instance, dictates various technical parameters. These include the resolution of graphics and alphanumeric information as well as the number of supported colors. Alphanumeric information encompasses alphabetic and numeric characters (letters and numbers). Other cards can serve as internal modems and, as covered later, video editing components.

Portable computers, described in the next section, similarly benefit from expansion products. In most cases, they are fitted with *PCMCIA* slots. Small credit card-sized cards, called PC cards, are inserted in the slots and can support data communications and other operations.

Data Storage

Different systems can be used to store your data once the computer is turned off. The most common are floppy, hard, and removable disk drives, all of which are governed by magnetic principles.[5]

Floppy disks were once the primary software distribution vehicle. Programs stored on one or more disks would subsequently be installed on the PC. However, floppy disks have largely been superceded in this application by CD-ROMs. See Chapter 10 for a discussion of CD-ROMs and other *optical media*.

Hard drives are the most common and popular data storage system. Contemporary drives can store gigabytes of data and support fast storage and retrieval operations. As covered in the multimedia and video production chapters, specialized systems are used in video editing and other, data-intensive tasks.

Inexpensive, removable magnetic storage systems became widely adopted in the latter half of the 1990s. Although they had existed for years, newer and cost-effective models, which could even potentially handle demanding video applications, were introduced.

In operation, a removable system functions much like a floppy. Once a cartridge is filled with data, it is replaced. But unlike a floppy, removable cartridges are true mass storage devices with fast data access times.

Monitors

A monitor is the computer's display component, and different standards exist in the PC world. These include the VGA (640 × 480 resolution) and Super VGA (800 × 600+ resolution) standards.

A key development in the PC market is the advancement of display technologies. The average monitor's size and corresponding

Figure 3.1

Contemporary software has simplified numerous operations. In this case, you can tap a PC-based Help system to quickly find information about a program's features and operation. Simply move the cursor to a term and click a mouse button— the linked information (the description) appears. (Software courtesy of Adobe Systems, Inc.; PageMill.)

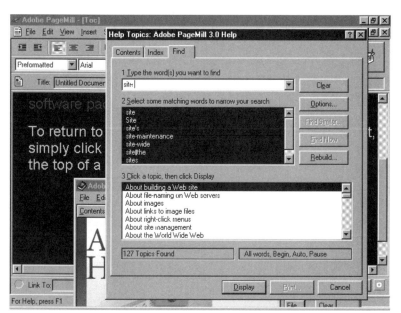

viewing area are increasing. So too is the resolution, that is, an on-screen image's apparent sharpness. Contemporary graphics cards and monitors are also easier on the eyes. They produce a higher resolution display with less flicker to help reduce eyestrain.

A graphics card may also speed up or accelerate certain software operations. It is also equipped with its own memory, which has an impact on the number of colors it may support and other performance characteristics.[6]

At this time, 24-bit systems have become an industry standard. The 24-bit figure refers to "a display standard in which the red, green, and blue dots that compose a pixel each carry 8 bits of information, allowing each pixel to represent one of 16.7 million colors."[7] Basically, instead of working with a limited palette or range of colors, 24-bit systems can gain access to millions of colors. (The actual number of on-screen colors, however, is lower than 16.7 million.) Visually, a picture is more vivid. When used with appropriate software, a landscape with clouds or other image may appear to be lifelike or photorealistic.

THE MICROCOMPUTER

Prior to the PC era, the computer industry was dominated by mainframe and minicomputer systems. Both may still be used by organizations with multiple users and extensive processing needs. But the PC is now a fixture in almost every market.

The PC's popularity can be partly traced to the introduction of the original, widely available, IBM microcomputer in 1981. People were already familiar with IBM products, and the corporation's entry in this field legitimized the PC in the eyes of the business community.

Other factors that contributed to the industry's growth over the years include

- The production of more advanced models;
- Powerful software that filled specific needs;[8]
- The PC's low cost in comparison with other systems, which helped extend

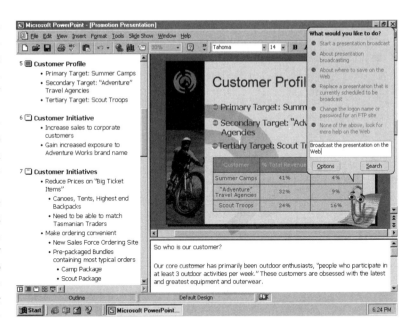

Figure 3.2
Presentation programs can be used to produce PC-based slide shows. A program may also support an Internet export feature that allows you to distribute/play the presentation via the Internet. (Courtesy of Microsoft Corporation; PowerPoint.)

computer processing to a broader user base; and
- Falling PC prices as PCs became more sophisticated.[9]

The computer industry is one of the few manufacturing areas where this type of price versus performance ratio prevails—to our benefit.

Portable PCs

Besides desktop PCs, manufacturers produce portable, battery-operated laptop and notebook computers. In one application, a portable PC can be used in the field by a reporter covering a story. After the story is written, it can be relayed to the newspaper via a modem and telephone line.

Contemporary portable PCs are typically equipped with high-quality color displays and an internal modem. Some units are fitted with removable drives and sport CD-ROM drives, enhanced sound capabilities, and other multimedia options. Another possible feature is a desktop docking station, which provides additional expansion capabilities.

In keeping with the downward or miniaturization trend in the PC world, portable computers are becoming smaller and lighter as their processing power increases. These systems have spawned even smaller computers, including handheld units of varying capability.

Although this miniaturization trend will continue, there are, at least for the near future, size constraints. These may have more to do with the human–PC interface than with the capability to design ever smaller computers. In general, as keyboard and screen sizes decrease, their effectiveness, at least for the operator, may decrease in kind. A partial solution may lie in adopting pen computing, speech recognition, and other, more advanced interfaces.

Human–Computer Interface

As stated, one consideration in using a computer is the human–computer interface: how we communicate with and control the computer. On one level, interfaces are the hardware tools we use. They range from keyboards to graphics tablets to virtual reality systems, in which we can actually become part of a computer-generated world.

Software is the other level, the driving force behind human–computer interfaces. A computer's operating system provides the basic link; other software families are discussed in later chapters.[10]

The most common interface is the keyboard. Alphanumeric and function keys are used for control and software operations. You can type a command, press two keys to magnify an on-screen graphic, or type a letter with a word processing program.[11]

A second category supplements and complements the keyboard. A mouse, for example, is a small device that is interfaced with the computer and sits on a desk. As you move the mouse, a cursor on the monitor's screen moves correspondingly. In one application, you use the mouse to select a command listed on a pull-down menu. You highlight the specific command, click a mouse button, and the command is carried out. A mouse also functions as a drawing tool for graphics software.

A more effective device for this task is the graphics tablet, a drawing pad typically interfaced with an electronic pen. You can literally sketch a picture in freehand, among other options. Some configurations provide very fine degrees of control when combined with specific software.

The touchpad, another interface, works much like a mouse. As implied by the name, a touchpad is a small, sensitive pad (physical area). You can use your finger with the touchpad to emulate mouse functions.

Another interface is the touch screen. Unlike a mouse or graphics tablet, you interact directly with the monitor. The touch screen enhances this process, eliminating the use of a manual tool. In a standard application, a menu with a list of commands appears on the screen. You select the command by touching the screen at the appropriate location. In a sample application, consider an interactive computer program for a museum. When you touch the museum's rooms, which are displayed on a monitor, the exhibits at each location are listed.

The touch screen is an intuitive interface because you do not have to learn how to use it. Simply touch what you want and the hardware and software complete the task.

Another human–computer interface uses a pen metaphor, pen-based computing. In one configuration, you hold a computer much as you would hold a clipboard. But instead of writing on paper, you write on a flat-

Figure 3.3

Human–computer interfaces allow us to explore new techniques, in this case, the potential to create an animated figure based on an individual's movements. (Courtesy of Polhemus, Inc.)

panel display with a pen. This interface can simplify routine jobs. The express mail/package industry, for one, uses electronic forms and pen-based systems. As deliveries are made, people working in the field can check off boxes on forms and fill in other information. The data are then stored and can be transferred to a central computer.

Contemporary systems can also recognize handwriting and special markings used for editing. Recognizing conventional cursive or nonprinted text is, however, more demanding.[12]

Future pen-based systems may also draw on the idea of an *electronic book*. It could be rapidly updated, annotated with your own notes, and integrated with graphics.[13] Like a touch screen, this type of interface is very intuitive because we would be using a familiar communication form, basically an electronic version of paper.

Additionally, small handheld computers have featured handwriting recognition. These compact pen-based devices marry computer technology with the convenience of using a pen and paper. They are typically used for storing addresses, short messages, and similar data.

COMPUTER SOFTWARE

For organizational purposes, software can be divided into two categories: general release programs, the focus of this section, and programs geared for specific communications applications. The latter include hypertext and multimedia authoring software.

Operating Systems and Graphical User Interfaces

The operating system (OS) is the most important piece of computer software. It controls the computer as well as specific data and file management functions.

MS-DOS, for example, emerged as a standard in the PC market through the widespread integration of IBM PCs.[14] MS-DOS supports a text-based interface. Keywords are typed to carry out various functions.

A graphical user interface (GUI), in contrast, uses a visual metaphor. Popularized by Apple and its Macintosh line, Apple helped establish what we now associate with a GUI:

- A pointing device, typically a mouse;
- On-screen menus;
- Windows that graphically display what the computer is doing;
- Icons that represent files, directories, etc.; and
- Dialog boxes, buttons, and other graphical widgets that let you tell the computer what to do and how to do it.[15]

A windowing operation generates multiple windows, one or more framed work spaces on the monitor's screen. Windows can be moved, resized, or removed. Windows can enhance file copying and other procedures, and you can switch between programs with the click of a button. A program is run or displayed in its own window.

An icon is a small picture, a pictorial representation of, for example, a disk drive. Instead of typing a command to display the drive's files, you click on the icon. Instead of typing a command to delete a file, you move the appropriate icon to another icon, typically a trash can.

Apple's popularization of the GUI carried over to other PC platforms. Different products have sported GUIs, including Microsoft Corporation's Windows family.[16] In fact, Apple was involved in a multiyear lawsuit with Microsoft over an intellectual property issue. Apple contended that Microsoft borrowed the look and feel of its GUI when Microsoft developed Windows.

Database and Spreadsheet Programs

A database program can organize a consultant's client list, the titles of a radio station's carts, and other information. Song titles, for instance, can be filed under a singer's name or even under the types of music a station plays.

Once stored, the information can be organized and manipulated. You may also be able to define and examine the associations

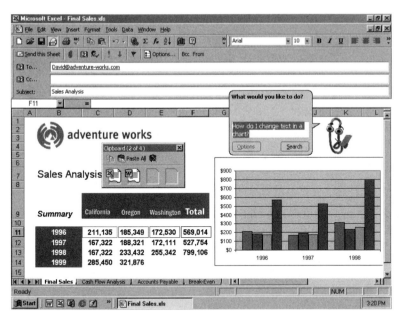

tion. You can now "see" the numbers instead of just rows and columns on a page.

Word Processing Programs

A word processing program is used to write letters, news stories, and many other types of documents. Some programs are also equipped with a mail merge option, which is used to merge a mailing or address list with a standard form letter.

A program has other functions that enhance the writing process. You can typically move, copy, and delete blocks of text, produce tables, and incorporate graphics, spreadsheet files, and other data on a page.

Most word processing programs can optionally save data in an international standardized code called the American Standard Code for Information Interchange (ASCII, pronounced "AH-skee"). Once saved, the information can be retrieved by other computers and programs. Thus, ASCII helps facilitate the exchange of information between computers and, ultimately, their human operators.

Word processing programs also use control codes, which are embedded command sequences, to trigger printer functions. Two control codes may signal the printer to start and then stop its underlining function.

A program may also be able to import and use another program's files through a conversion process. If this option is not available, the file can be saved as a text file (ASCII) and then retrieved, albeit with the loss of page formatting information.

Integrated and Communications Programs

An integrated program combines database, word processing, and other software modules in a single package. It can accommodate multiple applications, is usually cost effective, and supports data interchangeability. Although individual programs can match this function, you may have to buy more than one to complete the same tasks. But the individual programs are generally more comprehensive.

between information categories. For the radio station, these associations can include the sales staff, their clients, and appropriate sales figures.

A database may also accommodate graphics. Visual and textual information can be merged. In one application, a specification sheet that highlights a house's features may be integrated with a picture of the site. This configuration would enable a real estate company to maintain a written and pictorial database of houses on the market.

A spreadsheet program, in contrast, is primarily a financial tool. Data are entered via a table format, in columns and rows, and are tabulated and manipulated through a series of built-in mathematical and financial functions.

With a spreadsheet, you can rapidly finish jobs that would normally require hours to complete, as may be the case when it is used as a forecasting tool. As various figures on the spreadsheet are changed to reflect higher advertising rates, for example, all the pertinent figures are automatically recalculated.

A spreadsheet may also support a graphing capability. The data are portrayed as a line graph, pie chart, or other form. Viewing data in this fashion may make it easier to discover "hidden" relationships. The same graph may also create a more powerful presenta-

A communications program enables a PC to exchange information with other computers. The link for such operations is made through an RS-232 port. This standardized connection defines various technical characteristics that enable the computer to communicate with the outside world.

When a PC is connected to a modem, the communications software controls relay speeds, other technical parameters, and terminal emulation—a PC can emulate or function as a remote terminal.[17]

A standard terminal is generally designed to interact with mainframe and minicomputer systems. It consists of a keyboard and monitor. It is also less expensive than a PC because it is not equipped to process information internally.

Programming Languages

The application programs just described are created through programming languages. A programming language is considered a control language. It provides the computer with a set of instructions to perform a series of operations.

Common languages include BASIC, Pascal, COBOL, C, Java, and Ada. Each language has its own characteristics and is typically geared toward select applications.

Java and COBOL have been widely used with the Internet and business worlds, respectively. BASIC and Pascal are the first programming languages many people learn, and more esoteric languages, such as Lisp and Prolog, are the domain of the artificial intelligence community.

Ada, a language sponsored by the U.S. Department of Defense, is well suited for complex systems. Ada is named after Augusta Ada Byron, considered to be the world's first programmer.[18]

Graphics Programs

Graphics programs are used to create a company's logo, a rendition of the space shuttle, and other drawings. A graphic can also highlight next year's car model and can chart the U.S. population growth.

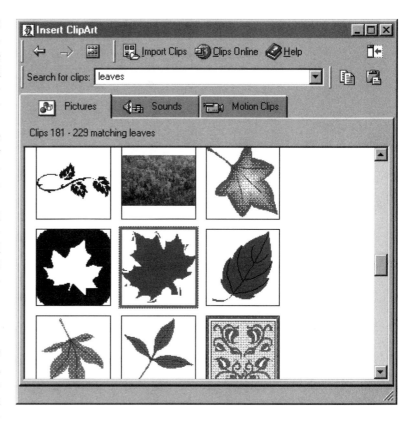

Figure 3.5
Samples of clip art or images that are available for Word, a leading word processing program. (Courtesy of Microsoft Corporation; Word.)

There are different types of graphics programs and formats (for example, TIFF and GIF), some of which are introduced in this section. Most programs can also import more than one format, and it is possible to share files between IBM and Mac platforms.[19]

Another trend is the convergence between program categories. A single program may now support multiple functions. Instead of using two or more of the programs listed in this section, a single program may suffice.

Paint Programs. A paint program addresses or manipulates the individual pixels on a screen. The pixels can be assigned specific colors and can be controlled to produce numerous effects. Applications include manipulating video-based images and computer art, in which an artist paints with an electronic rather than a traditional medium.

The paint program creates bitmapped graphics. Bitmapping refers, in part, to the computer's ability to manipulate the individual pixels that make up a graphic. It also describes the method by which the graphics information is stored. As succinctly stated by Gene Apperson and Rick Doherty, "The bit-mapped image is represented by a collection of pixel values stored in some orderly fashion. A fixed number of bits represents each pixel value. The display hardware interprets the bits to determine which color or gray levels to produce on the screen."[20] Thus, the values, which essentially are the pixels' colors, are coded, stored, and eventually retrieved and interpreted by the computer to create the graphic.

A paint program offers a selection of brush shapes and sizes for freehand drawing. Additional tools can be used to create geometric shapes and for other functions.

You can also control the palette of colors. The number of usable colors can vary from system to system, and you can alter a color to fit your project. One common method is to select a color and to change its red, green, and blue values by moving R-G-B slider bars.

Image Editing Programs.   An image editing program's primary function is that of an image editor or electronic darkroom. This software family can be viewed as the word processor of images. Just as you can edit and move text, a picture can be rotated, rescaled, and manipulated prior to its printing on paper and/or videotape.

For example, you can use special filters to either sharpen or blur an image (or select parts). In one operation, you can throw a distracting background out of focus. Programs also feature color editing and correction tools, paint modules, and possibly a hook to a scanner.

As described in Chapter 12, a scanner can digitize photographs and other still images, which are subsequently fed to a computer. With this software and hardware combination, a picture can be scanned and then retouched or altered.

Drawing Programs.   A drawing program does not address individual pixels. It treats a graphic as a series of individual geometric shapes or objects that can be manipulated and moved to other screen locations. Graphics are vector based, and picture information is expressed and stored mathematically, not as bitmaps.[21]

Drawing programs are typically used for creating ads and illustrations. They also have powerful text handling tools: Words can have a three-dimensional appearance and can follow designated nonlinear paths.

Like paint software, drawing programs have their own advantages and application areas in which they excel. For example, when a graphic produced by a drawing program is enlarged or reduced, the lines will remain sharp and well defined.[22]

Computer-Aided Design Programs.   A computer-aided design (CAD) program is similar to a drawing program in that it manipulates geometric shapes. But it has enhancements that make it a powerful architectural and industrial design tool.

There are two-dimensional (2-D) and three-dimensional (3-D) CAD programs. The latter can be used to create a 3-D view of a building. The image can then be rotated to show the building from different perspectives.

Initially, 3-D views were generally limited to wireframe drawings, images composed

Figure 3.6
An image editing program can support special effects. The two pictures on the right have been manipulated by the software. (Software courtesy of Image-In, Inc.; Image-In Color.)

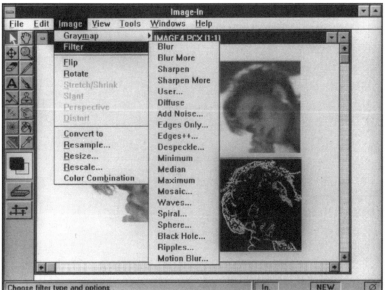

only of lines, with no solid appearance. Computer advances broke through these restrictions, and we can add physical surfaces and solid attributes. In 3-D, the building in our example would look more like a "real" building and would help the designers to better evaluate its physical characteristics.

A program may also have an animation feature. It may likewise be possible to tie into another program to reveal the area(s) of a designed part that may be subject to stress.

Finally, a car part or other information could be fed to a series of computer-controlled tools. At this point, an engineer's vision, a drawing on a computer monitor's screen, is transformed into the actual physical part by the tool. This function is an element of computer-aided manufacturing (CAM), an area closely allied to CAD systems. In many instances, CAD and CAM are linked in a CAD/CAM system.

Animation and 3-D Programs.   An animation can be described as a series of images that, when viewed in sequence, convey motion. We are all familiar with Saturday morning cartoons, such as Bugs Bunny and the X-Men. With computers, we can now tap animation techniques to produce our own projects.

Contemporary programs have also simplified this process. Some allow you to create predefined paths that a figure will follow. You select the number of frames, determine the starting point, and set up the path or direction(s) of the movement. The program will then draw the individual frames, out of the specified total, as the object progresses down the path.

A program may also support other functions. You may be able to speed up or slow down the animation at specific points for smoother transitions and enhanced realism. Or you may be able to define two distinct shapes and have the software draw the intermediate frames so one changes into the other (morphing).

Cycling the colors on a screen is another option. You select a range of colors, and then the on-screen colors cycle through the range.

This can create the illusion of motion while only using a single frame.

Animations can also be created in 3-D, adding realistic depth. A picture can also be wrapped around a 3-D object, and wood and other textures can be applied.

Programs may accommodate 24-bit images, and you typically have control over light sources and camera parameters. Light sources are individual lights that illuminate a scene. The number of lights can vary, as can their color, intensity, and type.[23] The camera parameters affect what we see on the monitor. A view could be changed from wide angle through telephoto and, in an animation, the camera itself could move.

Lifelike animations and still images can be created through ray-tracing, a rendering technique that "literally traces the paths of thousands of individual rays of light through a three-dimensional scene via computation."[24] Ray-tracing can produce very realistic images with accurate shadows and reflections.

The only penalty associated with ray-tracing is time. It may take hours to produce or render the final scene, depending on the PC's capability. But you can initially create the scene in a faster mode. When finished, the scene can be reviewed, changes can be made, and the final sequence can then be produced.

When the work is finished, it can be recorded on videotape. If the project exceeds the size of your usable memory, you can use a special setup to record the frames individually on a VCR as they are rendered.

A program may also save an animation as a series of individual frames (files). In one operation, they can then be imported by a nonlinear editing system, discussed in a later chapter, where they are processed and saved as an animation.[25]

Presentation   Programs.   Presentation programs can generate computer-based charts, graphs, and electronic slides. Templates are available to format the data or you can create your own designs. Interactive links may also be supported. When you activate an on-screen button, an action is triggered. This action may include playing a digitized audio and video clip.

If you have a series of on-screen graphics, you can control the length of time each image is displayed as well as transitional effects.[26] You can also incorporate audio and video clips and can typically export the presentation for replay on the Internet—the project is converted by the software for this environment. In this case, the same presentation can be used across multiple distribution venues.

Mapping Programs. A mapping program generates maps, on-screen graphics of geographical regions. The information can include a collection of U.S. street maps and specialized maps that serve as analytical tools when linked with the appropriate information. In one application, a map could highlight the demographic makeup of a state or locality.

There are also consumer programs, which can even be tied to databases. As you view a country's map, you may be able to retrieve climatic and other types of information. You may also be able to use a form of a mapping program for planning a road trip.

Another type of program may handle digital landscape data. You can produce and view accurate renditions of Mt. Saint Helens and other sites, and possibly even other worlds.

Visualization Programs. Scientific visualization is the "ability to simulate or model 3-D images of natural phenomena on high performance graphics computers."[27] Data are graphically represented, and this visual representation can provide insights into how our world and the universe function.

Instead of looking at pages of numbers, a scientist can view the data graphically. For example, while designing a new spaceplane, airflow and thermal characteristics could be observed.[28] Essentially, the visual image makes it easier to interpret the data because you actually see, in a sense, the data brought to life. The invisible is made visible.

Implications. To wrap up this discussion about graphics programs, it is appropriate to talk about creativity and computer systems. Certain advantages accrue when using computer-based technologies. In the graphics area, a PC with the appropriate software can help you transform an idea or vision into a reality, an actual product. It is the marriage between the conceptual and the concrete.

The same system can help you reach this goal even if you are not a graphic artist. You may have an idea for a poster but may not be adept at creating the 3-D letters your project demands. A computer with the appropriate software could help. In this case, the PC is functioning as a tool. You provide the guiding thought, and the PC helps to implement your idea.

But it is also important, regardless of your artistic ability, to at least have a basic grasp of underlying aesthetic principles relative to the job at hand. Otherwise, an ad's message, or other product, could be lost in a maze of words and graphics. In another example, you can use a PC-based system to create a video production. But unless you understand lighting, camera, and audio techniques, the final product may lack integrity.

A PC may be helpful, but it *does not* circumvent knowledge and aesthetics. A PC could extend your creative vision, but *you* still have to supply the imagination and skills.

Finally, this use of computers has raised questions. Does a computer dehumanize or actually enhance the creative process? Can traditional and computerized methods coexist?

PRINTERS AND LOCAL AREA NETWORKS

A computer system may have components other than those described thus far. These include the printer and a network that links two or more computers in a communications system.

Printers

A printer produces a paper or hardcopy of information. Three types of printers have dominated the general business and consumer markets, and there's a fourth category of important, yet less widely adopted, machines.

The first major category, the laser printer, can produce near-typeset-quality documents and graphics. A small laser is the heart of a printer's engine, the actual printing device. Other elements include a photoconductor drum, toner particles, which make up the image much like the toner particles of a copy machine, and the paper.

The laser printer helped trigger a publishing revolution. When combined with a PC and desktop publishing program, individuals and organizations had access to enhanced publishing tools. Even though the final copy did not equal a commercial publication, it was superior to the typical PC printer's output.

The second major category, the ink-jet printer, uses a reservoir of ink for printing. Its output can almost match a laser printer in certain areas, and ink-jet printers are cost effective and can support color.

The dot-matrix printer, the third category, was a favorite choice during the earlier years of personal computing. It produces alphanumerics and graphics through a matrix of closely spaced dots. They are still used, even though they have generally been supplanted by laser and ink-jet models.

The fourth category of printers is the plotter. A plotter uses a series of pens to create large-scale architectural and technical prints, as well as other line drawings. The output can range from plans for a sailboat to a new piece of machinery.

Color printers are also becoming increasingly popular. More software programs can handle color output, and cost-effective printers have been marketed to meet a growing demand.

Printers are discussed further in Chapter 12, the desktop publishing chapter.

Local Area Networks

In brief, a local area network (LAN) is a communications system that can tie together a group of offices or other defined physical area. It can link PCs so they can be used to share equipment and exchange information.

Printers, data storage drives, and other devices are included on the network, and program and data files can be tapped by the network's users.

A LAN's design and implementation is analogous, in certain respects, to a multiuser system, in which a central computer is typically connected to terminals. A terminal serves as a keyboard and a monitor, an interface device in this environment, since the central computer performs the processing tasks.[29]

A LAN, like a multiuser configuration, makes it possible to share system resources. The LAN has an advantage, though. Each PC has its own processing capability and may be able to operate independently if the network is rendered inoperable by equipment failure. If the central computer "goes down" in the typical multiuser environment, the whole system may come to a crashing halt.

In many cases, multiuser systems have been replaced by LANs. Cost-effective PCs and network interface cards, as well as a LAN's decentralized environment, have contributed to this development. A LAN also offers solid, administrative control, one of the traditional strengths of multiuser systems.

Operation

A LAN's topology, its physical layout, actually dictates how the system is tied together. The topology is usually based on a star, ring, or a bus design, and the information flow typically takes place over twisted-pair, coaxial, or fiber-optic cable.

Twisted-pair cable has the lowest information capacity, while coaxial cable is a shielded and superior relay line.[30] Fiber-optic cable is a newer contender, and one of its major strengths is a large channel capacity. As described in a later chapter, wireless LANs have also been created.

As the data are relayed through a network to the various PCs, the nodes, the data must be routed to the correct destinations. In this respect, a LAN can be considered a superhighway overseen by a traffic cop. Because multiple PCs are interconnected, the data relays and requests must be organized to facilitate the information flow.

Finally, an important element in this configuration is the server. The server is a computer that helps manage and control this flow. It also stores program and data files.

In a typical operation, you connect to the network by logging on, which may include using a password. At this point, you can gain access to the server's programs. Depending on the setup, you use either the server or your PC to store the data files. When finished, you disconnect from the network by logging off.[31]

It is also possible to create a peer-to-peer network. Instead of using a server, PCs send and receive files to and from each other. A peer-to-peer network also creates a more decentralized environment, and it can be less expensive to establish and operate.

Software and MAP

Software manufacturers have supported LANs with word processing and other standard programs. Software that can take advantage of a LAN's interconnectivity has also been introduced.

One option is *electronic mail* (e-mail), in which a message is electronically relayed to specified user(s). In another application, multiple users can collaborate on a project. In either case, this capability can enhance the

communication process within an organization and can save time and money.[32]

It is also important to note that the definition of a network extends beyond the configurations and applications described thus far. An example of a more novel use, which has provided an integrated work environment, has been the manufacturing automation protocol (MAP). MAP, a manufacturing and industrial-based system, has automated factory tasks and has linked equipment produced by different manufacturers.

Another example is the development of national and international electronic communications systems. LANs and individual PCs are islands unto themselves unless there is a mechanism to facilitate the exchange of information. A link that fulfills this goal is the Internet, a set of interconnected computer systems. By using the Internet as a data highway, you can communicate with colleagues and friends. You can also retrieve spacecraft images and tap into the vast information universe.[33] The Internet is discussed, in detail, in Chapter 18.[34]

CONCLUSION

Computer technology will exert an even greater influence on the communications field—and inevitably society—as more sophisticated computers are purchased by an expanding user base. Prices will continue to fall as technology continues to advance. This combination of factors will eventually contribute to the remaking of the computer world, especially at the PC level.

The refinement of technology has also progressed at a dizzying pace. The processing, graphics capabilities, and software sophistication of the new generation of computers represent a substantial design leap over earlier models. Eventually, the characteristics that separate the different computer classifications may blur—it may be difficult to distinguish one computer family from another. At this time, the PC probably matches the capabilities of computers, including the workstations pioneered by Apollo Computers and Sun

Figure 3.7
An example of a house created with a 3-D CAD program. Note the support for multiple views. (Courtesy of American Small Business Computers; DesignCad 3D.)

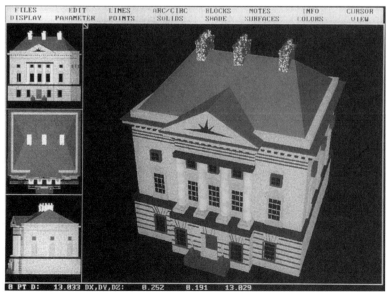

Microsystems, that were once the domain of engineering and design firms.

Apple, for one, took a step in this direction when it introduced a line of PCs equipped with RISC processors. When used with software that takes advantage of the processor's capabilities, computational tasks are greatly enhanced. Other computers support multiple processors that speed up operations, including PC-based video editing functions.

Finally, as stated in the opening, although this chapter focused on PCs, they are not the only players in the communication revolution. Other computer systems, including dedicated graphic stations, have played and will continue to play important roles in our future endeavors.

REFERENCES/NOTES

1. As defined by Anthony Ralston and Edwin D. Reilly, Jr., eds., *Encyclopedia of Computer Science and Engineering* (New York: Van Nostrand Reinhold Company, 1983), 969, a microprocessor is a "computer central processing unit (CPU) built as a single tiny semiconductor chip. . . . It contains the arithmetic and control logic circuitry necessary to perform the operations of a computer program." Microprocessors have been built into audio and video equipment for different functions.

2. This is due to their popularity. Other computer systems are also discussed when appropriate. These include mainframe and minicomputers.

3. Alan Freedman, *The Computer Glossary* (New York: AMACOM, 1991), 301.

4. There are also different types of RAM with different characteristics (for example, speed).

5. Another system, which employs optical technology, is described in a later chapter.

6. Special cards are also used for specialized applications (e.g., video editing).

7. "Glossary," *Publish* 6 (October 1991): 124.

8. These include a spreadsheet program as described in a later section.

9. For example, an ad in a December 1982 issue of *Byte* listed an Atari 800 computer, equipped with 16K of RAM, for $689.95, excluding its $469.95 disk drive or other major components. In 2000, even local computer stores were selling IBM PC compatibles, with 64MB of memory, hard and CD-ROM drives, and a color monitor, for well under $1000. See *Byte* 7 (December 1982): 583.

10. This includes VR in Chapter 22, Future Visions.

11. A keyboard—the way it feels and works—is also a personal matter.

12. Jon Udell, "Windows Meets the Pen," *Byte* 17 (June 1992): 159.

13. Laurie Flynn, "Is Pen Computing for Real?" *InfoWorld* 13 (November 11, 1991): 75. *Note:* Think about your own handwriting. You might even have trouble, at a later date, recognizing what you wrote on a note.

14. MS-DOS is Microsoft's DOS product for IBM PCs. Note also that a PC may be able to support multitasking and other advanced functions. Multitasking enables a computer to run more than one program simultaneously. Other operating systems have included UNIX.

15. Frank Hayes and Nick Baran, "A Guide to GUIs," *Byte* 14 (July 1989): 250.

16. Other GUIs have included IBM's Presentation Manager.

17. The modem also plays a role in this operation. This includes the data relay rate the modem can support.

18. Betty A. Toole, "Ada, Enchantress of Numbers," *Defense Science and Electronics* (Spring 1991): 32. *Note:* This issue has a number of articles about the Ada programming language and can be used to explore its roots.

19. There may, however, be some limitations.

20. Gene Apperson and Rick Doherty, "Displaying Images," in *CD-ROM Optical Publishing*, Vol. 2 (Redmond, Wash.: Microsoft Press, 1986), 134.

21. Corel Systems Corporation, Technical Reference (Ottawa, Ont.: Corel Systems Corporation, 1990), 18.

22. At times, it may be advantageous to convert a bitmapped image into a vector-based one. There is also some cross-compatibility between software in that a drawing program may, for instance, support bitmapped graphics.

23. The type may include spot and flood lights.

24. Impulse, Inc., Turbo Silver 3.0 User Manual, 1988, 17.

25. An advantage of the nonlinear editing station is that a single system can support editing and

animations. The disadvantage, though, is the possibility of image degradation when compared with a single-frame controller (for example, information compression on the part of the nonlinear system can reduce image quality).

26. The transitions between slides can range from dissolves to wipes.

27. Phil Neray, "Visualizing the World and Beyond," *Photonics Spectra* 25 (March 1991): 93.

28. Jim Martin, "Supercomputing: Visualization and the Integration of Graphics," *Defense Science* (August 1989): 32.

29. Multiuser systems are run by a powerful PC, or more typically, a mainframe or a minicomputer. A mainframe can be categorized as a physically large, powerful, and expensive computer that can accommodate numerous users through a multiuser environment. A mainframe is equipped with enhanced processing and memory systems physically built into a central console, the "main" frame. A minicomputer can be considered a scaled-down version of a mainframe. It can also support a multiuser environment, as can suitably equipped PCs, albeit to a lesser degree than a mainframe.

30. Because it is shielded, it is less susceptible to outside interference.

31. *Log on* and *log off* are two commands.

32. E-mail and other products, which allow more than one person to collaborate on a project, are part of a growing software category called *groupware*.

33. Space Digest, Bitnet, Ron Baalke, October 31, 1991, downloaded information. *Note:* If you're interested in these types of images, they are available from different Internet sites, including the Jet Propulsion Laboratory and the National Space Science Data Center.

34. The Internet was nominally geared for educational and research applications. But as discussed in a later chapter, business uses have soared.

SUGGESTED READINGS

Byte on CD-ROM. Compilations of Byte articles on CD-ROMs.

Glassner, Andrew S. "Ray Tracing for Realism." *Byte* 15 (December 1990): 263–271. A detailed examination of ray-tracing.

Hodges, Mark. "It Just Feels." *Computer Graphics World* 21 (October 1998): 48–56; Young, Jeffrey R. "Computer Devices Impart a Real Feel for the Work." *The Chronicle of Higher Education* XLV (February 26, 1999): A23–A24. Interesting looks at new human–computer interfaces.

Hodges, Mark. "Visualization Spoken Here." *Computer Graphics World* 21 (April 1998): 55–62; Mahoney, Diana Phillips. "Launching A Construction Simulation." *Computer Graphics World* 21 (August 1998): 60–62; Spohrer, Richard. "Making CAD Models Shine." *Computer Graphics World* 21 (January 1998): 51–56. Different computer graphic techniques, including CAD, and their applications.

Kincaid, John, and Patrick McGowan. "When Faced with the Office Wiring Decision, 'Let the Buyer Beware.'" *Communication News* (February 1991): 59–61; Quraishi, Jim. "The Technology of Connectivity." *Computer Shopper* 11 (May 1991): 187–198; Rosch, Winn L. "Net Gain." *Computer Shopper* 12 (April 1992): 534, 536–544. Overviews of LAN technology.

Mahoney, Diana Phillips. "The Picture of Uncertainty." *Computer Graphics World* 22 (November 1999): 44–50. Visualization and some applications.

Pournelle, Jerry. "User's Column." *Byte*. A column in the now discontinued *Byte* magazine. Jerry Pournelle is a well-known science fiction writer and computer expert. His column covered a range of topics for the computer user. As Pournelle might have said, this column was and still is "highly recommended."

Rivlin, Robert. *The Algorithmic Image*. Redmond, Wash.: Microsoft Press, 1986. A richly illustrated history of computer graphics and applications.

Webster, Ed, and Ron Jones. "Computer-Aided Design in Facilities and System Integration." *SMPTE Journal* 98 (May 1989): 378–384. The use of CAD software in the television industry.

GLOSSARY

Disk Drive: A data storage device.

Electronic Mail (e-mail): Electronic messages or mail that can be relayed all distances (for example, over a LAN or around the world).

Graphical User Interface (GUI): A visual, rather than text-based, interface.

Graphics Programs: The generic classification for computer graphics software. These range from paint to CAD software, and they excel at certain applications. Paint programs, for example, are used to create/manipulate bitmapped images. CAD programs are geared for architectural/technical drawings.

Human–Computer Interface: The tools we use to work with computers. They range from keyboards to touch screens.

Local Area Network (LAN): A dedicated data communications network. It can link computers, printers, and other peripherals for exchanging/sharing information, programs, and other resources.

Monitor: A computer's display component.

Multiuser System: A computer that can accommodate multiple users, typically through terminals. The computer provides central processing and data storage capabilities.

Personal Computer (PC): A computer typically designed to serve one user.

Photorealistic: A video display card/monitor that can produce realistic or lifelike images.

Printer: A device that produces a hardcopy of the computer's information.

Programming Language: A computer language used to create a computer program, the instructions that drive a computer to complete various tasks. Typical languages include C, Pascal, and Fortran.

Random Access Memory (RAM): A computer's working memory.

Server: A computer that manages and controls the flow of information through a network. It also stores program and data files.

4 Computer Technology: Legal/Health Issues, Y2K, and Artificial Intelligence

The previous chapter introduced us to computer technology. This chapter focuses on related topics. These include potential health problems, legal issues, and artificial intelligence. The latter has implications that cut across technical and philosophical fields.

COMPUTERS AND THE WORK/PLAY ENVIRONMENT

Monitors, Ergonomics, and Viruses

Computer technology may have an impact on our lives beyond its applications. Controversy has surrounded the use of monitors, which produce electromagnetic fields. There are concerns that emissions may play a role in miscarriages, birth defects, and cancer. But the data have been conflicting and inconclusive. The frequencies that concern us are very low frequency (VLF) and extremely low frequency (ELF) emissions. Much of the research has focused on the magnetic field, and studies have been conducted to determine its effect on the human body.

Until this issue is finally resolved, it may be wise to take some elementary precautions. You can buy a nonemission display, or as became the trend in the mid-1990s, a monitor that conforms to Sweden's strict emission standards.

You can also sit an arm's length from the monitor to further minimize potential exposure. But do not move closer to someone behind you who may be working with another computer. Magnetic fields are stronger from a monitor's back and sides. Finally, turn the monitor off when not in use.

Besides these potential effects, other safety and health factors should be considered. One of the most important is *ergonomic design*, the philosophy of developing equipment and systems around people, that is, making equipment conform to an operator's needs, and not the other way around. Desks, chairs, monitors, and keyboards are typical computer equipment influenced by ergonomic design.

Sound designs can help prevent some of the problems associated with working with computers. With carpal tunnel syndrome, your wrist can be damaged through repetitive keyboard motions. It can be alleviated,

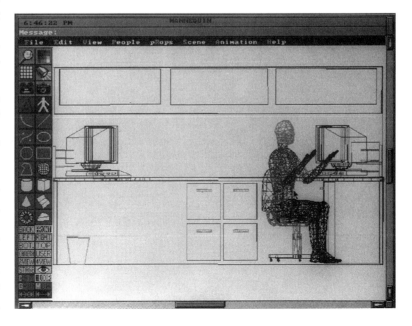

Figure 4.1
PC software can be used in ergonomic design. The same program may also create images suitable for desktop publishing and other projects. (Courtesy of Biomechanics Corp., HumanCad Division; Mannequin.)

or possibly even avoided, by ergonomically sound keyboard and desk designs.[1] Adopting good work habits can also help. Maintain a good sitting posture, and take a break every hour.

The latter can also reduce eye fatigue, as can proper light placement and intensity. A high-resolution display/board combination, which may also produce less flicker as a result of a high screen refresh rate, can likewise help.

Besides taking care of ourselves, it is important to protect information. As you probably know from the news, the number of computer viruses has been rising. A computer virus is a program that attaches itself, in a sense, to other programs. A virus may simply display a harmless message or, worse, cause valuable data to be lost.

Viruses can be spread by various means, including infected disks inadvertently shipped by commercial manufacturers. A virus may also remain dormant until triggered. In 1992, the well-publicized Michelangelo virus was triggered on Michelangelo's birthdate. Computers have internal clocks that keep track of the time and date. When the right date rolls around, if your system is infected, the virus could become active.[2]

What can you do to protect yourself? Don't exchange disks with other people. You should also use virus protection software. Once a virus is detected, it is neutralized. But since new viruses regularly pop up, it is crucial to update your program's virus database. As a final precaution, always back up your files. This would be a good practice even if viruses did not exist.

LEGAL ISSUES

The computer industry's spectacular growth has raised intellectual property and related legal issues. This section serves as an introduction, and some topics are further developed in other chapters.

Software Piracy

A challenge facing the industry is software piracy, the illegal copying and distribution of software. The United States is currently the world leader in software production, and international piracy has had an adverse economic impact on the software industry.

The piracy problem is a matter of intellectual property rights. A panel convened to examine the situation suggested that the U.S. government should increase its "antipiracy efforts" and "strengthen the enforcement of intellectual property rights abroad and in the United States."[3]

The last sentence is a key one because pirating is not limited to other countries. Although there have been crackdowns, led by software industry organizations, piracy is still rampant in the United States.

Piracy presents an almost impossible situation that points out a dilemma of the information age. The same tools that can create an information commodity can be used to steal it, in this case, software.[4] The situation is also exacerbated by the PC's ubiquitous presence at work and home.

Legislation can be enacted against piracy to protect intellectual property, but how do you enforce it? A few major "pirates" might be caught, but what of the thousands of individuals who may copy software either for sale or, more typically, for personal use?

Manufacturers have taken different steps in response. These have ranged from software protection schemes to prevent copying to prompting you to type in a specific word from a manual.[5]

The widespread use of local area networks (LANs) has not helped the situation. For LAN administrators, the number of potential users may make it harder to track and keep control of software, despite the release of other programs that address this need.

Another problem has been the language of licensing agreements. An agreement, which is made between the software company and you, can vary and can be unclear. This has caused confusion.

However, Borland International, a major software house, has adopted a clear-cut licensing agreement for its products:

This software is protected by both United States Copyright law and international treaty provisions.

Therefore, you must treat this software just like a book, with the following single exception . . . to make archival copies. By saying "just like a book," Borland means . . . this software may be used by any number of people and may be freely moved from one computer location to another, so long as there is no possibility of it being used at one location while it is being used at another. Just like a book that can't be read by two different people in two different places at the same time, neither can the software be used by two different people in two different places at the same time. (Unless, of course, Borland's copyright has been violated.)[6]

The language is not ambiguous and clearly points out your rights. It also cleared up the confusion with some agreements as to whether or not you could use a single copy of a program with your home and office PCs.

To sum up, software piracy is a major problem, especially in our information age, and manufacturers have taken steps to combat it. However, more can be done. As indicated by the panel mentioned earlier, the U.S. government can increase its efforts in this area.[7]

Education is one way. Intellectual property may not be viewed as a gold watch, money, or other "real property." A common perception is that you should not steal money, but it is all right to copy a disk.

Part of this problem is philosophical. While the legal system may safeguard intellectual property, the philosophical element may lack focus. We have to recognize and accept the philosophical basis behind the idea of ownership before we follow the legal guidelines. Otherwise, the only preventive measures are legal, and if they cannot be fully enforced, the situation will continue.

Finally, note that other types of pirating affect the communications industry. These include the illegal copying and distribution of movies and CDs and, as covered in the satellite chapter, pirating of pay television services.

Besides resulting in lost revenue, measured in billions of dollars, the issue led to a trade clash between the United States and China in 1996. China had been identified as a major offender of piracy laws.[8] On the flip side, a large videotape pirating operation was halted in New York City at approximately the same time. Both situations highlighted, again, the problem's wide scope.

L Is for Lawsuit

The computer industry has also been a fertile ground for lawsuits. The software arena has been especially hotly contested, and the focus has been on patent and copyright protections and violations.

In general, "a copyright provides long term conditional ownership rights in a specific expression of an underlying concept, without protecting the concept itself. A patent provides relatively short term . . . conditional ownership rights in an underlying concept (the patent's 'invention') without reference to the particular concept's embodiment."[9]

One of the most publicized lawsuits was the dispute between Apple Computer and the Microsoft Corporation for copyright violations.[10] Apple contended that Microsoft's Windows borrowed heavily from Apple's GUI in regard to the interface's "look and feel."[11]

The issue came to a head in 1992. At that time, most of Apple's claims were thrown out by U.S. District Judge Vaughn Walker in a series of rulings. This included Apple's claim that Windows was "substantially similar to the look and feel of the Macintosh user interface."[12] If the ruling had gone the other way, Apple could have had a hammerlock on GUI rights and, possibly, future developments.

With respect to patents, a 1981 Supreme Court decision, *Diamond v. Diehr,* opened the software patent floodgate. A patent can provide broader and wider protection than a copyright and, as such, is a valuable legal and financial commodity.

As of the late 1990s, companies began to file more patent applications, the U.S. patent office was overburdened, and the hardware end of the computer industry experienced its own share of lawsuits. The patent process is also laborious and expensive, which may shut out individuals and small businesses from this form of protection.

A pessimist might say this escalation of patent applications and potential litigation has a negative impact. If, for example, the developers of the first electronic spreadsheet had pursued and been granted a patent, they could have blocked the development of rival software products.[13] This action could have hampered the industry's growth.

An optimist might contend that legal protection can promote an industry's growth. There's a financial incentive for developers to continue their work since they know their investments can be safeguarded and potentially rewarded.[14]

Consequently, new products can be introduced, license agreements can be made with other companies, and the influx of new ideas in the marketplace might lead to other products, to the industry's benefit. It is, after all, a balancing act between protecting intellectual property while ensuring that a dynamic industry continues to grow.

Y2K

For want of a nail the shoe is lost,
For want of a shoe the horse is lost,
For want of a horse the rider is lost,
For want of the rider the battle is lost,
For want of the battle the war is lost,
For want of the war the nation is lost,
All for the want of a horseshoe nail.

—George Herbert[15]

As the millennium approached, the global community was faced with a potential time bomb—the Y2K bug. Early computer programmers worked with hardware and storage space limitations. To conserve resources, programmers adopted, and established as a norm, a shorthand method of designating dates. Instead of writing 1981, or any other such date, only the last two digits—81—were used. But what would happen in the year 2000? A computer could read the double zero (00) as 1900, thus potentially disrupting computer-based operations. In an information society, the results could be catastrophic.

Concerns ranged from air traffic control failures to massive power outages to banks losing financial data (for example, the amount of money in your savings account). The problem was exacerbated by embedded technology—hardware components in elevators, telecommunications equipment, and other systems that also relied on dates for operation.

From a systems perspective, other key factors included:

1. An enormous volume of date-sensitive transactions help drive our information society. What would happen if they were disrupted?

2. Programmers worked to solve and forestall potential problems. This included searching through computer programs and replacing old Y2K-sensitive code with new code. Programmers in older languages (for example, COBOL) also became in demand for this operation.

3. Estimates to fix the global problem ranged from 1 to 2 trillion dollars.[16] In one example, instead of spending money on software, some companies experimented with "setting the clock back so the [computer] program will think it is 1972 instead of 2000; the dates in 1972 fall on the same days of the week as in 2000."[17] This would be a temporary fix.

4. Organizations had to identify and replace affected components (for embedded technology). Banks, power companies, and other organizations also ran tests to evaluate potential problems and recently implemented Y2K corrections.

5. It was believed that some countries were not Y2K ready. In a related scenario, various European countries were switching their financial institutions to the "Euro dollar" standard during this general time period, thus diverting valuable programming resources. What were the potential international trade and monetary implications?

6. Legal issues surfaced. The Securities and Exchange Commission, for one, issued a statement about public companies and their obligation to disclose potential problems.[18] Companies were also afforded protection "from lawsuits based on wrong information they provide, as long as the disclosure is neither reckless nor fraudulent."[19] But according to one writer, "Damage claims could range from financial compensation for death or injuries from defective medical

equipment to irretrievable data and loss of business income."[20] In essence, the new millennium also promised to be a new boon for lawyers.

7. The Y2K problem could be likened to a row of dominos. Even if your company was Y2K ready, what would happen if a key supplier was not? It is also like an EMP—companies and countries were threatened by electronic paralysis.

8. People responded to the possible disruptions of services in different ways. Some individuals ignored the matter and believed the problem would be solved. Still others took a middle-of-the-road approach. They believed there would be minor disruptions, much like the inconveniences caused by a hurricane or other storm.

A third group, though, looked back to the Cold War era and began to stockpile food, water, and other essential supplies. They believed services would be cut off, and unprepared national and global economies would be shattered.

The reality of the situation? The year 2000 dawned with only minor disruptions. In fact, part of this chapter was written shortly after the Y2K problems were supposed to hit. So for better or worse, the chapter was finished.

The smooth transition to the year 2000 also surprised a lot of people. As they watched New Year's parties televised from around the world, they partly expected to see the lights go out in different countries at the stoke of midnight. It did not happen.

This led to some charges that the potential problems were actually exaggerated. The media were also blamed for "hyping" the situation. It made good news. This was countered by others who believed the investment in time and money actually paid off. In some cases, it led to the modernization and enhancement of computer operations, thus paying long-term dividends for this investment.[21]

Regardless of the viewpoint, it is interesting to note that all this chaos was caused by the placement of two numbers—seemingly insignificant—but with monumental, cascading implications. Much like the nail in the poem that opened this section.

ARTIFICIAL INTELLIGENCE

This section of the chapter examines artificial intelligence (AI), another computer area relevant to our discussion of information and communications technologies.

The AI field is dedicated, in part, to developing computer-based systems that seemingly duplicate the most important of human traits, the ability to think or reason. This section serves as a brief introduction to the field and related topics.[22] More detailed information can be found at the end of the chapter in the Suggested Readings section.

Natural Language Processing

Natural language processing can simplify the communication between humans and computers. For example, you may not have to learn specific command sequences because the computer is able to "understand" you. For a database program, a query for information can be phrased in an ordinary sentence.[23]

This type of interface could be valuable for inexperienced users. It could also be extended to other programs to simplify operations and to make computer technology more accessible and transparent.

Speech Recognition

Speech recognition, a subset of natural language processing, allows a computer to recognize human speech or words. You verbally instruct the computer to perform an operation instead of inputting the instructions with a keyboard. The system will initially digitize your voice, and then it must recognize various words before the instructions are carried out.

For example, while working with a microscope, you could issue a command to a computer. Or instead of using a keyboard with your favorite software, simply speak. This same capability could also be incorporated into consumer electronic products and other devices. Regardless of the application, this type of operation can simplify your work. It is also a very natural interface because it employs one of our most common communication tools, speech.

To complete the circle, the computer could answer you. Sophisticated systems have been developed at the Massachusetts Institute of Technology's Media Lab, one of the leading institutions in speech research.[24] At the PC level, a speech synthesis card can be used with special software, enabling a blind PC user to, for example, type on a keyboard, and have the words subsequently vocalized. A program may even be compatible with a GUI whereby menu selections are vocalized and on-screen icons are described.[25]

This capability can be linked with an optical character recognition (OCR) system, as examined in the desktop publishing chapter, Chapter 12. An OCR configuration can recognize text from printed documents. The information can subsequently be reproduced in computer-generated speech, making the material available to visually impaired individuals.[26]

When this operation is viewed with the entire spectrum of speech research, it is an important achievement. Even at this developmental stage, our productivity is boosted, the human–computer interface becomes increasingly transparent, and working with computers turns into a natural process. These tools can also help individuals to communicate their thoughts and ideas in a manner that would have been impossible to achieve a few short years ago.

Expert Systems

An expert system is a computer-based advisor. It is a computer program that can help in the medical, manufacturing, and other fields.

The heart of an expert system is knowledge derived from human experts and includes "rules of thumb," or the experts' real-world experiences. Additional information sources could encompass books and other documents.

This expertise can then be retrieved during a consultation. The computer leads you by posing different questions. Based on your answers, in conjunction with a series of internal rules, essentially the stored knowledge, the computer "reaches" a decision.[27]

In one application, an expert system was designed to help service bureaus. The goal? To better define potential target audiences for their clients, companies primarily engaged in direct marketing sales through catalog and mail-order transactions.[28]

In another application, the Foundation Bergonie, a French research center, developed a system that would "help doctors in general hospitals in the management of breast cancer patients."[29] The expert system provided doctors with a level of advice that would not normally have been available.

In other examples, expert systems have been used to promote aquaculture as well as by credit card companies and the government. Consequently, expert systems have emerged as valuable information tools. They can

- Help fill an information gap,
- Support a field where there may be too few human experts, and
- Preserve, in a sense, the knowledge and experience that would otherwise be lost when a human expert dies.

Figure 4.2

A screen shot of an expert system. The window on the left shows the code. The window on the right shows one element of this expert system in action: a question and list of possible answers. (Screen shot by permission of Knowledge Garden, Inc.; KnowledgePro Gold for Windows.)

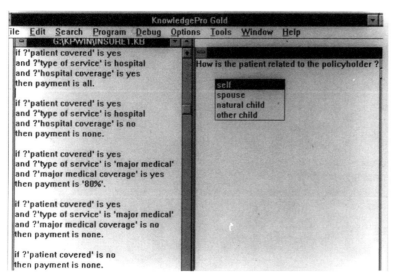

Yet despite their advantages, expert systems are not infallible. Their capabilities are limited by the quality of the information, the rules that govern the system, and by other criteria.

If a novel situation is encountered, an expert system's "help" may also be limited. A human expert, in contrast, could more readily adapt to a new set of conditions.

There's also the problem of working with a machine. A human expert is typically better equipped to ask the right questions to define a client's situation. Thus, the human expert may arrive at a superior solution.

Another potential trouble area has been litigation. If a medical expert system made a mistake, who would be liable? This question, the high cost for liability insurance, and other factors, have held up product releases.[30] Other fields could be similarly affected, and in an article devoted to this topic, two pages outlined potential risk areas for expert system developers and designers.[31]

Finally, expert systems have been joined by a related field, neural networks. In brief, "neural network technology, which attempts to simulate electronically the way the brain processes information through its network of interconnecting neurons, is used to solve tasks that have stymied traditional computing approaches."[32] A network can, for example, be trained: It learns by example. A neural network can also work with incomplete data, much like humans, but unlike traditional computer systems.[33] These characteristics make a neural network well suited for OCR because it could recognize characters that were degraded and not perfectly defined.[34]

Neural networks have been trained to adjust a telescope to improve its performance, to handicap horse races, to make stock market predictions, and to recognize a face, even if the expression changes.[35] In keeping with the communications industry, research has also been conducted to tap a neural network as part of a sophisticated tool for measuring television audiences.

Computer Vision

Computer vision can be described as the field in which a picture, produced by a camera, is digitized and analyzed by a computer. In this setup, the camera and computer serve as machine equivalents of the human eye and brain, respectively.

Figure 4.3

Demeter. Demonstrates the "automation of terrestrial agricultural operations through the application of space robotics technology." (Courtesy of NASA/Space Telerobotics Program.)

During this operation, objects in a scene can be identified through template matching and other procedures.[36] In template matching, stored representations are compared to the objects in the image. The computer identifies the various objects when the templates match.

In one application, circuit boards can be examined for defects. If such a board is identified, it could be removed from the assembly line.[37]

Various technologies have also been harnessed to develop autonomous robotic vehicles. An autonomous vehicle would also process visual information, and in operation, could function without direct human intervention.

An example of an ambitious project is an autonomous rover for space exploration. This vehicle, equipped with a robotic vision system, would be activated on its arrival on Mars or some other world. It would explore the planetary body's surface, conduct experiments, and relay pictures back to Earth.

The vision system would provide the onboard computer with information about the surrounding terrain. Based on an analysis and interpretation of the information, the computer could identify and avoid a potentially dangerous obstacle (for example, a crater) without waiting for human intervention.[38]

This vehicle would be born from the union of hardware components and a sophisticated AI program that could process picture information. The integration of computer software and hardware may also serve as a model for vehicles built for use on Earth. These include experimental vehicles that already exist and those that may be placed in production.

Philosophical Implications

I.   A robot may not injure a human being, or through inaction, allow a human being to come to harm.

II. A robot must obey the orders given it by human beings, except where such orders would conflict with the first law.

III. A robot must protect its own existence, as long as such protection does not conflict with the first or second law.

—Isaac Asimov, *The Robots of Dawn*[39]

The development and integration of AI systems in society has philosophical implications. Some AI opponents believe, for example, that the new generation of intelligent machines will eliminate millions of jobs. There is also the fear that AI systems, including robots, diminish us as humans. Other AI-based systems have also raised the specter of machines wreaking havoc, as presented in *The Terminator, Creation of the Humanoids, Colossus: The Forbin Project*, and other movies. Although these concerns may be legitimate, AI proponents have offered counterarguments:

1. The tools of the AI field are just that, tools. An autonomous vehicle will not, for example, necessarily replace the human exploration of the planets, just as expert systems did not replace doctors.

Figure 4.4

A screen shot from the ExperTest(R) general-purpose tester. The ExperTest has been used as a test development bed for PC motherboards, among other applications. By tapping artificial intelligence technology, the system can deliver an enhanced diagnostic capability, much like a human expert. (Courtesy of Array Analysis.)

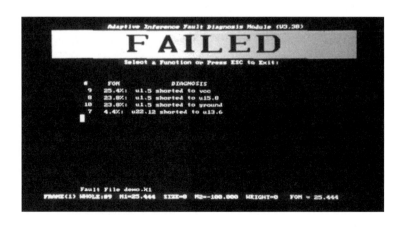

2.   Although AI systems could lead to mass job displacement, it could be argued that the technologies are not at fault. Rather, we have failed to cushion the impact of this development through job training programs and other actions. Expert systems and other AI products are also generally designed to complement and not replace humans.

3.   Instead of diminishing our humanity, AI systems may enhance our lives. Cancer patients have benefited, and speech recognition and synthesis systems can provide individuals with better control over their environments.

4.   The last concern was partly addressed by Isaac Asimov in his "Three Laws of Robotics" quoted earlier. The laws cut across scientific and science fiction boundaries and could help guide us as we start implementing advanced AI products.

To wrap up this discussion, it is also important to remember that while the arguments presented by both camps may have some merit, we are ultimately responsible for AI systems and their impact in our lives. If used correctly, AI and other technologies may continue to enhance the human exploration of our own world, of other worlds, and, just as important, the exploration of the human condition.

REFERENCES/NOTES

1.   This includes using a special wrist support.

2.   "Information Sheet on the Michelangelo Virus," compiled by J. M. Allen Creations/ Michael A. Hotz, February 17, 1992.

3.   Gary M. Hoffman, *Curbing International Piracy of Intellectual Property* (Washington, D.C.: The Annenberg Washington Program, 1989), 7. *Note:* The panel was not solely concerned with software issues.

4.   Ibid, 10.

5.   Without the manual, you would not have the word and the program would not run. Sim-City, a computer simulation program, used a creative variation. If you didn't respond correctly, the game played, but your city was hit by natural disasters, including earthquakes that shook the on-screen landscape.

6.   Borland International, *Turbo Prolog Reference Guide* (Scotts Valley, Calif.: Borland International, 1988), C2.

7.   See pp. 19–24 of Hoffman, *Curbing International Piracy,* for a detailed discussion of the panel's recommendations.

8.   For more details, see John V. Pavlik, *New Media Technology* (Needham Heights, Mass.: Allyn & Bacon, 1996), 288–290.

9.   Steve Gibson, "U.S. Patent Office's Softening Opens Floodgates for Lawsuits," *InfoWorld* 14 (August 17, 1992): 36.

10.   Thanks to A. Jason Mirabito, a partner in the patent law firm of Wolf, Greenfield, and Sacks, for his suggestions for this section.

11.   Beth Freedman, "Look-and-Feel Lawsuit Expected to Go to Trial," *PC Week* 9 (February 24, 1992): 168. *Note:* The case also involved the Hewlett-Packard Company and its NewWave product.

12.   Jane Morrissey, "Ruling Dashes Apple's Interface Hopes," *PC Week* 9 (August 17, 1992): 117.

13.   Brian Kahin, "Software Patents: Franchising the Information Infrastructure," *Change* 21 (May/June 1989): 24. This is a sidebar in Steven W. Gilbert and Peter Lyman, "Intellectual Property in the Information Age," *Change* 21 (May/June 1989): 23–28.

14.   "Patents: Protecting Intellectual Property," *OE Reports* 95 (November 1991): 1. *Note:* This interview with a patent lawyer provides a good overview of patent law and what can and cannot be patented.

15.   Taken from *For Want of a Nail* by Robert Sobel (London: Greenhill Books, 1997), foreword.

16.   Barnaby J. Feder and Andrew Pollack, "Computers and 2000: Race for Security," *New York Times* CXlVIII (December 27, 1998): 22.

17.   Ibid.

18.   Jeff Jinnett, "Legal Issues Concerning the Year 2000 Computer Problem: An Awareness Article of the Private Sector," downloaded March 16, 1999.

19.   Feder and Pollack, "Computers and 2000," 23.

20.   Elizabeth McCarthy, "Short Fuse," *Sacramento Business Journal,* dated January 4, 1999, downloaded March 16, 1999.

21.   Mel Duvall, "Y2K Payoff: Systems Poised for New Projects," *Inter@ctive Week* 7 (January 10, 2000): 8.

22.   For an introduction to AI's important elements, see Barry A. McConnell and Nancy J.

McConnell, "A Starter's Guide to Artificial Intelligence," *Collegiate Microcomputer* 6 (August 1988): 243; and John Gilmore, "Artificial Intelligence in the Modern World," *OE Reports* (May 1987): 4A.

23. The Q&A program used such an interface. It also had the capability to add words to its vocabulary.

24. Stewart Brand, *The Media Lab* (New York: Penguin Books, 1988), 51

25. Richard S. Schwerdtfeger, "Making the GUI Talk," *Byte* 16 (December 1991): 118. *Note:* Even though speech synthesis may not be considered an AI element, it is included here for organizational purposes.

26. Joseph J. Lazzaro, "Opening Doors for the Disabled," *Byte* 15 (August 1990): 258. *Note:* The article also provides an excellent overview of PC-based systems for the blind, deaf, and motor disabled.

27. The representation of knowledge by rules is only one expert system development tool. An expert system can also be created with a conventional programming language. A rule can take the form of:

> IF the car doesn't start AND
> the lights do not turn on
> THEN the battery needs charging.

This example is very simplistic, and in a real-world situation, multiple rules would be employed.

28. Persoft Inc., "Separating Fact from Fiction about More/2," News from Persoft, press release.

29. Texas Instruments, "French Expert System Aids in Cancer Treatment," Personal Consultant Series Applications, product information release.

30. Edward Warner, "Expert Systems and the Law," *High Technology Business* (October 1988): 32.

31. G. Steven Tuthill, "Legal Liabilities and Expert Systems," *AI Expert* 6 (March 1991): 46–47.

32. Michael G. Buffa, "Neural Network Technology Comes to Imaging," *Advanced Imaging* 3 (November 1988): 47.

33. Gary Entsminger, "Neural-Networking Creativity," *AI Expert* 6 (May 1991): 19.

34. Larry Schmitt, "Neural Networks for OCR," *Photonics Spectra* 25 (October 1990): 114.

35. Maureen Caudill, *Neural Networks Primer* (San Francisco: Miller Freeman Publications, 1989), 4. See also Andrew Stevenson, "Bookshelf," *PC AI* 7 (March/April 1993): 30, 38, 57, 58, for an in-depth review of *In Our Own Image,* by Maureen Caudill (Oxford: Oxford University Press, 1992), a book relevant to this overall discussion.

36. Louis E. Frenzel, Jr., *Crash Course in Artificial Intelligence and Expert Systems* (Indianapolis, Ind.: Howard W. Sams & Co., 1987), 201.

37. Ibid, 205.

38. A neural network could be very appropriate in this type of situation, especially since it could be trained to avoid numerous obstacles and deal with novel situations.

39. Isaac Asimov, *The Robots of Dawn* (New York: Doubleday and Company, Inc., 1983), back cover.

SUGGESTED READINGS

Benfer, Robert A., and Louanna Furbee. "Knowledge Acquisition in the Peruvian Andes." *AI Expert* 6 (November 1991): 22–27; Keyes, Jessica. "AI in the Big Six." *AI Expert* 5 (May 1990): 37–42; Shafer, Dan. "Ask the Expert." *PC AI* 3 (November/December 1989): 40, 49. The first two articles examine different expert system applications; the third examines some of the differences between expert systems and neural networks.

Blackwell, Mike, and Susan Verrecchia. "Mobile Robot." *Advanced Imaging* 2 (November 1987): A18–A21; Caudill, Maureen. "Driving Solo." *AI Expert* 6 (September 1991): 26–30. A look at autonomous vehicles.

The January 10, 2000, issue of *Inter@ctive Week* carried a series of Y2K articles and various implications. Two include Brown, Doug. "Y2K Overseas: What Went Right," 9; and Babcock, Charles. "Spending on Y2K: Waste of Time or Prudent Investment," 8–9.

Cleveland, Harlan. "How Can 'Intellectual Property' Be Protected." *Change* 21 (May/June 1989): 10–11; Fisher, Francis Drummer. "The Electronic Lumberyard and Builders' Rights." *Change* 21 (May/June 1989): 13–21; Hoffman, Gary M. *Curbing International Piracy of Intellectual Property.* Washington, D.C.: The Annenberg Washington Program, 1989; Middleton, Kent R., and Bill F. Chamberlin. "Intellectual Property," in *The Law of Public Communication.* New York: Longman Publishing Group, 1991, 240–277. Earlier looks about intellectual property.

Dyson, Esther. "Intellectual Property on the Net." Downloaded from EFF web site, November 1998; "The Digital Millennium Copyright Act." The UCLA Online Institute for Cyberspace Law and Policy; Samuelson, Pamela. "Digital Media and the Law (Legally Speaking)." *Communications of the ACM* 34 (October 1991). Intellectual property issues. The Digital Millennium Copyright Act is covered in a later chapter.

Frenzel, Louis E., Jr. *Crash Course in Artificial Intelligence and Expert Systems.* Indianapolis, Ind.: Howard W. Sams & Co., 1987. An guide to AI technology and applications.

Kavoussi, Dr. Louis R. "NY Doctors Use Robotics in Baltimore Surgery." *Communications Industries Report* 15 (October 1998): 23. An example of using robotics in telemedicine; that is, from a distance.

Markowitz, Judith. "Talking to Machines." *Byte* 20 (December 1995): 97–104. Speech recognition technology/products.

NASA. "Remote Agent Experiment." Downloaded from NASA web site May 1999; Rhea, John. "Report from Washington and Elsewhere." *Military & Aerospace Electronics* 9 (October 1998): 8, 19. AI and space applications.

Public Service Center, U.S. Patents and Trademark Office. "Patents in Brief." Information sheet, February 1991. An overview of patents and filing procedures.

Smith, J. C. "The Charles Green Lecture: Machine Intelligence and Legal Reasoning." *Chicago-Kent Law Review* 73 (1998). Downloaded from Lexis. An interesting study that explores the possible ties between expert systems and the legal field.

Stanley, Jeannette, and Sylvia Luedeking, eds. *Introduction to Neural Networks.* Sierra Madre, Calif.: California Scientific Software, 1989. An easy-to-understand introduction to neural network systems (accompanied software).

GLOSSARY

Artificial Intelligence (AI): The field dedicated, in part, to developing machines that can seemingly think.

Computer Virus: A program that "attaches" itself to other programs. It may display a harmless message or cause data to be lost.

Computer Vision: The field that duplicates human vision through a computer and a video camera system.

Copyright and Patent: A copyright provides protection in the expression of a concept. A patent protects the underlying concept.

Ergonomic Design: The philosophy of developing equipment and systems around people, making equipment conform to an operator's needs, and not the other way around.

Expert Systems and Neural Networks: An expert system manipulates knowledge and can serve as an in-house expert in a specific field. It combines "book information" with the knowledge and experience of human experts. A neural network, for its part, simulates the way the brain processes information through its network of interconnecting neurons.

Natural Language Processing: Natural language processing focuses on simplifying the human-computer interface. Rather than using special keywords to initiate computer functions, you use conventional word sequences.

Optical Character Recognition (OCR): After a document is scanned, an OCR system can recognize the text.

Software Piracy: The illegal copying and distribution of software.

Speech Recognition: A subset of natural language processing, it allows a computer to recognize human speech or words.

Y2K: A potential, adverse computer problem caused by using shorthand notation to represent a year (e.g., 98 for 1998) in computer programming. Massive financial and personnel time investments helped rectify the situation.

5

The Magic Light: Fiber-Optic Systems

The concept of harnessing light as a modern communications tool stretches back to the late nineteenth century. Alexander Graham Bell, the inventor of the telephone, patented an invention in 1880 that used light to transmit sound.[1] Bell's invention, the photophone, employed sunlight and a special light-sensitive device in the receiver to relay and subsequently reproduce a human voice.

It wasn't until the twentieth century, though, that the idea for such a tool was transformed into a practical communications system. Two developments helped bring about this advance: the perfection of the laser and the manufacturing of hair-thin glass lines called fiber-optic (FO) lines. When combined, they helped create a lightwave communications system, a system in which modulated beams of light are used to carry or transmit information.

LIGHT-EMITTING DIODES AND LASER DIODES

FO technology has contributed to the development of high-capacity communications lines. A transmission is conducted with optical energy, beams of light, produced by a transmitter equipped with either a light-emitting diode (LED) or a laser diode (LD).

The light is then confined to and carried by a highly pure glass fiber.

Both LEDs and LDs are used in different communications configurations. An LED is less expensive and has generally supported lower volume, short-distance relays. An LD, like an LED, is a semiconductor, but in the form of a laser on a chip. It is a small, powerful, and rugged semiconductor laser that is well suited for high volume and medium- to long-distance relays. The LD family has also served optical storage systems.

THE FIBER-OPTIC TRANSMISSION

In an FO transmission, a beam of light, an optical signal, serves as the information-carrying vehicle. Both analog and digital information are supported.

In operation, the light is launched or fed into the fiber. The fiber itself is composed of two layers, the cladding and the core. Due to their different physical properties, light can travel down the fiber by a process called total internal reflection. In essence, the light travels through the fiber via a series of reflections that take place where the cladding and core meet, the cladding-core interface. When the light reaches the end of the line, it is picked

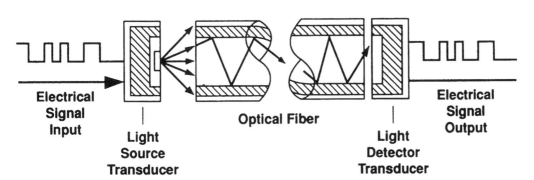

Figure 5.1

How a fiber optic system works. The electrical signal is converted into an optical signal, and back, following the relay. (Courtesy of Corning Inc.)

up by a light-sensitive receiver, and after a series of steps, the original signal is reproduced.

To sum up, a video camera's output or other such signal is converted into an optical signal in a FO system. It is subsequently transmitted down the line and converted back following its reception.

Finally, the fiber, which may be made out of plastic in short-distance runs, is covered by a protective layer or jacket. This layer insulates the fiber from sharp objects and other hazards, and it can range from a light protective coating to an armored surface designed for military applications. The now-protected fiber is called an FO cable, and it consists of one or more fiber strands within the single cable enclosure.

Advantages

An FO line has distinct advantages as a communications channel:

1.   In comparison with some other communications systems, light—the information-carrying vehicle in an FO system—can accommodate an enormous volume of information. Transmissions are in the gigabit-plus range (billions of bits per second), and a single 0.75-inch fiber cable can replace 20 conventional 3.5-inch coaxial cables.[2]

2.   FO lines are immune to electromagnetic and radio interference. Because light is used to convey the information, adjacent communications lines cannot adversely affect the transmission. An FO line can also be installed in a potentially explosive environment where gas fumes may build up over time.

3.   An FO line offers a higher degree of data security than conventional systems. It is more difficult to tap, and FO lines do not radiate. A signal cannot be picked up by instruments unless the line is physically compromised.[3]

4.   Information can be relayed a great distance without repeaters. A new generation of LDs and complementary fiber, as well as sensitive receivers, has made it possible to relay signals more than 100 kilometers without repeaters. In practical terms, if fewer repeaters are used, building and maintenance costs are reduced.

5.   An FO line is a valuable asset where space is at a premium, such as a building's duct space for carrying cable. Because fiber is comparatively narrow, it can usually fit in a space that may preclude the use of conventional cable.

Disadvantages

Some factors do, however, limit an FO line's effectiveness. Like other communications systems, there may be a loss of signal strength, which in this case, may be caused by physical and material properties and impurities.[4] In another example, dispersion can affect the volume of information a line can accommodate in a given time period, that is, its channel capacity.

The latter can be addressed by using single-mode rather than multimode fiber. The single-mode fiber is constructed with a very narrow core, and the light essentially travels straight down the fiber in a single path, thus preventing the smearing of pulses that comprise the optical signal.[5] A single-mode fiber can also accommodate a high information rate and has been the backbone of the telephone industry and other high-capacity long-distance systems.

Fiber is also harder to splice than conventional lines, and the fiber ends must be accurately mated to ensure a clean transmission. But special connectors, among other devices and techniques, have facilitated this process.

Another element is the current state of the communications industry. Despite the inroads made by FO technology, the overall system is still dominated, in certain settings, by a copper standard. This includes the local telephone industry.

FO components can also be more expensive. An FO cable hookup, for example, may be more expensive than a conventional one. The actual FO line may likewise cost more in specific applications, even though this price differential has diminished.

Thus, FO technology is somewhat analogous to digital technology. It has been integrated in the current communications structure.

The prices for LDs and other components are also dropping, and the growing information torrent our communications systems must handle may mandate the use of such high-capacity channels. They may be joined by fiber/coaxial hybrid systems. Note also that by using compression techniques, the traditional copper plant is still very much alive and well.

APPLICATIONS

An FO line is an attractive medium for the video production industry. Its lightweight design and transmission characteristics can make it a valuable in-field production tool. Several hundred feet of line, for instance, may weigh only several pounds, and a camera/FO combination can be used at a greater distance from a remote van than a conventional configuration.[6]

Fiber's information capacity also makes it an ideal candidate for an all-digital television studio. The digitization of broadcast-quality signals can generate millions of bits per second, and an FO system could handle this information volume. The line would also fit in a studio's duct space, which is usually crammed with other cables, and its freedom from electromagnetic and radio-frequency interference would be valuable assets in this environment.

Besides these roles, FO systems have been used to create efficient and high-capacity LANs. Businesses, hospitals, schools, and other organizations are also designing and implementing their own FO lines that can accommodate computer data as well as voice and video.

The telephone and cable industries also have a vital interest in this technology. Long-distance carriers have developed extensive FO networks, and eventually, fiber may be extended to all our homes. Cable companies, for their part, are converting their systems to fiber or, as covered in the next chapter, fiber/copper hybrid systems.

Fiber can also provide us with high-speed data highways. As we generate not only more information, but more complex information with each passing year, the communications infrastructure may be hard pressed to handle this data flow. But FO technology will continue to play a key role in solving this problem. It can also support an array of new television and information options.

Underwater Lines

Beyond landline configurations, AT&T headed an international consortium that developed the first transatlantic undersea FO link between the United States and Europe. The system, TAT-8, is more than 3000 nautical miles in length and was the first transoceanic FO cable. TAT-8 was designed to handle a mix of information. When inaugurated, it had an estimated lifetime in excess of 20 years.[7] Even though fiber had already been used in long-distance land and short-distance undersea operations, TAT-8 was the first of a new class of cables. Its installation was preceded by extensive deep-water experiments and trials conducted in the early 1980s to demonstrate the project's feasibility. Once completed, the findings confirmed the designers' expectations: The FO cable and repeater relayed the information within acceptable error rates and transmission losses. This held true even when the system was subjected to the ocean's cold temperature, tremendous water pressure, and other extreme environmental conditions. The cable also survived the physical strain imposed by the laying and recovery operations.

Like other new applications, though, FO lines have had some problems. Certain TAT-8 users, for instance, experienced some outages from cable damage caused by fishing trawlers.[8]

FO Lines and Satellites

At first glance, undersea cables and long-distance landlines may appear to be obsolete in light of satellite communication. Yet an FO system has a wide channel capacity, an extended lifetime, and can be cost effective for long-distance applications.

An FO line also has certain other advantages. A satellite transmission may be susceptible to atmospheric conditions, and a slight time delay is inherent in satellite traffic.[9] An FO link is not subject to these constraints.

Fiber may also be a more secure conduit for sensitive material. A satellite's broad transmission area makes it possible for unauthorized individuals to receive its signal. Although it could be encrypted for protection, the encryption scheme could be broken.

Fiber is also well suited for intracity relays and point-to-point communication (e.g.,

Figure 5.2
A medical application of fiber optic technology. (Courtesy of KMI Corp., Newport, Rhode Island, USA.)

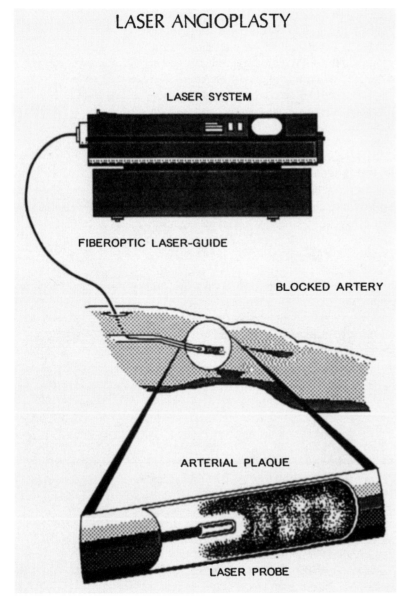

New York to Washington, D.C.). It has been used for television network news, by the telephone industry, and in teleconferencing applications. Various processing techniques have also enhanced an FO system's relay.

A satellite, for its part, can support a flexible point-to-multipoint operation that can readily accommodate additional receiving sites. This may not be the case with an FO system where a special line may have to be laid to reach the new location.

Thus, FO lines and satellites actually supplement and complement each other on national and international levels. They also have their own particular strengths, and both will continue to support our communications needs.

Other Applications

FO technology has also been adopted for other applications. In the medical field, an FO line can be used in certain types of laser surgery. The fiber can act as a vehicle to transport the intense beam of light to the operating site. In a related application, fibers serve as a visual inspection tool in the form of a fiberscope. This device "consists of two bundles of optical fibers. One, the illuminating bundle, carries light to the tissues, and the other, the imaging bundle, transmits the image to the observer."[10] Doctors can use the device to peer inside the human body.

Scientists have also experimented with and have used FO sensors to monitor the physical condition of different structures. In one application, fibers have been embedded in composite materials, which can be used to create items ranging from airplane parts to, potentially, a space station. Through testing, it would be possible to determine if a structure was subject to damage and undue stress.[11]

Although the fiber may not be relaying computer data, it is still providing observers with information. As discussed in the book's opening, this is one of the key features of the communication revolution. Information will take new forms and shapes, and in this setting, the data about a structure's physical condition does, in fact, convey an intelligence.

All the applications described thus far only touch on FO technology's different roles. It has even been tapped by the aerospace industry to supplement and to potentially replace fly-by-wire systems with fly-by-light configurations. In this set-up, conventional, heavier wiring would be replaced by fiber for command-and-control operations. The fiber would save weight, and the freedom from electromagnetic interference would lend itself to this environment.[12] The automobile industry has also explored this option.

CONCLUSION

FO lines have had an impact on our communications system. On one front, they have already supported sophisticated national and international telephone connections. On another, they promise to change the way we view television and, potentially, the way we receive and use information.

The growth of FO lines can also be viewed as an evolutionary rather than a revolutionary development. For example, the infrastructure is modernized as FO lines and components are integrated, but the current system is not immediately abandoned. This follows the pattern of digital and other technologies and applications.

In a related example, transoceanic FO cables may be complemented and supplemented by global systems. In Project Oxygen, for instance, the design called for a global system that would "have a maximum throughput of 2.56 terabits (2560 gigabits) per second on [its] undersea cable segments."[13] Besides this high-capacity capability, the system would support a sophisticated network design that would provide users with enhanced connection options.

FO capabilities will also improve. One driving force is the marriage between optical and electronic components (optoelectronics) and the use of light alone. In one example, optical amplifiers can improve an FO system's performance. The signal could remain in the optical domain for longer distances, thus reducing the number of regenerative repeaters.[14]

A similar development is taking place in the switching field, where an optical configuration would be advantageous.[15] In one application, data traffic may be managed by microelectromechanical systems (MEMS). One such system could consist of "tiny moving mirrors [which] are used to deflect beams of light" to different fibers. Mounted on an integrated circuit, an individual mirror may be "smaller than the diameter of a human hair."[16] In a related application, light may also be used to develop newer, faster computers.

In sum, these developments are exciting— and we have only just begun to tap light's potential as a communications tool. Its role in our information society cannot and should not be underestimated.

REFERENCES/NOTES

1. Richard S. Shuford, "An Introduction to Fiber Optics," Byte 9 (December 1984): 121.

2. Arthur Parsons, "Why Light Pulses Are Replacing Electrical Pulses in Creating Higher-Speed Transmission Systems," Communications News (August 1987): 25.

3. "Optical Fiber Technology: Providing Solutions to Military Requirements," Guidelines (1991): 6.

4. The latter factor, though, has been greatly reduced.

5. Ronald Ohlhaber and David Watson, "Fiber Optic Technology and Applications," Electronic Imaging 2 (August 1983): 29.

6. Richard Cerny, "Fiber-Optic Systems Improve Broadcast ENG/EFP Results," Communications News (April 1983): 60.

7. R. E. Wagner, S. M. Abbott, R. F. Gleason, et al., "Lightwave Undersea Cable System," reprint of a paper presented at the IEEE International Conference on Communications, June 13–17, 1982, Philadelphia, Penn., 7D.6.5.

8. Barton Crockett, "Problems Plaguing Undersea Fiber Links Raise Concern Among Users," Network World 20 (August 21, 1989): 5.

9. As described in Chapter 7, a communications satellite is typically placed in an orbit some 22,000 miles above the earth. The round-trip to and from the satellite results in the slight time delay.

10. Abraham Katzir, "Optical Fibers in Medicine," Scientific American (May 1989): 120. Note: A fiberscope may also be incorporated in an endoscope, which provides physicians with remote access to the body regions under observation. The article presents an excellent overview of this field.

11. Richard Mack, "Fiber Sensors Provide Key for Monitoring Stresses in Composite Materials," Laser Focus/Electro-Optics 23 (May 1987): 122.

12. Luis Figueroa, C. S. Hong, Glen E. Miller, et al., "Photonics Technology for Aerospace Applications," Photonics Spectra 25 (July 1991): 117.

13. "Frequently Asked Questions About Project OXYGEN™; downloaded from Project Oxygen's web site February 2000.

14. Robert G. Winch, Telecommunication Transmission Systems (New York: McGraw-Hill, 1998), 456.

15. Paul R. Prucnal and Philippe A. Perrier, "Self-Routing Photonic Switching with Optically Processed Control," Optical Engineering 29 (March 1990): 181.

16. Robert Pease, "Microscopic Mirrors May Manage Future Optical Networks," Lightwave 16 (May 1999): 1.

SUGGESTED READINGS

Belanger, Alain. "Broadband Video Goes Mainstream." Lightwave 15 (November 1998): 49–52, 56; Cerny, Richard. "Using Fiber in the Field." Broadcast Engineering 38 (January 1996): 52–60; 64. Coverage of fiber-based video applications; the first article also examines fiber systems in other countries; the second article also covers video/field production applications.

Branst, Lee. "Optical Switches are Coming—But When." Lightwave 15 (July 1998): 63–64, 68; Hecht, Jeff. "Optical Switching Promises Cure for Telecommunications Logjam." Laser Focus World 34 (September 1998): 69–72. Optical switching developments.

Chinnock, Chris. "Miniature Fiber Endoscopes Offer a Clear View Inside the Body." Laser Focus World 31 (August 1995): 91–94; Katzir, Abraham. "Optical Fibers in Medical Applications." Laser Focus/Electro Optics 22 (May 1986): 94–110. Fiber-optic technology and medical applications.

Corning Incorporated, Opto-Electronics Group. Just the Facts. Corning, N.Y.: Corning, Inc., 1992; Higgins, Thomas V. "Light Speeds Communications." Laser Focus World 31 (August 1995): 67–74; Snell, Geoff. "An Introduction to Fiber Optics and Broadcasting." SMPTE Journal 105 (January 1996): 4–7. Excellent overviews of FO technology and applications.

Heath Company. Heathkit Educational Systems—Fiber Optics. Benton Harbor, Mich.: Heath Company, 1986. A hands-on introduction to fiber-optic technology and systems.

Hecht, Jeff. "Optical Computers." High Technology 7 (February 1987): 44–49; "Optical Computer: Is Concept Becoming Reality?" OE Reports 75 (March 1990): 1–2; Silvernail, Lauren P. "Optical Computing: Does Its Promise Justify the Present Hype?" Photonics Spectra 24 (September 1990): 127–129. A look at optical computing.

Hutchins, Don, and S. J. Campanella. "Is Fiber a Threat to Satellite-Based Nets?" Network World 4 (March 23, 1987): 26–27. A fiber versus satellite debate.

Keller, John. "ARPA to Fund Fly-by-Light." Military & Aerospace Electronics 5 (September 1994): 1, 10; Thompson, Dan and Carlos Bedoya. "Optical Fiber Finally Takes Off." Photonics Spectra 29 (April 1995): 82–90. Newer, fly-by-light systems (aerospace uses).

Kotelly, George. "OFC '96 Assesses Latest Fiber-Optic Technology, Components and Networks." Lightwave (February 1996): 6–14, 28. A comprehensive look at optical developments, including optical processing.

Marshall, Larry R. "Blue-Green Lasers Plumb the Mysteries of the Deep." Laser Focus World 29 (April 1993): 185–197; Tozer, Bryan A. "Telecommunications Systems Minimize Laser Risks." Photonics Spectra 33 (February 1999): 124–126. Interesting looks at lasers and underwater imaging (for example, active imaging and imaging classes) and laser safety issues, respectively.

Moore, Emery L., and Ramon P. De Paula. "Inertial Sensing." Advanced Imaging 2 (November 1987): A48–A50; "Fiber Optics for Astronomy." Sky and Telescope (December 1989): 569–570. Two nonmainstream FO applications.

Pease, Robert. "Project Oxygen Poised for Construction Under New Strategy." Lightwave 15 (July 1998): 1, 28, 31. The development of a world-spanning fiber optic system.

GLOSSARY

Fiber-Optic Line (FO): A hair-thin glass fiber that can handle information ranging from voice to video. An FO line has a wide channel capacity and a superior transmission capability.

Laser Diode (LD): One of the more versatile communications tools. As an FO system's light source, LDs have supported high-speed and long-distance relays.

Light-Emitting Diode (LED): Another FO system's light source. LEDs have generally supported short- and medium-haul relays.

Single-Mode Fiber: An efficient and high-speed, high-capacity FO line.

6 The Cable and Telephone Industries and Your Home

The fiber-optic lines described in the last chapter may serve as the backbone of digital-based entertainment and information systems. They could feature interactive and conventional television programming, stock market quotes, and additional services. These offerings could be supported by cable, telephone (telcos), and possibly satellite companies.

VIDEO-ON-DEMAND AND VIDEO DIALTONE

A goal of the communications industry is to provide viewers with more control over programming choices, whether it is a television show or a new optional service. Two terms have been associated with this development, video-on-demand (VOD) and video dialtone (VDT).[1]

In brief, VOD, which comes in different flavors that sport different features, can be thought of as enhanced pay-per-view (PPV). With PPV, you are charged a set fee to view a movie, concert, or special event.

VOD expands this capability by supporting a more diverse and tailored programming pool. In one configuration, VOD could function much like a video rental store. You select the programming you want to see when you want to see it. There could also be a library of movies, old television series, and other programming. Interactive controls, which would function much like those on a conventional VCR, could also be featured.[2]

VDT, on the other hand, was initially designed to allow telcos to compete in the video marketplace. Options included VOD programming.

Since the 1980s, the broadcasting and cable industries have not relished the idea of the telcos competing in their backyard. At that time, the telcos were blocked from participating in these and related fields.

As part of the ongoing evaluation of the consent decree that led to the divestiture of AT&T, Judge Harold Greene, the architect of this decision, barred the seven regional holding companies from offering various information services. The companies were created after the breakup of AT&T, and the restrictions were adopted to prevent the telephone industry from dominating the new, emerging information infrastructure.

But political pressures and other factors led to a relaxation of the original decision in the late 1980s and early 1990s. This opened the door for the telcos to support information services, such as electronic Yellow Pages and, potentially, to enter the video market.[3]

The next major step was VDT. In 1992, the FCC modified its rules to allow telephone companies to compete in the video arena. One provision called for the support of a "basic platform that will deliver video programming and potentially other services to end-users" on a common carrier, that is, a nondiscriminatory basis.[4] In essence, the telcos could now carry video programming, with certain restrictions.[5]

It was hoped that VDT would help fuel the growth of an advanced communications system. Ultimately, consumers would be able to tap an array of new services easily and transparently.

VOD and VDT also reflect a fallout of the new communications technologies. As described in other chapters, rather than serving the mass audience, individual needs could now be satisfied. In one sense, the communication revolution could be viewed as a personal revolution.

The next important development was the 1996 Telecom Act.[6] For this discussion, four points merit our attention:

1. Support was provided for the proliferation of set-top boxes described in the next section.

2. Cable companies could enter the telephony business.

3. Existing VDT policies and rules were repealed.

4. Rules for an Open Video System (OVS) were outlined. They could be considered an extension to the original VDT concept.[7]

The last provision was designed to help relax the guidelines that restricted the broad telco entry into the video field. Telcos were also granted the right to be regulated as OVSs, allowing them to "deliver video without incurring common carrier regulations,"[8] among other features.

If a telco opted for this framework, it would be free of certain regulations, but would have to provide capacity to other program providers, among other restrictions. It was a mix of regulatory relaxation and new requirements.[9]

THE ENTERTAINMENT-INFORMATION MERGER

Qube

The idea of providing customers with more control over their viewing choices did not begin in the 1990s. The Qube cable television service, launched in Columbus, Ohio, in the 1970s, let subscribers tailor their cable service to match their viewing requirements.

Its most unique characteristic was an interactive capability. After a speech delivered by then-President Jimmy Carter, for instance, Qube subscribers participated in a televised electronic survey. The subscribers registered their answers to questions via keypads and the results were tabulated.[10]

Similar viewer response programs were produced in other areas, and the system had the capability to support a PPV option, information services, and electronic transactions. The latter could encompass shopping and banking at home. Even though Qube later abandoned its ambitious plans, it actually helped define the concept of an *integrated entertainment and information utility*.

Implications

For our purpose, an integrated entertainment and information utility implies that a single company may provide you with entertainment and information services. You may use your television for conventional viewing, VOD, retrieving an electronic television viewing guide, and other activities. The same delivery mechanism may also support an Internet hookup. These broad service categories can be grouped under the *interactive television* umbrella.

This type of system could become a reality through cable, telephone, or other links. For example, experimental VOD operations were launched, and during the early to mid-1990s, the cable and telephone industries rushed in to support different levels of this application. At the same time, they experimented with new information delivery methods.

It is also important to stress digital technology's role in this endeavor. It will help supply the software, the movies, and the television programming to potential customers. To provide these services, digital video servers, essentially high-speed and high-capacity information processing/storage devices, have been used.

Another element is the convergence factor. Numerous companies, as well as technologies and applications, will have an impact on this infrastructure's growth. For example, new technological developments and a competitive marketplace have played a role in inter- and intra-industry mergers. Software companies have also joined with broadcast entities, and cable and computer companies have becomes partners with television equipment manufacturers.

This sector of the communications industry is a complex mix of services and companies. The rest of our discussion focuses on potential players, the key issues that must be addressed, and the questions that are, as of this writing, still unanswered.

Issues and Questions

1. True or real-time VOD and other interactive services require enhanced converter boxes. In fact, the new *set-top boxes* resemble computers in terms of their sophistication and user interfaces. Various computer companies have also contributed to this development. The planned operations call for boxes that could handle data, digital video, and other information.

This market also received a boost from new cable rules. Instead of being locked into a single box offered by, for example, your cable company, you could go to a local retailer and purchase a unit that fits your needs. The more you pay, the higher the level of sophistication the unit may support.[11] This philosophy also offers consumers more flexibility in another way. A box could be upgraded to offer new or enhanced services—you are not tied to a single system.

2. A cable company must have the channel capacity to support new entertainment and information services. In one setup, a fiber/coaxial hybrid system could be constructed. Fiber could form the trunk line with coaxial cable being used in other parts of the system. In another configuration, fiber could be directed to a curbside terminal where multiple subscribers would then be connected by standard cable.[12]

The hybrid systems would benefit from fiber's transmission and channel capacity advantages and the cost effectiveness of retaining the current technology base for individual subscribers. In one implementation, a hybrid system initially provided subscribers with 150 channels.[13]

3. Even though it appears the trend is for the eventual implementation of all-fiber systems, the timetable is still unknown. The cost for a full switchover would be expensive, and as described, fiber/coaxial hybrids and/or data compression may prove suitable, at least in the interim.

Digital compression could also increase the number of channels a cable system could support. Consequently, instead of building a new FO plant, a coaxial plant may be able to handle more advanced services.

For telephone companies, digital compression may make it possible to use existing telephone lines to deliver video programming. If successful, the process could provide the telcos with immediate access to most homes in the United States.[14]

But this switchover, be it based on telephone or even cable technologies and systems, may mandate the adoption of either new set-top boxes or other conversion devices. The cost can vary, and to achieve a wide market penetration, it must be somewhat competitive with current pricing structures.

4. The development of an integrated entertainment and information utility hinges on the growth of information services. This could mean videoconferencing, support for multimedia presentations, and Internet access.

As described in Chapter 18, the Internet is an ever-widening information pool that

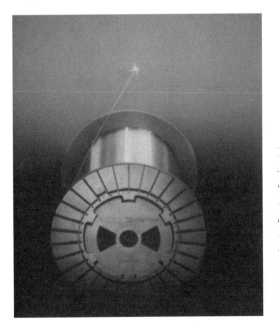

Figure 6.1

An advantage of a fiber-optic line is its great information-carrying capacity. New developments in this field, led by companies such as Corning, have revolutionized the way we use and relay information. (Courtesy of Corning, Inc.)

subscribers can tap via various communications channels. It is also one of the unknown factors in our discussion. In brief, the goal of many organizations is to provide subscribers with an Internet link. Satellite companies have developed high-speed data delivery systems while cable companies have supported sites with cable modems.

By using a satellite- or cable-based relay, you can receive data at a rate that far outstrips a conventional telephone connection. It is also possible to receive video and other information.

Telephone companies have also engaged in system upgrades to match their competitors' advantages. These have ranged from the extension of their FO networks to the adoption of DSL and other schemes.[15]

But despite these advances, questions remain: Will a company, which is your Internet service provider, also support your television entertainment needs? By extension, will an Internet "hook" convince subscribers to switch to a full-service company that may offer a high-speed Internet link, television programming, and other communications and information options? Will we generally continue to use multiple companies for television, telephone, and information service deliveries?

5.   Other new and emerging technologies may help shape this system. In one example, high-definition television (HDTV) systems can deliver enhanced pictures and sounds to our homes. When combined with VOD, your living room could be converted into a movie theater. HDTV has also promoted the merger between television and computer-based technologies. In essence, could your computer serve as your television viewing device?

6.   What are the television and satellite industries' roles in this endeavor? Television stations may be positioned to offer consumers multiple entertainment channels, data, and information services. Satellite operators, on the other hand, have already surpassed the cable and telephone industries in certain respects. DirecTV has relayed digitally compressed television programming to its subscribers, and it could support an Internet link.

The satellite industry also has two major advantages. First, it may be easier to upgrade a service. Once a subscriber owns a dish, technical changes may be a matter of software changes and set-top box enhancements. Cable and telephone companies may have to additionally retool their physical plants.

Second, like DirecTV, a satellite-based company does not have to be tied to an existing, and possibly antiquated, communications infrastructure. Thus, it could take advantage of new technological developments. Cable and telephone companies that have serviced a given region for years, and may have a sizable investment in current facilities, may not have the same luxury.

7.   Another factor may be the initiation of more cooperative ventures between different industries.[16]

8.   Technically, how many years will it take for the necessary technologies to reach smaller markets? Are the changes inevitable, as is the case with most industries where a physical plant is modernized over time, or will some companies balk at the additional expense?

A wild card in this mix has been the 1996 Telecom Act. One provision called for updating the universal service concept. Previously, the goal was to provide all Americans, including those who lived in rural areas, with a basic telephone service. This concept was extended to advanced telecommunications services.[17] But even if advanced services are universally available, when will they be universally adopted (if at all)? Although you can provide access, what if someone cannot afford the prerequisite hardware/software?

9.   Although VOD or other entertainment and information services may be quickly adopted by specific demographic groups, the question about consumer acceptance is still unanswered. How much money are people willing to spend for such services? Can they be sustained by targeting select subscriber groups via narrowcasting?

10.   The concept of a VOD system has been extended to other environments. In one case, VCRs used for playing movies in certain hotels have been replaced by a hybrid

digital/analog system. A similar or all-digital system could also be used for corporate training, linking information kiosks in departments stores, and delivering college lectures.[18] What future applications could be accommodated? Will they help create a niche market for VOD if the general consumer market fails to materialize?

The telephone industry's overall role in this endeavor is still not clear as of this writing.

CONCLUSION

The communications industry is undergoing a series of rapid changes. As indicated, companies, and individuals must contend with a complex mix of current and potential services. We may also be able to gain access to VOD and the Internet through our cable, telephone, or satellite connection. But there are still questions, especially for the consumer market.

Will people be willing to spend additional money for these services? If broad consumer acceptance is not achieved, should projects be abandoned? Will it take time for people to accept and use these services, that is, will it be a slow, but sure, growth pattern? Or should companies focus more on niche markets?

REFERENCES/NOTES

1. VDT, used in the context of computer systems, is a video display terminal. For this specific discussion, though, it represents video dialtone.
2. Richard L. Worsnop, "Pay-Per-View TV," *CQ Researcher* 1 (October 4, 1991): 743.
3. Steve Higgins, "NYNEX, Pacific Bell Tout Electronic Information Services," *PC Week* 9 (March 9, 1992): 67.

Figure 6.2

Three of the players that want to deliver information and possibly entertainment programming to your home: satellite, cable, and telephone companies.

4. FCC, CC Docket 87-266, July 16, 1992, Local Telephone Companies to Be Allowed to Offer Video Dialtone Services; Repeal of Statutory Telco-Cable Prohibition Recommended to Congress, downloaded from CompuServe, October 1992. *Note:* There had been a call for a VDT-type option for a number of years.

5. Harry C. Martin, "Telcos Offer Video Services," *Broadcast Engineering* 34 (September 1992): 8.

6. The Telecom Act is discussed in various chapters of this book.

7. Chris McConnell, "FCC Begins Work on Telecom Act," *Broadcasting & Cable* (February 19, 1996): 15.

8. Chris McConnell, "Cable to Fight FCC's OVS Rules," *Broadcasting & Cable* (June 10, 1996): 16.

9. Michael Katz, "OVS: Other Door to Cable Entry," *Broadcasting & Cable* (February 12, 1996): 54.

10. Edward Meadows, "Why TV Sets Do More in Columbus, Ohio," *Fortune* (October 6, 1980): 67.

11. Robert Kapler, "Engaging in the Set-Top War," *TV Technology* 16 (May 18, 1998): 1.

12. Larry Aiello, Jr., "Bring Fiber Home," *Guidelines* (1992): 7.

13. "TW's Plan for Queens: 150 Channels, 40 of PPV," *Broadcasting* (March 11, 1991): 21. *Note:* Besides replacing lines, other costs for an all-fiber operation would include the necessary optoelectronic components. Conventional cable also has a very wide channel capacity, especially when used for short distances. This factor helps make hybrid systems feasible.

14. "Telcos Eye Twisted Pair Video," *TV Technology* 10 (August 1992): 1.

15. Please see Chapters 2 and 18. They also cover high-speed communications systems/Internet access issues.

16. In one operation, New York Telephone and Liberty Cable Television, a wireless operator in Manhattan, developed a plan to deliver programming. See Rich Brown, "New York Connects to Video Dialtone," *Broadcasting* (October 12, 1992): 38.

17. Dan Carney, Congressional Quarterly Staff Writer, Highlights of the Sweeping Revision of Telecommunications Laws (S652-H Report 104–458), cleared by Congress on February 1; downloaded from AOL, February 11, 1996.

18. Claire Tristram, "Stream On: Video Servers in the Real World," *NewMedia* 5 (April 1995): 50.

SUGGESTED READINGS

Anderson, Lessley. "Interactive TV's New Approach." *The Industry Standard* 2 (October 11, 1999): 60, 62; Ducey, Richard V. "A New Digital Marketplace in Information and Entertainment Services: Organizing Around Connectivity and Interoperability." National Association of Broadcasters, downloaded from www.nab.org, November 1999. Excellent overviews of interactive television services and various issues.

"Breakup of the Bell System." *Communications News* (September 1984): 98–99; Tanzillo, Kevin. "Flood Gates Open." *Communications News* (January 1989): 48–51, 65. The first article describes the divestiture of AT&T; the second looks at the industry 5 years later.

Broadcasting & Cable magazine has chronicled the telcos' entry in the video market. Sample articles, prior to the 1996 Telecom Act include these: "Bell Atlantic Sees Television in Its Future." May 8, 1989, 30; "Telco's Army Poised for Assault on TV Entry." October 3, 1988, 3–47.

Brown, Peter. "Digital Set-Top Boxes—Slow to Shine." *Digital Television* 2 (May 1999): 1, 28, 19; Healey, John. "All Set for Advanced Set-Tops?" *Broadcasting & Cable* (April 27, 1998): 48–54; McGarvey, Joe. "Competition Heats Up Early Digital Set-Top Market." *Inter@ctive Week* 2 (January 16, 1995): 26–27. An overview of set-top boxes and related issues.

FCC. CS Docket No. 96-46. March 11, 1996. Implementation of Section 302 of the Telecommunications Act of 1966. Open Video Systems. Everything you want to know about OVS. The introduction also provides a good overview of video dialtone and its repeal by the 1996 Telecom Act.

Telecommunications Act of 1996, 104th Congress, 2nd Session, January 3, 1996. Sections of the act have an impact on some of this chapter's applications.

GLOSSARY

Entertainment and Information Utility: A company that will deliver entertainment programming and information services. A precursor of such a system was Qube, a cable system in Columbus, Ohio.

Open Video System (OVS): As an alternative to traditional cable systems, the OVS model allows telcos to deliver video without incurring certain regulations.

Telecommunications Act of 1996 (1996 Telecom Act): Comprehensive legislation that has influenced the communications industry.

Two-way Cable: An interactive, two-way cable system.

Video Dialtone (VDT): VDT permitted telcos to participate in the video marketplace. It was superseded by the 1996 Telecom Act.

Video-on-Demand (VOD): VOD can be viewed as enhanced pay-per-view. At one level, you could gain access to and view a movie or program from a central library at your convenience.

Video Server: A high-speed and high-capacity information processing/storage device geared for a video relay. Applications include VOD and commercial playbacks (by a television station).

7

Satellites: Operations and Applications

Satellite communication has become a part of everyday life. We make international telephone calls as easily as local calls down the block. We also see elections in England, tennis matches in France, and other international events with the same regularity as domestic affairs.

This capability to exchange information is made possible by satellites. For those of us who grew up before the space age, satellite-based communication is the culmination of a dream. It stretches back to an era when the term *satellite* was merely an idea conceived by a few inspired individuals. These pioneers included authors such as Arthur C. Clarke, who in 1945 fostered the idea of a worldwide satellite system. This idea has subsequently blossomed into a sophisticated satellite network that spans the globe.

The first generation of satellites was fairly primitive compared to contemporary spacecraft, and they embodied active and passive designs. A passive satellite, such as Echo I launched in 1960, was not equipped with a two-way transmission system. Echo was a large aluminized-mylar balloon that functioned as a reflector. After the satellite was placed in a low Earth orbit, signals relayed to Echo reflected or bounced off its surface and returned to Earth.

In contrast with the Echo series, the Telstar I active communications satellite, launched in 1962, carried receiving and transmitting equipment. It was an active participant in the reception-transmission process. As the satellite received a signal from a ground or Earth station, a communications complex that transmitted and/or received satellite signals, it relayed its own signal to Earth. Telstar also paved the way for today's communications spacecraft because it created the world's first international satellite television link.

SATELLITE TECHNOLOGY

Satellite Fundamentals

Geostationary Orbits. Telstar, Echo, and other earlier satellites were placed in low Earth orbits. In this position, a satellite traveled at such a great rate of speed that it was visible to a ground station for only a limited time each day. The satellite appeared from below the horizon, raced across the sky, and then disappeared below the opposite horizon. Because the ground station was cut off from the now invisible satellite, another station had to be activated to maintain the communications link; otherwise, it would have been necessary to use a series of satellites to create a continuous satellite-based relay. The latter would have entailed the development of a complex Earth- and space-based network.

This problem was solved in 1963 and 1964 through the launching of the Syncom satellites. Rather than racing across the sky, the spacecraft appeared to be stationary or fixed. Today's communications satellites have generally followed suit and are placed in *geostationary* orbital positions or slots. Simply stated, a satellite in a geostationary orbital position appears to be fixed over one portion of the Earth. At an altitude of 22,300 miles above the equator, a satellite travels at the same speed at which the Earth rotates, and its motion is synchronized with the Earth's rotation. Even though the satellite is moving at an enormous rate of speed, it is stationary in the sky relative to an observer on the Earth.

The primary value of a satellite in this orbit is its ability to maintain communication with ground stations in its coverage area. The orbital slot also simplifies this link: Once a station's antenna is aligned, it may only have to be repositioned to a significant degree when

Figure 7.1

Satellites play a major role in the communication revolution. This artist's illustration depicts a satellite that delivers television programs directly to viewers, as described in this chapter. (Courtesy of Hughes Space and Communications Company.)

Uplinks and Downlinks. According to the FCC, an uplink is the "transmission power that carries a signal . . . from its Earth station source up to a satellite"; a downlink ". . . includes the satellite itself, the receiving Earth station, and the signal transmitted downward between the two.[1] To simplify our discussion, the uplink refers to the transmission from the Earth station to the satellite, and the downlink is the transmission from the satellite to the Earth station. This two-way information stream is made possibly by special equipment. The station relays the signal via an antenna or dish and a transmitter that produces a high-frequency microwave signal.

The communications satellite, for its part, operates as a repeater in the sky. After a signal is received, the satellite relays a signal back to Earth. This is analogous to an Earth-based or terrestrial repeater, but now it is located more than 22,000 miles above the Earth. The satellite

1. receives a signal,
2. the signal is amplified,
3. the satellite changes the signal's frequency to avoid interference between the uplink and downlink, and
4. the signal is relayed to Earth where it is received by one or more Earth stations.

To create this link, the satellite uses transponders, which conduct the two-way relays. A communications satellite carries multiple transponders.

As illustrated by the Intelsat family, the number of transponders per satellite class has increased over the years. For example, the original Intelsat satellite, Early Bird, was equipped with 2 transponders that supported either a single television channel or 240 voice (telephone) circuits. The later Intelsat IV satellites, launched in the early to mid-1970s, carried 12 transponders that generally accommodated 4000 voice circuits and two television channels.

The newer Intelsat VIII spacecraft are equipped with 44 transponders. Intelsat VIII can accommodate, on the average, "22,500 two-way telephone circuits and three televi-

contact is established with a different satellite. Prior to this time, a ground station's antenna had to physically track a satellite as it moved across the sky.

Based on these principles, three satellites placed in equidistant positions around the Earth can create a worldwide communications system. This concept was the basis of Arthur Clarke's original vision of a globe-spanning communications network.

Finally, note that although geostationary slots are important, low and medium Earth orbits are still used. These may range from remote sensing (discussed later in this chapter) to specialized communications satellites (Chapter 9).

sion channels."[2] Using digital technology, a satellite could potentially handle more than 100,000 simultaneous two-way telephone circuits.

Intelsat VIII is also a hybrid satellite. It supports both the C- and Ku-bands. As will be covered, C- and Ku-band satellites employ the C- and Ku-band communications frequencies, respectively. Consequently, while some satellites employ only one band, other satellites support both, enhancing a satellite's communication capabilities.

Parts of a Satellite

Satellite Antennas. An important factor that influences the satellite communications network is a satellite's antenna design. A satellite's transmission is focused and falls on a specific region of the Earth. This reception area, the footprint, can vary depending on the satellite and its projected applications. The footprints can range from global to spot beams, in descending order of coverage.[3]

A global beam provides the most coverage and is used for international relays. A spot beam falls on a narrowly defined geographical zone, making it particularly effective for major metropolitan areas. Because this signal is concentrated in a relatively small area, it is also stronger than one distributed to the entire country for cable television relays. A smaller and less expensive dish could subsequently be used.

Satellite Spacing and Antennas. Geostationary communications satellites had traditionally been spaced four degrees or well over 1000 miles apart in the orbital arc to create buffer zones. The zones reduce the chance of cross-interference during transmissions. If eliminated, an uplink, that, for example, was not tightly focused, could unintentionally spill over and affect another satellite.

In the 1980s, the FCC decided to situate communications satellites closer together to open up additional orbital slots. This action was complemented by specifications for Earth-based antennas that satisfied the technical demands imposed by this arrangement.[4]

The plan was devised to address the orbital scarcity problem. As stated, a geostationary slot is an advantageous orbital position. But the number of available slots is finite. The region of space that supports this type of orbit is confined to a narrow belt above the Earth.

A political consideration also plays a role in the allocation process. A nation cannot launch and place a satellite in any slot it chooses because slots are assigned on an international basis through the auspices of the International Telecommunications Union.

The FCC's program was implemented to maximize the United States' orbital allocations. It was also a reflection of the increased pressure to launch more satellites, which demanded the availability of additional orbital assignments.

Power System. A satellite's electrical power is supplied through the conversion of sunlight into electricity by solar cells and ancillary equipment. Cylindrically shaped satellites (spin-stabilized) are covered with the cells, whereas a three-axis stabilized satellite uses wings or extended solar panels.[5] A communications satellite is also equipped with a battery system.[6]

Besides solar cells, satellites use another power source during their 12+-year approximate lifetime. A satellite is equipped with external thrusters and a fuel supply. The thrusters, when activated by ground station controllers, emit small jets of gas to help maintain the satellite's station, which is its position in its slot. Even though a satellite in a geostationary orbit appears to be fixed, it actually moves slightly or drifts. The thrusters correct this drifting.

Once the fuel is expended, however, the satellite may be lost to ground operators.[7] The satellite's drifting can no longer be corrected, and the satellite may become inoperable even though its other systems may still be functional. In view of this situation, a plan has been devised to extend a satellite's life: Let the satellite drift in a controlled fashion. Because the thrusters are now fired less frequently, the remaining fuel can be stretched.

The system's primary drawback is antenna modifications: The satellite must be tracked. But recent designs have made this process simpler and less expensive.[8] The upgrade would also be compensated by the satellite's longer life.

Finally, while communications satellites rely on solar energy, spacecraft designed for deep space and long-term missions have used radioisotope thermoelectric generators (RTGs), that is, nuclear power. The RTGs have extended lifetimes and have provided power on missions where a spacecraft may be too far from the sun to tap its energy as a power source.[9]

Transmission Methods.   A satellite's information capacity is limited by different factors, as are other communications systems. For a satellite, these include the number of transponders and the power supplied to the transmission system. A typical 36-MHz transponder carried by a C-band satellite, for example, can accommodate a television channel and a number of voice/data channels or subcarriers.

Like terrestrial systems, satellite relays can be either analog or digital. Various transmission schemes have also been implemented, including multiple-access systems. Multiple ground stations can gain access to and "share" a satellite's transponder.[10]

Satellite traffic is also increasingly digital, and processing techniques have decreased the bandwidth requirement and lowered transmission costs. An important implication of this development, digital compression, is making satellite communication available to a broader user group. By employing compression, an organization can use a portion of a transponder for a video relay.[11] Although the receiving sites must be equipped to handle this information, the transmission costs are reduced and the satellite can carry additional channels.

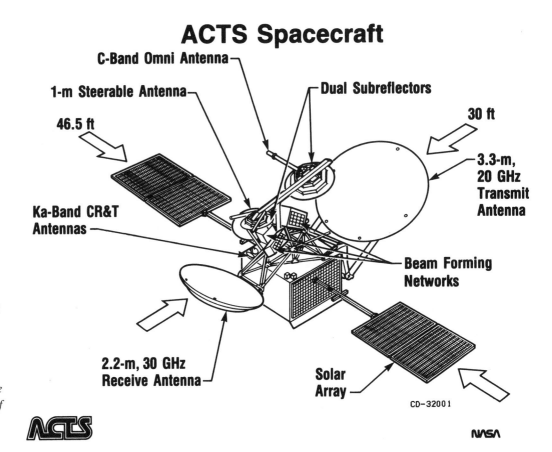

Figure 7.2

NASA's ACTS spacecraft showing, as originally envisioned, its various components. This advanced communications vehicle is described in the next chapter. (Courtesy of NASA; Lewis Research Center.)

Supported applications include those outside of the traditional cable and broadcast industries. In education, a school could start a distance learning network, because satellite time would be less expensive. Or a school could produce a series of satellite-distributed educational programs. If properly implemented, it could be an efficient and cost-effective way to share resources and to support students.

Note also, as indicated in previous and later chapters, that compression is not limited to the satellite industry. It is a powerful tool that has reshaped whole segments of the communications industry. These include the teleconferencing and video production markets.[12]

C-Band and Ku-Band. Until the early 1980s, transmissions in the commercial satellite communications field were primarily conducted in the 4/6 gigahertz (GHz) range, the C-band. The 4/6 notation indicates the downlink and the uplink, respectively. Ground stations that receive their signals are generally equipped with 10-foot to about 15-foot-diameter dishes.

Besides C-band spacecraft, a newer generation of satellite that employs a higher frequency range, the K-band, has become operational. The first class of this type of spacecraft uses the Ku-band (12/14 GHz). This has an advantage. C-band satellite transmissions have been limited in power "to avoid interference with terrestrial microwave systems."[13] A Ku-band satellite is not similarly restricted, and the power of its downlink can be increased. This higher power also translates into smaller receiving dishes and points out a generalization between a satellite's transmission and a dish's size. As the power increases, the dish's size can decrease.

The Ku-band also offers a user more flexibility. A dish's smaller size and a Ku-band system's freedom from terrestrial operations simplifies finding a suitable dish site. C-band systems are not afforded this same luxury. The possible joint interference and a dish's size may make it harder to find a location.[14]

Very Small Aperture Terminals. Ku-band technology has spurred the growth of a satellite system that employs very small aperture terminals (VSATs). A VSAT is a compact dish mated with the necessary electronic hardware to create a cost-effective communications system composed of a few or numerous sites.

VSAT technology has grown in popularity. The small dish can be mounted in a confined area, and a complete station can be cost effective. A VSAT setup also supports various network configurations, information, and can even service geographically remote sites.

A VSAT's multipoint distribution capability also highlights a satellite's key strength. Unlike point-to-point configurations, where a dedicated line may have to be laid to connect a new site, with a VSAT, you set up a dish and its companion electronics. It also allows an organization to create its own communications network.

In one example, Wal-Mart tapped this technology to process credit card purchase transactions. It is faster and less expensive than traditional systems.[15] VSATs have also been used to distribute press pictures to news organizations and to deliver video to K-mart stores.[16] In essence, VSATs help large and small organizations take advantage of satellite technology.

There are, however, some disadvantages of Ku-band systems. This includes a greater susceptibility to interference created by rain. A severe rainstorm could disrupt a transmission to varying degrees.[17]

Finally, it should be noted that the *Ka-band* for satellites has more recently been adopted. A satellite that helped pioneer this new communications tool is described in the next chapter.

SATELLITE SERVICES

General Services

The U.S. communications satellite fleet consists of government and commercial spacecraft. Government satellites are used for military and nonmilitary applications. These range from the creation of a worldwide communications net for military installations to serving as a testbed for new technologies.

On the commercial front, Western Union's Westar I satellite, launched in 1974, was the country's first commercial satellite designed to serve domestic communication needs. The FCC, which oversees the private element of the satellite fleet, also helped spur the industry's growth through its open-skies policy. Articulated in the early 1970s, the policy promoted the commercialization of outer space with regard to commercial satellite communication.

A company that plans to launch and operate a commercial communications satellite must also seek FCC approval. Other details must then be resolved, including contracting with a launch agency and obtaining insurance against possible losses. The latter could include the destruction of the launch vehicle (and the satellite) and the satellite's failure to activate once it reaches its assigned station.

The satellite's emergence as a dominant player in the communications field is a reflection of its general reliability and its

- Wide channel capacity,
- Ability to handle a mixed bag of information, and
- Ability to simultaneously reach multiple sites, including areas not serviced by terrestrial lines.

Organizations have also been able to tap this tool by leasing satellite channel space or by contracting for a turnkey communications system for a specified fee and time period. Another option calls for leasing transponder time for occasional use.

Intelsat, Inmarsat, and Comsat. The international market has been dominated by the International Telecommunications Satellite Organization (Intelsat). Founded in 1964 and composed of more than 100 nations, Intelsat is the world's largest satellite consortium. Intelsat owns and operates a fleet of satellites that supports the international distribution of television and telephone signals and other services.

The Communications Satellite Corporation (Comsat) has served as the U.S. representative to Intelsat. This has made Comsat a powerful and far-reaching satellite organization that provides an array of services.

But storm clouds have appeared on the horizon. Independent organizations, such as PanAmSat, have stepped forward to compete with Intelsat in the international market. While other satellite networks have served various regions of the Earth, they are generally not viewed as Intelsat's competitors. The new private systems would, however, be direct competitors for satellite communication traffic.

This development has been fueled by the rapid integration of satellite technology and equipment in the world market. When Intelsat was founded, the satellite industry was in its infancy, and the consortium was partly established to foster international satellite communication for developed and developing nations. Today, the tools of satellite technology have reached a level of maturity and cost effectiveness where it is possible to launch private operations.

Another organization, the International Maritime Satellite Organization (Inmarsat), extended satellite communication to ships at sea, oil drilling rigs, and even remote land sites. Ships, for example, have established satellite links with land bases through stabilized antennas.

Teleports. Satellites can work with terrestrial communications systems. A company's data may be transmitted over a high-speed landline prior to an uplink and after the downlink. Consequently, satellite and terrestrial systems can be interdependent. The satellite transmission can serve as the long-distance connection, while terrestrial lines provide the intracity hookup.

This integration of systems has been exemplified by New York City's teleport. The Port Authority, Western Union, and Merrill Lynch joined forces to develop a sophisticated communications center on Staten Island. A series of ground stations tied the teleport to national and international satellites and high-speed lines provided the local connection.

Teleports will also help spur the satellite industry's growth. Organizations could share satellite facilities to reduce each participant's financial burden. This could make satellite communication more economically attractive for organizations that have not yet entered the field.

Similarly, a company's geographical location may preclude the building of an Earth station.[18] The establishment of a nearby teleport would provide the company with the capability to establish a satellite link.

Satellites and the Broadcast and Cable Industries

Satellite-distributed television programming is the backbone of the U.S. cable television industry. HBO and other companies uplink programming to a satellite, where it is subsequently downlinked and received by cable companies. The programming is then locally distributed to individual subscribers.[19]

WPIX in New York City, WTBS in Atlanta, and other independent television stations have also joined the satellite revolution. They have used satellites to distribute programming much the same way as HBO. The television networks have also turned toward satellite communication. NBC, for example, the first Ku-band network, established a national satellite system that linked affiliate television stations. The same technology also helped give birth to new networks by providing organizations with a national distribution vehicle. These developments, and others, follow up on the success enjoyed by the Public Broadcasting System, a pioneering satellite organization.

Satellite Newsgathering.   Satellite communication has revolutionized another facet of the television industry: television news. Besides using satellites for story distribution, television stations can participate in satellite newsgathering (SNG), a newer production form.

Wider satellite availability, lower costs, and portable equipment have prompted stations to create their own remote setups. A station buys either a van or truck and a portable satellite dish. This rig is taken on the road, and an uplink to a satellite is established when the reporters reach the story's site. The transmission is subsequently picked up by the home television station. This capability makes it possible for the station to conduct relays from distant sites. For example, a station from Seattle, Washington, could send its satellite rig to Washington, D.C., to establish a link between Seattle's congressional representatives and their constituents. Thus, a station can provide its viewers with information that may have been previously unavailable.

On the international front, flyaway systems have extended satellite communication to regions where standard satellite links may not be available or accessible. Accommodated by a commercial airliner, the system is stored in trunks and reassembled on arrival. The Cable News Network (CNN) has pioneered the use of such systems to cover world events. These include 1989's Tiananmen Square student occupation in Beijing, China, and Operation Desert Storm, 1991's Persian Gulf conflict.[20]

Satellite technology also made Operation Desert Storm the first "real-time" war.[21] People witnessed missile attacks and other events as they actually occurred. Satellite

Figure 7.3

The Harris S-15 SNG Mobile System, a satellite newsgathering vehicle. (Courtesy of Harris Corp.)

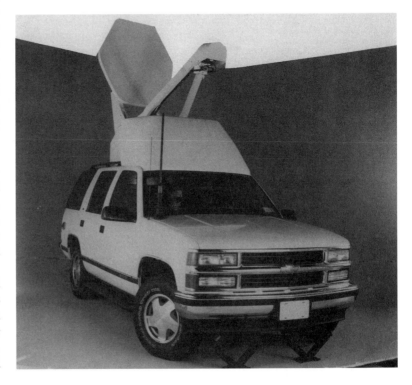

links also provided reporters with the ability to relay voice, video, and computer data (news stories) to their home offices in a timely fashion.

This immediacy triggered negative and positive U.S. responses. In the most publicized case, Peter Arnett, a CNN reporter, continued to file stories from Baghdad, Iraq, during the war. He was criticized for this action by the Victory Committee, a coalition that included the Accuracy in Media organization.[22] The criticism centered on his reporting while under Iraqi censorship and that stories were used for pro-Iraqi propaganda. Similarly, though, severe restrictions were placed on the press by the U.S. government and its allies.

Just as important, while satellites and other communications tools made it possible to monitor this conflict to an unprecedented degree, new technologies did not guarantee good reporting. As stated in a report that explored the media's role, "technology cannot be an end in and of itself in making sense of war-related events. That, as always, remains the job of the journalists themselves, with the new technology facilitating but not replacing the task."[23]

Remote Sensing and the News. Pioneered in part by Mark Brender and ABC, the news media have adopted another satellite-based system, remote sensing satellites, to support their reports. A remote sensing satellite is a sophisticated spacecraft equipped with high-resolution cameras and an array of scientific instruments. Instead of being locked in a geostationary orbit, the satellite can cover the Earth in successive orbits. The pattern is then repeated.

Remote sensing satellites were originally designed to examine and explore the Earth. They have helped document the Earth's physical characteristics as well as environmental changes, such as the impact of deforestation and pollution.

The United States and France have led the world in this field through their Landsat and SPOT satellites, respectively, and news organizations have subsequently tapped this

tool. For example, this type of satellite was an invaluable resource in the aftermath of the Chernobyl nuclear reactor accident. Satellite pictures of the site were obtained and released by the media. The pictures highlighted the facility's damage and helped prevent a potential cover-up.

ABC News also used remote sensing images in a special program televised in July 1987. The pictures documented various facets of the Iran–Iraq war. The same region was also under scrutiny during Operation Desert Storm.

Despite this service's benefits, all governments, including the U.S. government, were not happy with this newfound capability. Because the media could order a photograph of a region of the Earth covered by the spacecraft, military maneuvers and other situations, which a government might want to keep hidden, could be revealed. In the United States, the government could have imposed restrictions on private, domestic remote sensing licenses, via ambiguous licensing procedures. If implemented, the media's access to specific images may have been limited.[24]

This stance, and others, triggered a reaction from the Radio-Television News Directors Association (RTNDA) and other media groups. It was argued that such restrictions violated First Amendment rights, and the government's concern over potential national security breaches was unfounded. The press had generally been responsible in the use of sensitive information in the past and would follow suit with the satellite pictures.

Besides this initial government intervention, media organizations were and still are faced with other problems that may hinder this investigative tool's effectiveness. In one case, a satellite may not be in the correct orbital position to deliver a requested picture immediately. This delay could hamper the coverage of fast-breaking events.

Nevertheless, the remote sensing field holds great promise for continued news coverage. U.S. government restrictions have generally been relaxed, and new domestic and international satellites have been developed.

Figure 7.4
Satellites have produced high-resolution images of the earth. This 1-meter resolution black-and-white image of New York City was collected October 11, 1999, by Space Imaging's IKONOS satellite. It features Lower Manhattan, including the World Trade Center and the Brooklyn Bridge. (Courtesy of Space Imaging.)

This includes the deployment of Ikonos in 1999, the "first commercial imaging satellite."[25] It can produce high-resolution images, and this class of satellite should reduce the response time for pictures. Simplified image manipulation software has also contributed to enhanced operations. ABC News, for one, employed such a system during Operation Desert Storm to generate graphics for its programs.

Finally, note that the U.S. government still plays a pivotal role in this field. In one example, the Clinton administration released a policy statement in 1994 about remote sensing licenses and the export of pertinent technologies. In part, the policy indicated that a satellite operator had to "keep records so the U.S. government can know who has purchased what data, and it authorizes the government to restrict the flow of data to protect national security interests during a crisis or a war."[26] Consequently, even though restrictions were somewhat relaxed, they could still be implemented. In 1996, for instance, tighter controls on imaging, which some categorized as vague and a First Amendment violation, were passed.[27] These issues are further discussed in the chapter's conclusion.

Satellites and Scrambling.   Independent television stations and cable services (e.g., HBO) are distributed by satellite to U.S. cable companies. Once the signals are received, they are routed to individual subscribers. This distribution system is efficient, but creates a problem for cable companies and program providers. Other satellite dish owners could bypass the local cable company to gain access to this programming.

The situation developed into a serious problem by the mid-1980s. The Television Receive-Only (TVRO) industry, partly composed of companies that manufacture and supply Earth stations, experienced a rapid growth of its consumer business. This was a result of the FCC's 1979 deregulation of receive-only stations.

Consumer-based TVRO systems, consisting of small backyard receive-only dishes and the complementary electronic components, were purchased by more than a million Americans. These configurations, which can be called home satellite dishes (HSDs), sprouted up across the country, and people were able to watch pay television and other programming for free.

Eventually, though, the VideoCipher encryption or scrambling scheme, which would theoretically prevent this free viewing, was developed.[28] A television signal was rendered unintelligible by the system unless you had access to a special descrambling device. For cable companies, each site would be equipped with a descrambling unit so their subscribers would continue to receive uninterrupted programming.

Full-time scrambling was inaugurated in the mid-1980s by HBO, and other services joined the bandwagon. This development created a furor in the TVRO industry. Sales for HSD systems declined. Dish owners had two alternatives: Either buy or lease a descrambling device and pay a monthly program subscription fee, or avoid the monthly fee and use an illegally altered decoder. Illegal decoders became available when the supposedly unbreakable scrambling system was cracked.

One individual became so angry at this situation that he illegally interrupted HBO's programming on April 27, 1986, and relayed his own antiscrambling and antisubscription fee message:

Good evening HBO
From Captain Midnight
$12.95/month
No way!
Showtime/Movie Channel beware

This satellite pirate, Captain Midnight, was eventually identified as a part-time employee at a teleport and was subsequently convicted.[29] Captain Midnight used the teleport's facilities to override HBO's signal, and the satellite distributed his signal instead.

Each group of players—HSD owners and the TVRO industry on one side and the cable and television industries on the other—presented its own arguments to support its position. Scrambling opponents indicated that many backyard dishes were purchased by individuals who were not served by cable companies and over-the-air broadcast stations. Their only recourse was to buy an HSD system, and this expensive solution became even more expensive with the introduction of the scrambling device and subscription fees.

The television and cable industries, for their part, asserted their property rights. Illegally received signals were pirated signals. They also stated that some individuals purchased a dish to avoid paying a monthly cable bill.

The solution to this problem lay in a compromise. Subscription prices were lowered, and HSD owners were generally treated more equitably, on a par with standard cable subscribers, in terms of monthly fees. These initiatives, among others, helped to somewhat defuse the situation, but the problem with illegal descramblers continued.

As with computer software, the philosophical basis behind the idea of property rights, in this situation, satellite-distributed programming, has to be accepted. Otherwise, legal and technical measures must be adopted. But in the technical arena, even new protection schemes have the potential to be broken.

The General Instrument Corporation, VideoCipher's parent company, had campaigned against illegal descramblers for years. In one example, an enhanced decoder was introduced in the early 1990s.[30] But estimates indicate that at one time, a quarter of the descramblers in the marketplace were illegally receiving programming (pirating signals).[31]

Finally, the controversy between these two groups has clouded another important issue, the vulnerability of the commercial satellite fleet. If Captain Midnight could disrupt HBO's transmission, other individuals with access to the proper facilities could follow suit. Even though this group's size may be somewhat limited, the situation could change as new facilities, both permanent and portable, are brought on line.

Beyond television programming, financial data and other information vital to the world community is exchanged daily. An ongoing disruption of these services would be disastrous. Ultimately, the same technology that advanced our communications system could harm us unless precautions are taken.

In response, the FCC indicated that various security systems had to be implemented to protect the integrity of satellite transmissions. In 1991, the FCC took a step in this direction by adopting an automatic transmitter identification system (ATIS). Designed for video broadcasts, "ATIS repeats the name and phone number of the broadcaster in Morse code on a subcarrier just beyond the audio spectrum, making signal identification easy."[32] The system was implemented so accidental and intentional interference sources could be identified.

Direct Broadcast Satellites

As conceived, a direct broadcast satellite (DBS) was a class of spacecraft with a very powerful transmission system, on the order of 150 to 200 watts of power. A satellite would operate in the K-band and would bypass television stations and cable companies to relay programming directly to consumers. The original concept has, however, evolved over the years. The vision of very high power satellites, which could only support a limited number of channels, has shifted in the United States.

Nevertheless, two DBS concepts were and still are important: a dish's size and programming choices. Thanks to a powerful transmission and to technical advancements, a dish less than 2 feet in diameter can be used. This is in contrast to the typical 7-foot or more C-band HSD configuration.

Dish size is a vital concern for DBS companies since a small dish is unobtrusive, compact, fairly inexpensive, and easy to set up. It is also permanently aligned toward a specific satellite, unlike an HSD dish, which can be moved.

Viewing choices are also important. A DBS company can provide subscribers with movies as well as television and sports programming. Current systems, as covered in a later section, can also compete with cable operations in the delivery of a mixed bag of program options.

Figure 7.5
The DSS(R) satellite receiving system: 18-inch antenna, a decoder box, and remote control. The system is used to receive programming from DirecTV®. (Courtesy of DirecTV.)

Early History. During the early 1980s, various companies floated DBS proposals, some of which shifted over time. For example, the Satellite Television Corporation (STC), a subsidiary of Comsat, planned to use four satellites to cover the United States.[33] Because each satellite would target only a sector of the country (for example, the Eastern time zone), its "focused" signal would help make it possible to use the smaller receiving dish. But in one modification, this geographical zone was extended so the entire country could be serviced in an accelerated time frame.[34] Proposed DBS systems were also somewhat handicapped by a limited channel capacity. Weight and power demands had an impact on the number of transponders and channels a satellite could support.[35] Consequently, five- or six-channel offerings were not uncommon.

Regardless of the scheme, none of the high-power DBS ventures became operational. Different factors contributed to this situation:

1.   The development of a national system demanded a large capital investment. Beyond the millions of dollars to build, launch, and maintain the satellites, a terrestrial support network had to be created. The latter ranged from local sales and repair offices to an advertising campaign to program licensing fees. For some organizations, the investment was too high for an untested and potentially risky business.[36]

2.   The rapid expansion of the TVRO/HSD and VCR industries exacerbated this situation since the consumer market was already served by these applications. In fact, more than 40 million households were already equipped with VCRs by the mid-1980s.[37] This was an unfortunate development for DBS companies since movies were slated to be a staple feature.

3.   As stated, subscribers would have received only a limited number of channels. Although this may have been acceptable to consumers who lived in areas with few programming options, would consumers in areas served by cable follow suit?

4.   The DBS industry could not sustain a sufficient level of financial support. It was also dealt a severe blow in the 1980s when STC suspended its plans. Other companies, including CBS, had previously bowed out of the field. Consequently, a high-power DBS system did not materialize in the United States.

In contrast, a low-power service was actually created by United Satellite Communications, Inc. (USCI). Instead of constructing a fleet of expensive and untested high-power satellites, a more proven medium-power Ku-band satellite was used. Launched in 1983, USCI offered subscribers five channels of entertainment programming. Future options tentatively included specialized information services and bilingual programming.

But despite the advantages of using a less expensive spacecraft and beating high-power systems to the punch, financial pressures forced USCI to close its doors in 1985.

The Digital Option.   Other DBS ventures followed suit, including one proposed by NBC, Hughes Communications Inc., Cablevision Systems Corporation, and the News Corporation Limited. The plan collapsed, though, in the early 1990s.[38]

Nevertheless, as exemplified by DirecTV, a new generation of DBS systems finally became a reality around the same era. Initiated in 1994 and owned by the Hughes Electronics Corporation, DirecTV had already attracted more than a million subscribers by 1996.[39]

Digital technology and processing techniques have been tapped to make DirecTV an efficient and comprehensive service. The several channel limitation has been eliminated and enhanced audio and video signals can be relayed.

As of this writing, high-power Ku-band satellites deliver well over 100 digital channels. Movies, pay-per-view options, and standard programming fare are supported, and this programming depth, as well as comparable pricing structure, makes DirecTV directly competitive with cable systems.[40]

High-definition television (HDTV) relays could also be featured. HDTV provides for a superior television signal that could be viewed on new, wide-screen sets.[41]

Unlike the broadcast industry, which would require a retooling of its infrastructure for HDTV, satellite-based systems may be more easily implemented. This may give DirecTV and similar services a leg up on broadcasters.

As described in the Internet chapter, you may also be able to opt for a high-speed data relay. Thus, a satellite service could emerge as an integrated entertainment and information utility.

One drawback, however, as of this writing, is a limited two-way capability. Although a satellite can sustain a high-speed feed to subscribers, the return loop has been conducted via telephone lines. Under current technological conditions, this generally precludes the adoption of teleconferencing and other interactive ventures, even though it is satisfactory for many tasks.[42] This includes sending requests for specific information that would subsequently be relayed by satellite.

Summary. The attraction of a DBS system is a powerful one that fuels continued interest in the field. Consumers can receive an array of programs, including ones that may not be otherwise available. The prerequisite technology base has also matured since the 1980s, making DBS operations more feasible.

It also appears that DBS operations have become fully integrated in the U.S. communications infrastructure. They can support individuals in rural areas who do not have broadcast or cable options and can offer traditional cable subscribers another choice.

It will be interesting to watch the overall communications field as satellite, cable, and telephone companies compete for subscribers. As the latter two industries upgrade their physical plants, they will be better positioned to contend with DBS systems.

Note also that on the international front, other countries are well versed in DBS technology and continue to draft plans for sophisticated systems. Japan and various European nations are the major contenders in this field.

Figure 7.6
Satellites have played a crucial role in monitoring the weather. This shot shows Hurricane Fran as viewed by the GOES-8 satellite. (Courtesy of NASA.)

DirecTV-type services were also targeting Latin America during the late 1990s. In one case, Brazil represented a marketing prize. It held a good portion of Latin America's total households, while only "2.5 per-cent [were] reached via cable television lines. . . ."[43]

This same environment would also offer satellite companies new challenges. In the United States, telephone lines are used for the return loop and for relaying billing information. In countries where universal telephone service did not exist, other options had to be adopted.[44]

Finally, like other fields, the DBS industry had to contend with auctions. Bids for available DBS slots surpassed several hundred million dollars.[45] Similar auctions were held for spectrum allocations for personal communications services, discussed in Chapter 9, and broadcasters were concerned that proposed HDTV channels would also be affected by this new trend.

CONCLUSION

Satellites helped transform our communications system. In one example, distance is less

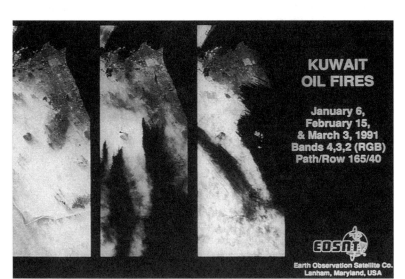

Figure 7.7
*Satellite systems have en-
hanced our communica-
tions capabilities and our
ability to cover world
events. In this photo, the
impact of the Kuwait oil
fires (e.g., smoke), set
during operation Desert
Storm, is evident in this
sequence of photos.
(Courtesy of the Earth
Observation Satellite
Company, Lanham,
Maryland, USA.)*

of a factor than it was in the past. Satellites enable us to relay information rapidly around the world and to view news and human events, such as the 1989 crisis at Tiananmen Square, in real time. Other applications support remote sensing and the delivery of entertainment programming through DBS companies.

In essence, satellites have provided us with a new set of communications tools. It is also important to remember that we have only been using these tools for a relatively short time. We have only just begun to tap their potential and, as introduced in Chapter 9, their potential to promote a personal communication revolution and ultimately their potential to better explore our own world. But this optimistic view could be dampened, in the remote sensing arena, by government mandates. As described, although some restrictions have been lifted, the industry is still subject to ambiguous rulings.

Barbara Cochran, RTNDA president, wrote an article about this scenario, "Fighting the Feds Over Shutter Control." (Copyright by Radio-Television News Directors Association. Reprinted with permission of the RTNDA.)[46] Her words clearly define this issue, and just as important, the power inherent in the images viz a viz a free press.

Proposed government restrictions on the use of commercial satellite images are unconstitutional.

History was made two months ago when the first image from a U.S.-owned commercial satellite was beamed back to earth and released to the general public. That image was remarkable because, although it was made by a satellite orbiting 423 miles above Earth, it showed detail around the Washington Monument so clearly that you could pick out automobiles on the street.

The technology is nifty, but why should you care? Because these images can help television journalists tell stories with a graphic reality never before possible and can provide images from places where cameras are forbidden. Satellite imagery can help you report on major storm systems, ecological change or urban sprawl in your community. Soon you will be able to give viewers a "fly-through" of your city or the surrounding area. With satellite imagery, producers will be able to show terrain in forbidden territory or a gravesite indicating a massacre.

There's one more reason you should care: The U.S. government has adopted a policy for these commercial satellites that would allow the government to cut off access to images whenever the State Department or Pentagon deems it necessary. The policy is unconstitutional, a violation of the First Amendment right of the press to publish or broadcast without government interference, and RTNDA is leading the fight against this policy.

Until recently, government satellites collected pictures from space, which were released to the public at the government's discretion. For years, journalists have used weather satellite imagery in their newscasts to inform viewers and even save lives of those threatened by hurricanes and tropical storms. Now Earth imagery is available from Russia, France, and India. When these images are combined with computer technology, producers can create much more realistic 3-D graphics that replace artists' sketches and give the viewer the sense of "flying through" a landscape.

Dan Dubno of CBS News says the network has used satellite imagery in recent months to report on "everything from the crisis in Kosovo to the Gulf War; North Korean nuclear development; the atomic bomb tests in Pakistan and India; the assault on Osama bin Laden's hideaway in Afghanistan; the 1999 hurricane season; El Nino; and the shuttle launch."

This evolution in journalists' use of satellite imagery brings enormous benefits. Stories can be more accurate and truthful and can give the public access to geographic areas that are politically inaccessible or too expensive to get to.

Now, with the advent of more commercial satellite companies that will create a business out of making these images widely available, journalists will have more opportunities than ever to use this imagery.

But the choices will exist only to the extent that the U.S. government permits the imagery to be made available. And contained in the government policy for licensing of commercial satellites is some disturbing language on "shutter control."

Shutter control is the term for cutting off imaging over a given geographic area for a given period of time. If a satellite is really nothing more than a camera in the sky, government exercise of shutter control constitutes prior restraint of publication of images.

U.S. constitutional law puts a heavy burden on the government if it wants to prevent publication. The First Amendment ensures that the press is free of government control and restraint. The Supreme Court has insisted that the government must take its case to a court of law and present evidence to show why publication should be forbidden before the fact. In the case of government arguing that publication would endanger national security, the court has said the government must show that publication presents a clear and present danger to the national security.

The policy for commercial satellite licensing is much broader and more vague than constitutional law prescribes. In addition, the policy puts the power to decide to exercise shutter control solely in the hands of the executive branch, leaving out the judiciary entirely. The Secretary of Commerce may invoke shutter control after being informed by either the Secretary of State or Defense that a period exists when national security, international obligations, or foreign policy interests may be compromised.

No judicial review. No presentation of evidence in court. No "clear and present danger" test. Rather, the policy makes the executive branch of government both judge and jury in the matter. And the causes that could trigger shutter control are so broad as to be meaningless.

These conditions for shutter control were included in new rules proposed by the federal government in 1998. RTNDA and the National Association of Broadcasters filed comments strongly objecting to the rules. I also met personally with Secretary of Commerce William Daley to explain our position. So far, the government has not issued a final version.

But in 1998 the dispute was theoretical. No satellite had been successfully launched, no

company was providing imagery to the commercial market. That all changed on September 24, when IKONOS successfully launched and three weeks later when it began distributing images with a resolution of one meter, or approximately three feet. Now, journalists and other users have access to imagery with much more detail than ever before of countries, weather systems, ecological change and a host of other subjects. Other companies with licenses have plans to launch in the future.

So the question is now a practical one. What if the conflict in Kosovo were going on now, with severe restrictions on reporting and deep involvement of U.S. military forces? What would the U.S. government do? Would it prevent the commercial satellite from photographing Kosovo and the surrounding area? Would it prove a national security impact, or would the reasons be "foreign policy interests"? And how would news organizations react? Would they try to win the right to argue the case with the executive branch in court?

At RTNDA, we believe this issue is essential to preserving the right to broadcast material in the public interest without prior restraint. A test of the government's role in shutter control is almost certain to occur in the near future. When it does, RTNDA will be in the forefront on the issue, protecting the ability of journalists to use the best tools available—including satellite imagery—to tell stories that are accurate, independent and in the public interest.

REFERENCES/NOTES

1. *Telecommunications: A Glossary of Telecommunications Terms* (Washington, D.C.: Federal Communications Commission, April 1987), 4, 14.
2. Intelsat web site; downloaded January 2000.
3. Mark Long, *World Satellite Almanac* (Boise, Idaho: Comm Tek Publishing Company, 1985), 73.
4. For example, to produce a more focused signal.
5. *Stabilization* refers to the way a satellite maintains its stability while in orbit. A spin-stabilized satellite rapidly rotates around an axis while the antenna is situated on a despun platform so it continues to point at the Earth. Three-axis-stabilized spacecraft use gyros to maintain their positions. Other systems also

play a role in this process. NASA has used a three-axis-stabilized concept for many of its outer space probes. See Andrew F. Inglis, *Satellite Technology* (Boston: Focal Press, 1991), 32; and P. R. K. Chetty, *Satellite Technology and Its Implications* (Blue Ridge Summit, Penn.: TAB Books, 1991), 174–179, for more detailed information.

6. It can be used during eclipses.

7. Another vital element in the transmission between a satellite and a ground station is telemetry data. This is essentially housekeeping data that are relayed by the satellite to indicate its current operational status. In addition, the ground station can uplink commands, such as the aforementioned activation of the thrusters and a possible command to move the satellite to a different orbit. Thus, depending on its design, a spacecraft is not necessarily locked into a position once it achieves its assigned slot.

8. Phil Dubs, "How to 'Recycle' a Dying Bird," *TV Technology* 12 (May 1994): 20.

9. A controversy erupted about RTGs prior to the launching of the *Galileo* spacecraft. Designed to investigate Jupiter and its moons during an extended mission, *Galileo*'s systems are powered by RTGs. Protesters sought to block its launch legally. There was a concern over the possible contamination of the Earth if the RTGs' fuel was scattered in a launch disaster. NASA and space advocates replied the nuclear fuel was in protective containers. Even if there was an explosion, the fuel would not be scattered. Ultimately, the spacecraft was launched in October 1989. Robert Nichols provides an excellent overview of this issue in his article "Showdown at Pad 39-B," *Ad Astra* 1 (November 1989): 8–15. Other, subsequent spacecraft raised similar concerns.

10. Long, *World Satellite Almanac,* 68. See also Chetty, *Satellite Technology and Its Implications,* 402–403, for details. *Note:* Two such systems have been frequency- and time-division multiple access (FDMA/TDMA). They also created a more efficient communications operation.

11. Conversation with Scott Bergstrom, Ph.D., director, Technology-Based Instruction Research Lab, Center for Aerospace Sciences, University of North Dakota, August 6, 1992. For additional information, see Peter Lambert, "Digital Compression; Now Arriving on the Fast Track," *Broadcasting* (July 27, 1992): 40–46.

12. Other areas are also discussed at appropriate points in the book.

13. Andrew F. Inglis, *Satellite Technology* (Boston: Focal Press, 1991), 30. *Note:* They share a frequency range.

14. Satellite Communication Research, *Satellite Earth Station Use in Business and Education* (Tulsa, Okla.: Satellite Communication Research), 18.

15. George Lawton, "Deploying VSATs for Specialized Business Applications," *Telecommunications* 28 (June 1994): 28.

16. Ibid, 30.

17. Long, *World Satellite Almanac,* 99. *Note:* At these higher frequencies, the signal can be weakened, that is, absorbed or scattered by raindrops. This is analogous to the way the light from a car's headlights is dispersed and reduced in intensity by fog.

18. For example, a downtown district.

19. The placement of HBO on satellite also helped spur the growth of urban cable systems.

20. Marc S. Axelrod, "Transmitting Live from Beijing," *InView,* Summer 1989.

21. John Pavlik and Mark Thalhimer, "The Charge of the E Mail Brigade: News Technology Comes of Age," in *The Media at War: The Press and the Persian Gulf Conflict* (New York: Gannett Foundation Media Center, 1991), 35.

22. "Group Launches Campaign to 'Pull Plug' on CNN's Arnett," *Broadcasting* (February 18, 1991): 61.

23. Pavlik and Thalhimer, "The Charge of the E-Mail Brigade," 37.

24. Jay Peterzell, "Eye in the Sky," *Columbia Journalism Review* (September/October 1987): 46.

25. Karen Anderson, "Eagle Eye in the Sky," *Broadcasting & Cable* (October 12, 1999): 72. *Note:* In one case, a 10-meter resolution limitation was lifted (Ikonos has a 1-meter capability). Other countries have also gotten on the remote sensing bandwagon, including Russia for high-resolution images. See The Commercial Space Act of 1997, H.R. 1702, 105th Congress, 2nd session and the Land Remote Sensing Policy Act of 1992 for information about U.S. policy issues.

26. U.S. Congress, Office of Technology Assessment, "Civilian Satellite Remote Sensing: A Strategic Approach," OTA-ISS-607 (Washington, D.C.: U.S. Government Printing Office, September 1994), 114.

27. "RTNDA Protests Imaging Satellite Constraints," *Broadcasting & Cable* (August 12, 1996): 84.

28. HBO provided the impetus for the development of the VideoCipher system, originally developed by M/A-Com, Inc. This concept was described in a 1983 HBO brochure, "Satellite Security," for its affiliates.

29. William Sheets and Rudolf Graf, "The Raid on HBO," *Radio-Electronics* 10 (October 1986): 49.

30. Peter Lambert, "Countdown to Renewable Security," *Broadcasting* (July 27, 1992): 56.

31. Gary M. Hoffman, *Curbing International Piracy of Intellectual Property* (Washington, D.C.: The Annenberg Washington Program, 1989), 11.

32. Hughes Communications, Inc. "Staying Clean," *Uplink* (Spring 1992): 6.

33. David L. Price, "The Satellite," *COMSAT* 11 (1983): 14.

34. "STC Asks for Modifications in DBS Plans," *Broadcasting* (July 23, 1984): 99.

35. Andrew F. Inglis, "Direct Broadcast Satellites," *Satellite TV* (October 1983): 33.

36. "DBS Ranks Cut in Half," *Broadcasting* (October 15, 1984): 75.

37. "Commerce Department Sees Bright Future for Advertising," *Broadcasting* (January 12, 1987): 70.

38. "USSB, Hughes Revive DBS in $100 Million+ Deal," *Broadcasting* (June 10, 1991): 36.

39. "The Growing World of Satellite TV," *Cable & Broadcasting* (February 5, 1996): 59. *Note:* Stanley Hubbard has been one of the most vocal DBS advocates in the United States and started such a service.

40. Hughes Communications, Inc., "DirecTV," information flyer.

41. See Chapter 15 for details.

42. A cable system may be similarly affected.

43. William H. Boyer, "Across the Americas, 1996 Is the Year When DBS Consumers Benefit from More Choices," *Satellite Communications* 20 (April 1996): 24.

44. Ibid, 26. *Note:* One option was to use local institutions, such as banks, as billing centers.

45. Rich Brown, "DBS Auctions Yield $735 Million," *Broadcasting & Cable* (January 29, 1996): 6.

46. Barbara Cochran, RTNDA president, "Fighting the Feds Over Shutter Control." Copyright by Radio-Television News Directors Association.

SUGGESTED READINGS

Boyer, William. "Across the Americas, 1996 Is the Year When DBS Consumers Benefit from More Choices." *Satellite Communications* 20 (April 1996): 23–30; "The Growing World of Satellite TV," *Broadcasting & Cable* (February 5, 1996): 59. DBS growth and its potential in the Americas.

Brender, Mark E. "Remote Sensing and the First Amendment." *Space Policy* (November 1987): 293–297. An excellent review of remote sensing, journalism, and First Amendment issues, prior to the Clinton administration.

Chetty, P. R. K. *Satellite Technology and Its Applications*. Blue Ridge Summit, Penn.: TAB Professional and Reference Books, 1991. Satellites and their applications. This book is especially strong in its coverage of satellite design.

"Commercializing Spy-Satellite Technology Should Be a Boon to Photonics Industry." *Photonics Spectra* 28 (April 1994): 50–51. Imaging systems on satellites and government regulations.

Communications News. From late 1987 through 1988, *Communications News* ran a series of articles devoted to VSATs. These articles provide an interesting perspective of this field. Includes David Wilkerson, "VSAT Technology for Today and the Future—Part 3: Use Private Networks or Leased Services?" (November 1987): 60–63.

Dorr, Les, Jr. "PanAmSat Takes on a Giant." *Space World* W-12-276 (December 1986): 14–17; "Anselmo, Landman Team Up to Tackle Intelsat." *Broadcasting* (August 5, 1991): 48. Early look at private, international satellite networks.

Dubs, Phil. "How to 'Recycle' A Dying Bird." *TV Technology* 12 (May 1994): 20. An interesting look at how to extend a satellite's useful lifetime.

"EOSAT Operations Underway." *The Photogrammetric Coyote* 9 (March 1986): 8, 17; "SPOT to Fly in October." *The Photogrammetric Coyote* 8 (September 1985): 2,4; NASA. "The Landsat Satellites: Unique National Assets." FS-1999(03)-004-GSFC (downloaded from www.nasa.org, November 1999). Two early viewpoints and one contemporary look at remote sensing satellites.

Flanagan, Patrick. "VSAT: A Market and Technology Overview." *Telecommunications* 27 (March 1993): 19–24; Lawton, George. "Deploying VSATs for Specialized Business Applications." *Telecommunications* 28 (June 1994): 28–32;

Picasso, Gino. "VSAT's: Continuing Improvements for a Workhorse Technology." *Telecommunications* 32 (September 1998): 56–57; Sweitzer, John. "The VSAT Ka-band Configuration." *Satellite Communications* 23 (October 1999): 42–44, 48. VSAT developments, applications, and the players.

Frieden, Robert M. "Satellites in the Global Information Infrastructure: Opportunities and Handicaps." *Telecommunications* 30 (February 1996): 29–33; Hartshorn, David. "Conjuring the Cure for Asian Flu." *Satellite Communications* 23 (January 1999): 24–31; Kirvan, Paul. "Bargaining for Satellite Communications Services." *Communications News* (April 1993): 49; Schober, Eckart. "Breaking Open the Private Network Market in Latin America." *Satellite Communications* 22 (October 1998): 34–37. The satellite's role in a worldwide information infrastructure and satellite use costs.

GAO. "Telecommunications: Competitive Impact of Restructuring the International Satellite Organizations." GAO/RCED-96-204, Letter Report, July 7, 1996. A look, in part, at the potential impact of restructuring Intelsat.

Inglis, Andrew F. *Satellite Technology: An Introduction.* Boston: Focal Press, 1991. An excellent guide to satellite technology and applications.

Niekamp, Raymond A. "Satellite Newsgathering and Its Effect on Network-Affiliate Relations." An AEJMC Convention paper, Radio-TV Journalism Division, August 1990. Network affiliates and "their policies in sharing news video with their networks and other stations."

"RTNDA Protests Imaging Satellite Constraints." *Broadcasting & Cable* (August 12, 1996): 84. Domestic remote sensing restrictions and security concerns (on the part of the U.S. government.

U.S. Congress, Office of Technology Assessment. "Civilian Satellite Remote Sensing: A Strategic Approach," OTA-ISS-607. Washington, D.C.: U.S. Government Printing Office, September 1994. A comprehensive review of remote sensing policy/applications.

GLOSSARY

Active Satellite: A satellite equipped to receive signals and to relay its own signal back to Earth.

C-band: A satellite communication frequency band and satellite class. Commercial C-band satellites are the older of the contemporary communications satellite fleet.

Communications Satellite Corporation (Comsat): The U.S. representative to Intelsat.

Direct Broadcast Satellite (DBS): A powerful communications satellite that delivers movies and other offerings to subscribers equipped with compact satellite dishes.

Earth Station: An Earth station establishes a communication link with a satellite. Some Earth stations, also called ground stations, transmit and receive signals; others only receive signals.

Footprint: The shape of a satellite transmission's reception area on the Earth.

Geostationary Orbit: A desirable orbital position/slot for a communications satellite. The satellite's motion is synchronized with the Earth's rotation and appears, to ground observers, to be stationary. This has technical advantages for maintaining a communications link.

International Maritime Satellite Organization (Inmarsat): An international organization that extended satellite communication to ships at sea and other remote sites.

International Telecommunications Satellite Organization (Intelsat): An international satellite consortium. Intelsat supports a broad range of satellite services.

Ku-band: A newer satellite communication band and class. Ku-band satellites also support more powerful downlinks and have news (media) applications.

Orbital Spacing: Buffer zones physically separate the satellites to help eliminate interference.

Passive Satellite: A satellite that does not relay its own signal back to Earth.

Remote Sensing Satellite: A remote sensing satellite scans and explores the Earth with different instruments, including cameras. Images can highlight the Earth's physical characteristics, such as wetland acreage losses. The media have also used these satellites to cover potentially inaccessible regions for news coverage.

Satellite Newsgathering (SNG): The process of using small, transportable satellite dishes to directly relay news stories from almost anywhere in the field.

Scrambling: A process in which a satellite's signal is rendered unintelligible. The receiving site is equipped with a decoder to return the signal to its original state.

Teleport: A satellite dish farm.

Transponder: The heart of a satellite's communications system that acts like a repeater in the sky.

Very Small Aperture Terminal (VSAT): A small satellite dish and the complementary electronic components. A VSAT system can be cost effective.

8 Satellites: New Developments, Launch Vehicles, and Space Law

The previous chapter covered satellite fundamentals, ranging from basic operations to communications applications. This chapter focuses on complementary topics: future technologies and launch vehicles. The latter is critical. Without cost-effective launch vehicles, new satellites may never literally leave the ground. We conclude with a quick look at relevant space law and a history of the U.S. space initiative.

FUTURE SATELLITE TECHNOLOGY

The plans for the next generation of communications satellite have been drafted. The spacecraft will improve on current designs and will carry sophisticated on-board switching and processing equipment. These intelligent satellites will direct the flow of communications signals, which will help streamline the ground network and the establishment of communications links. Such satellites will reduce the cost to create, run, and maintain our communications system.

Advanced Communications Technology Satellite

This new generation of spacecraft is exemplified by NASA's experimental Advanced Communications Technology Satellite (ACTS). Launched in 1993, ACTS operates in the Ka-band. Its features range from those just described to a sophisticated transmission system that can support "fixed beams and

hopping spot beams that can be used to service traffic needs on a dynamic basis. A hopping spot beam . . . sends/receives information and then the beam electronically 'hops' to a second location. . . ."[1] It can respond to user demands and traffic needs.

ACTS can sustain a high data rate, smaller receiving antennas can be used, and it is a flexible communications system. Tests have also been conducted to gauge its performance. Three representative examples include ISDN experiments, military and medical applications, and the capability to quickly restore communications services when terrestrial links are disrupted.[2] ACTS has also paved the way for other Ka-band satellites, and it may serve as a blueprint for spacecraft that can support personal, satellite-based relays. We will eventually communicate via satellite with small, handheld devices.

ACTS was also originally designed to accommodate an experimental optical communications system. Devised by the military to produce a secure relay, it was later scrapped from the mission.

Smallsats, Space Platforms, Space Weather, and Space Debris

Besides an ACTS-type spacecraft, other developments have and will continue to advance satellite technology. These include the launching of *smallsats* and space platforms.

Smallsats are small, cost-effective satellites. They can be used for remote sensing, for creating personal communications networks, and for other applications.[3] The bottom-line figure for countries and organizations is that satellite technology has become more affordable. A smallsat is less expensive than a conventional satellite and can be designed and assembled in an accelerated time frame.

These factors are crucial for new and developing applications. As described in Chapter 9, satellite-based personal communication could be supported, in one configuration, by a satellite series or constellation placed in low Earth orbit.

Basically, if you have to build and launch a number of satellites to start a service, you cannot spend two or more years to manufacture a single spacecraft. This is where smallsats step in. Instead of building customized satellites, you reuse existing technologies and products. Modular satellite systems, which can accommodate different space-based applications, are also employed.[4]

The process is somewhat analogous to mass production, and the trade-off is that the smallsat may not be as sophisticated or capable of handling as many tasks as its larger and more costly counterpart. But this may not be the design goal in creating a smallsat in the first place.

A space platform, in contrast, would be a large structure placed in orbit. It would route a high volume of information while occupying only a single slot. A single platform could potentially replace several contemporary spacecraft.

It is important to note, however, that some scientists believe various members of the newer generation of satellites may be more susceptible to space weather—for our discussion, adverse solar activity that could damage satellites.[5] As we move toward more mass-produced systems, which may employ commercial, off-the shelf components, they may not be properly shielded from radiation. Other solar activities may also harm "traditional" satellites as well. In fact, the National Oceanic and Atmospheric Administration (NOAA) has developed a "space weather scale" that tracks the potential damage that could be caused by solar-originating "geomagnetic storms."[6]

The upshot of these effects? If communications satellites are damaged, our communications system may be affected, and by extension, certain communications capabilities.

Finally, besides natural phenomena, satellites, particularly in low and medium orbits, face another hazard. The Earth is surrounded by a "cloud" composed of millions of pieces of debris. This space junk, which can vary from paint particles to radioactive droplets leaked from other satellites, poses new challenges for satellite designers.[7] Although the possibility for a fatal impact is slight, precautions will have to be taken. Organizations have also drafted plans to produce less "polluting" spacecraft and rockets.[8] Much like environmental conditions on certain regions of the Earth, our lack of foresight has had unforeseen consequences. This time, it is in space.

LAUNCH VEHICLES

The growth of the satellite system will be fueled by new launch vehicles and organizations. In the past, companies and most nations signed with NASA for this job, placing a heavy demand on launch vehicles and facilities. This situation looked as if it was going to be altered in the 1980s. The development of NASA's space shuttle and the entry Arianespace, a private consortium, promised to facilitate satellite launch operations.

The Space Shuttle

The space shuttle is the world's first refurbishable manned or piloted spacecraft. After a mission, the shuttle returns to Earth and is refurbished for its next flight. The space shuttle can carry a self-contained laboratory, scientific experiments, and satellites in its hold, the cargo bay. In its latter role, the shuttle initially carries a satellite to a low Earth orbit where it is subsequently released. If the satellite's final destination is a geostationary orbit, an attached rocket booster propels the satellite to a specified altitude, and it eventually reaches a preassigned orbital slot after maneuvers.

The shuttle's original promise and premise was to make space accessible and operations cost effective. Yet it was plagued by mechanical and structural problems. This was partly a reflection of the spacecraft's heritage:

- As a piloted vehicle, it had to support a crew.
- As a multipurpose vehicle, it had to support a range of missions.
- It was refurbishable.

Consequently, the shuttle was a complex spacecraft. While it proved successful in many ways, its complex design and other factors, including refurbishing delays, led to setbacks. The shuttle program was also hampered by a somewhat burdensome organizational hierarchy.

The *Challenger* Explosion

In February 1984, the space shuttle's credibility as a launch vehicle received a blow. After its release, the Westar VI satellite's booster malfunctioned, and the satellite was stuck in a useless orbit. Indonesia's Palapa-B2, the mission's second satellite, suffered a similar fate.[9] A little less than 2 years later, the world was shocked by the explosion of the space shuttle *Challenger*. The entire crew was lost in the most devastating tragedy in NASA's history.

In the wake of the *Challenger* disaster, and a report released by an investigatory commission, President Ronald Reagan announced that NASA would generally withdraw from the commercial satellite launch industry. The frequency of future flights would also be scaled down, and the shuttle's primary role would be to support scientific and military missions.

This directive reflected the president's attitude toward the government's role in private enterprises and the realization that an overly ambitious launch schedule contributed to *Challenger*'s destruction. As stated by the commission, "The nation's reliance on the Shuttle as its principal space launch capability created a relentless pressure on NASA to increase the flight rate."[10] This pressure played a role in the decision to launch

Challenger under adverse weather conditions. Other contributing elements, which led to *Challenger*'s destruction, were design flaws in the shuttle's rocket boosters, as well as possible flaws in the spacecraft's overall design and the booster refurbishing process.[11]

Various companies subsequently stepped forward to fill the void in the commercial launch industry. Even though there was some prior activity, the list of interested parties has grown at an accelerated rate.[12]

Companies adopted existing rockets and developed new unpiloted rockets, expendable launch vehicles (ELVs). An example of the latter was Orbital Sciences Corporation's *Pegasus*. Instead of a typical ground launch, *Pegasus* was carried by a jet, released, and then proceeded on its own power.

This class of ELV was developed to lift small payloads into low Earth orbits. They also complemented smallsats since they were less expensive than conventional ELVs.[13] But depending on the circumstances, a smallsat

Figure 8.1

The Challenger *at liftoff. (Courtesy of NSSDC.)*

Figure 8.2
Pegasus *launch vehicle.*
(Courtesy of Orbital
Sciences Corp.)

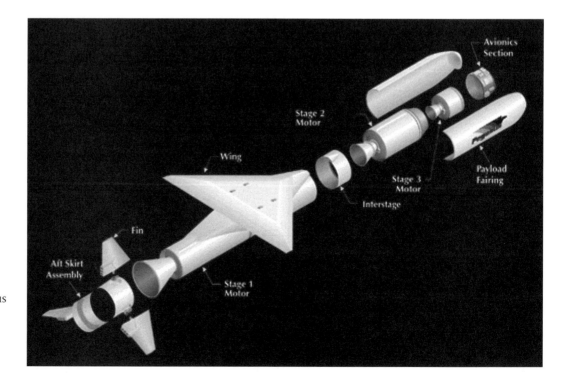

Figure 8.3
Diagram of the Pegasus
launch vehicle, high-
lighting its features.
(Courtesy of Orbital
Sciences Corp.)

could even hitch a ride on a conventional ELV, as part of another payload.[14]

These innovations will help fuel the small-sat industry. An organization may have access to a smallsat, but without a cost-effective ELV, it may not be able to launch it. Launch costs are also factors in the creation and maintenance of satellite constellations.

The future U.S. satellite launch fleet will include the space shuttle and ELVs. The commercial sector will use ELVs while the government will also rely on the shuttle.

NASA has helped support this industry by leasing its facilities to private companies and through other programs. This support has been timely in light of the stiff competition American companies will continue to face in the international satellite launch market.

This mixed fleet has also provided the United States with a more balanced launch capability. When the shuttle program was in full swing, ELVs were delegated to a secondary role. But they have reemerged from the background.

The space shuttle, for its part, will continue to fulfill the role for which it is best suited, that of a special utility and research vehicle. An example of the former was a dramatic and televised salvage operation.

In 1992, the shuttle *Endeavor* rendezvoused with an Intelsat VI satellite stuck in a low orbit. After some difficulties, astronauts retrieved the satellite, brought it into the shuttle's bay, and attached a motor for a subsequent boost to its final geostationary position.

Besides contributing to the knowledge base for recovery missions, some of the shortcomings of simulations were revealed. A simulation, which attempts to duplicate the conditions of an actual event, was inaccurate in this case. An alternate plan had to be devised and implemented.

This type of practical experience in dealing with novel situations is important for the future of extravehicular space-based activities. It also highlighted, at this stage of our technology base, the value of the human presence in space. Humans, unlike current robotic devices, can adapt to unique circumstances.[15]

Arianespace

Arianespace is a private commercial enterprise and an offshoot of another European organization, the European Space Agency (ESA). Arianespace was created in response to NASA's earlier domination of the satellite launch industry.[16] Arianespace has aggressively promoted and marketed ELVs and a sophisticated launch and support operation. Its rockets have a flexible payload capability and can accommodate heavy payloads.

Arianespace has also maintained a competitive price structure, and its launch site in Kourou, French Guyana, is particularly well situated to place satellites in geostationary

Figure 8.4

The Ariane 42P, with two solid strap-on boosters. (Courtesy of Arianespace.)

and other orbital positions. These factors and others contributed to its growing share of the international launch market when NASA was still a participant in the field.

Despite its successes, Arianespace has suffered some of its own setbacks. Satellites have been destroyed by rocket failures, and the organization must face a host of new and potential competitors, including China, private U.S. companies, and Japan.

The Current and Future State of the Satellite Launch Industry

The commercial launch industry, as indicated, experienced a series of upheavals during the mid- to late 1980s. While it is true that NASA had left the field, companies and other nations have filled the void. The increased competition triggered by NASA's decision may actually make it easier, in the long run, for an organization to launch a satellite.

However, on a bleaker note for the United States, its preeminent position in the satellite manufacturing field is eroding. During the late 1980s to early 1990s, 36 communications satellites were manufactured by the United States and 23 were built by Europe and Japan.[17] Prior to this time, the United States had dominated the industry.

This erosion was somewhat mirrored by other U.S. space initiatives. Mixed signals about the viability of a proposed space station were sent to the public and international community. There were other mishaps, including the loss of Mars probes and a technical problem with the *Galileo*'s Jupiter mission.[18] Nevertheless, there were and are some positive signs. These range from the ACTS mission to enhanced satellite manufacturing techniques (for example, smallsats). Congress also recognized the necessity of supporting a strong satellite and launch industry. As stated, "this industry contributes to the U.S. economy, strengthens U.S. scientific interests, and supports foreign policy and security interests."[19]

New Ventures

The United States and other countries have also tried to make space more accessible and affordable. One proposed initiative was the National AeroSpace Plane Program (NASP) with its experimental X-30 vehicle. The NASP was slated to pave the way for aerospace planes that could take off and land on conventional runways, attain a low Earth orbit, and be reusable instead of refurbishable.[20]

Much like an airline, the term *reusable* implies a quick turnaround time. Unlike the shuttle, a vehicle could be prepared for its next flight without major refurbishing. This capability would save time and money. Projected applications included retrieving low orbit satellites and servicing a space station.[21] Another spinoff was more down to earth. New airliners based on this concept could have carried passengers between, in one route, Los Angeles and Sydney, Australia, in 2.5 rather than 13.5 hours.[22]

Another program will develop a reusable launch system through a single-stage to orbit rocket. The Reusable Launch Vehicle Technology Program's objective is to create "technologies and new operational concepts that can radically reduce the cost of access to space. The program will combine ground and flight demonstrations. An important aspect . . . will be the use of experimental flight vehicles—the X-33 and X-34—to verify full-up systems performance. . . ."[23]

Consequently, the goal is to replace current launch vehicles, where practicable, with cost-effective reusable systems. The Delta-Clipper Experimental (DC-X) rocket already tested key concepts at the White Sands Missile range in the early to mid-1990s.[24]

In July 1996, NASA actually awarded Lockheed Martin a contract to build and fly an experimental X-33. Launched vertically and landing horizontally, its VentureStar promised to "cut the cost of a pound of payload to orbit from $10,000 to $1,000," the magical number that would help make space realistically affordable.[25] But despite a similar projection for the shuttle, it is hoped that technological advances may make the X-33's promise a reality.

Xs made another appearance in 1996 in the guise of the X-prize. Borrowing a tactic from aviation's early days when monetary

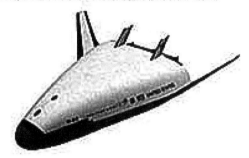

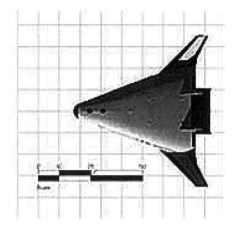

Figure 8.5
*Lockheed Martin's X-33
from different views.
(Courtesy of NASA.)*

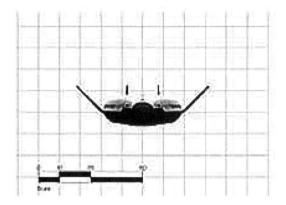

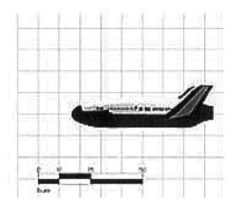

prizes were offered to advance flying, the modern X-prize seeks to "accelerate the development of low-cost, reusable vehicles and thereby jump-start the creation of a space tourism industry."[26] Finally, as has been indicated, satellite launches are not infallible. Accidents do occur, and a launch vehicle and accompanying satellite(s) can be lost. Future launches may also be postponed until an accident's underlying cause has been determined. This waiting period may be unacceptable to an organization with a tight timetable. Thus, an organization may sign contracts with multiple vendors to launch its satellites. If one ELV is "grounded," other satellites could continue to be launched.[27] It is hoped, though, that as new ELVs come on line, the number of accidents will diminish and the decades-old dream of making space access affordable will be realized—not only for organizations and governments, but also for the rest of us.

SPACE EXPLORATION

In closing this chapter, it is appropriate to examine an area related to satellite communication: space exploration. Both fields coincide to a certain extent, and space probes are sophisticated communications tools in their own right. More important, developments in this field have had an impact on the information and communications industries.

As discussed in Chapter 16, NASA helped pioneer image processing techniques to enhance and correct pictures transmitted by outer space probes. Similar techniques have been applied on Earth in desktop publishing, in the medical field, and in other applications. Remote sensing satellites, originally designed to explore the Earth, have also been used by the media. Consequently, cross-fertilization can occur between what can be called outer and inner space operations. As such, outer space developments at least merit an overview.

Figure 8.6
*The Hubble Space Tele-
scope being refurbished
during a shuttle mission.
(Courtesy of NSSDC.)*

This section also provides a brief history of
NASA and highlights some of the forces that
have shaped and continue to shape the space
program. These include social and political
issues and pressures. The section concludes
with an overview of legal implications governing space-based activities.

History
The 1950s witnessed the birth of the modern
era of space exploration.[28] A milestone was

the inauguration of NASA on October 1,
1958, as the successor to the National Advisory Committee for Aeronautics (NACA).
NACA, founded in 1915, helped advance the
nation's aeronautical industry through research and related activities. The new agency
was given the same mandate and oversight
of the civilian space program.

Some of the space program's major events
and influencing factors are as follows:

1.   The Soviet Union launches the first artificial satellite, *Sputnik 1*, on October 4, 1957.
2.   On April 12, 1961, Soviet cosmonaut
Yuri Gagarin becomes the first human in
space. The U.S. space program centers about
Project Mercury and its seven astronauts.
3.   President John F. Kennedy commits
the nation to landing an astronaut on the
moon before the end of the decade (Project
Apollo).
4.   On January 27, 1967, a fire in the
Apollo command module kills three astronauts.[29] A Russian cosmonaut loses his life in
the same year.[30]
5.   On July 20, 1969, Neil Armstrong and
Edwin ("Buzz") Aldrin of *Apollo 11* become
the first humans to walk on the moon, while
their comrade, Michael Collins, orbits overhead.
6.   The Apollo program comes to a halt
after *Apollo 17* in December 1972, owing to
financial considerations and the changing
U.S. social and political climate (for example,
the Vietnam War).
7.   The 1970s and early 1980s witness
other missions, including the Apollo-Soyuz
Test Project (1975), in which a Soviet and
U.S. spacecraft dock in orbit, and the development of the space shuttle.
8.   Budgetary constraints and the loss of
the space shuttle *Challenger* contribute to the
dearth of planetary missions from the late
1970s until the late 1980s.[31] The potential
lack of appropriations for ACTS and other
projects could have similarly derailed advanced satellite technologies.
9.   The *Galileo* probe investigates Jupiter
and its moons in the 1990s and the Hubble
Space Telescope continues its exploration of

the universe from a low Earth orbit. Both systems, though, have suffered from performance problems. A rover, controlled from Earth, explores Mars.

10.   Space probes explore the Solar System, including Mercury, Venus, Mars, Jupiter, Saturn, Uranus, Neptune, asteroids, the Moon, and the Sun. Only Pluto remains to be visited by a probe.

11.   The Earth is explored by different satellites. For example, Landsat satellites image the Earth for geological and, more recently, for media applications.

Legal Implications

As our satellites' capabilities increase and we develop commercial space-based enterprises, legal issues become more important. Relevant topics already discussed are orbital assignments, signal piracy, and the role of Intelsat and competing organizations in the international arena. Additional subjects include the following:

- The U.N.'s concern about the international free flow of information and its balance; not solely from the developed to the developing world.
- The international dissemination of data by remote sensing satellites.
- The U.S. media's use of such images and the impact of government restrictions and regulations.
- The sociopolitical impact of DBS relays if the signals spill over to a neighboring country.[32]
- The "upward extent" of a country's national sovereignty. Is it 100 miles, or could it be lower or even higher?[33] How high is high? What are the political implications for satellites with respect to their orbital slots?
- Ownership/property rights in space. Various legal questions remain as to property rights on the Moon and other celestial bodies.[34] In one case, would a company or consortium invest hundreds of millions of dollars to develop a mining operation if its rights were not clearly defined?
- Individual rights for space explorers and/or colonists. The U.S. Constitution Bicentennial Committee covered this issue in a subcommittee. An outcome, the Declaration of First Principles for the Governance of Outer Space Societies, declared that the U.S. Constitution should also apply to "individuals living in outer space societies under United States jurisdiction."[35] The document's drafters believed that individual rights, such as freedom of speech, assembly, and *media* and *communications* [my emphasis], are fundamental principles that would extend to U.S. space societies, balanced against the unique environment afforded by outer space.[36]

Finally, there are other legal issues that are beyond the scope of this book, such as licensing policies and procedures for communications satellites.[37] For the realm of outer space, there's a growing body of space law. It is a fascinating field, and one that will continue to evolve as we begin to take our first outward steps in space.

REFERENCES/NOTES

1.   NASA, "Advanced Communications Technology Satellite (ACTS) Hardware," Information Sheet, 3.

2.   Frank Gedney and Frank Gargione, "ACTS: New Services for Communications," *Satellite Communications* (September 1994): 48.

3.   Brian J. Horais, "Small Satellites Prove Capable for Low-Cost Imaging," *Laser Focus World* 27 (September 1991): 148.

4.   Amy Cosper, "Crank Up the Assembly Line," *Satellite Communications* (February 1996): 29. *Note:* NASA has adopted a similar philosophy for some of its outer space missions.

5.   Sten Odenwald, "Solar Storms: The Silent Menace," *Sky and Telescope* 99 (March 2000): 54. *Note:* The article provides an excellent overview of solar storms and their potential impact, including terrestrial implications.

6.   Ibid, 55.

7.   William J. Broad, "Radioactive Debris in Space Threatens Satellites in Use," *The New York Times* (February 26, 1995): 12.

8. Leonard Davis, "Lethal Litter," *Satellite Communications* (January 1996): 24.

9. Following this mission, the satellites were subsequently recovered and returned to Earth via a shuttle.

10. William P. Rogers, Neil Armstrong, David C. Acheson, et al., *Report of the Presidential Commission on the Space Shuttle Challenger Accident* (Washington, D.C.: U.S. Government Printing Office, 1986), 201.

11. Yale Jay Lubkin, "What Really Happened," *Defense Science* 9 (October/November 1990): 10.

12. The Reagan administration had been a proponent of the commercialization of outer space, especially in the area of the launch industry. See Edward Ridley Finch, Jr., and Amanda Lee Moore, *AstroBusiness* (Stamford, Conn.: Walden Book Company, 1984), 56–63, for more information.

13. A *Pegasus* launch would have cost approximately $7 to $10 million versus millions of more dollars for a standard ELV.

14. Rick Fleeter, "The Smallsat Invasion," *Satellite Communications* (November 1994): 29.

15. Note that salvage missions of this nature are limited, at least at this time, to low Earth orbits.

16. The ESA and its member states support a wide range of space activities (for example, ELV developmental work and space exploration). See the special advertising supplement to *NASA Tech Briefs* 15 (June 1991) and *Ad Astra* 3 (December 1991): 18–40, for more information.

17. R. T. Gedney, "Foreign Competition in Communications Satellites Is Real," *ACTS Quarterly* 91/1 (February 1992): 1.

18. Despite the mission's successful deployment of a probe into Jupiter's atmosphere, an antenna mishap reduced its transmission rate (for example, of photos).

19. The Commercial Space Act of 1997, H.R. 1702, 105th Congress, 2nd session.

20. U.S. General Accounting Office, National Aero-Space Plane; A Technology and Demonstration Program to Build the X-30, GAO/NSIAD-88-122, April 1988, 50.

21. See Jim Martin, "Creating the Platform of the Future; NASP," *Defense Science* (September 1988): 55, 57, 60, for more information about the NASP program, including potential military applications. NASP-based vehicles would not eliminate ELVs or possibly even the shuttle.

This is one lesson we should have learned from the shuttle program: One vehicle may not be able to perform all functions equally well.

22. U.S. General Accounting Office, National Aero-Space Plane; A Technology Development and Demonstration Program to Build the X-30, GAO/NSIAD-88-122, April 1988, 14. Note: The X-30 is also a hypersonic flight vehicle. As stated in the same report: "Hypersonic is that speed which is five times or more the speed of sound in air (761.5 mph at sea level)." As envisioned, the X-30 could reach hypersonic speeds up to Mach 25.

23. NASA Facts OnLine, Marshall Space Flight Center, "The Reusable Launch Vehicle Technology Program."

24. A witness to a DC-X flight (launch/hovering/flight periods), indicated "the landing was the way God and Robert Heinlein intended." For details, see Marianne J. Dyson, "A Sign in the Heavens," *Ad Astra* 5 (November/December 1993): 17.

25. "Lockheed Martin Selected to Build the X-33," NASA Press Release, July 2, 1996.

26. X-Prize brochure, 1996.

27. James M. Gifford, "Going Up," *Satellite Communications* 20 (February 1996): 33.

28. Parts of this section are taken from Michael Mirabito, "Space Program," in *The Reader's Companion to American History* (New York: Houghton Mifflin Company, 1991), 1013–1014. Houghton Mifflin kindly extended permission for its use.

29. Virgil Grissom, Edward White, and Roger Chaffee.

30. Vladimir Komarov. See J. K. Davies, *Space Exploration* (New York: Chambers, 1992), 193. Three other cosmonauts, Georgi Dobrovolsky, Vladislav Volkov, and Victor Patseysev, died in 1971.

31. Dick Scobee, Mike Smith, Ellison Onizuka, Judy Resnik, Ron McNair, Gregory Jarvis, and Christa MacAuliffe.

32. Stephen Gorove, "The 1980 Session of the U.N. Committee of the Peaceful Uses of Outer Space: Highlights of Positions on Outstanding Legal Issues," *Journal of Space Law* 8 (Spring/Fall 1980): 179.

33. S. Houston Lay and Howard J. Taubenfeld, *The Law Relating to Activities of Man in Space* (Chicago: The University of Chicago Press, 1970), 41.

34. Ty S. Twibell, "Legal Restraints on the Commercialization and Development of Outer

Space," *University of Missouri at Kansas City Law Review* (Spring, 1997), downloaded from Lexis.

35. Nathan C. Goldman, "Space Colonies: Rights in Space, Obligations to Earth," in Jill Steele Meyer, ed., *Proceedings of the Seventh Annual International Space Development Conference* (San Diego: Univelt, Inc., 1991), 220.

36. Rights versus the space environment include the right to bear arms, an important U.S. concept. But in space, where a weapon could physically compromise the integrity of a colony's protective shielding, this issue becomes more complex. The same question applies to the freedom of assembly and the press, among others. This general concept has also been used as the plot in various works, including Robert A. Heinlein's science fiction book, *The Moon Is a Harsh Mistress* (New York: Berkeley Publishing Corp., 1968). For more information, see William F. Wu, "Taking Liberties in Space," *Ad Astra* 3 (November 1991): 36.

37. For additional information, see Carl J. Cangelosi, "Satellites: Regulatory Summary," in Andrew F. Inglis, ed. *Electronic Communications Handbook* (New York: McGraw-Hill Book Company, 1988), 6.1–6.9.

SUGGESTED READINGS

Abutaha, Ali F. *The Space Shuttle: A Basic Problem.* This videotape was produced by George Washington University, Washington, D.C. This is a taped lecture conducted by Ali Abutaha for George Washington's Continuing Engineering Education program. The tape covers shuttle design problems discovered by Abutaha after the Challenger disaster.

Banke, Jim. "Ticket to Ride." *Ad Astra* 8 (January/February 1996): 24–26; Diamandis, Dr. Peter. "The 'X' Prize. *Ad Astra* 7 (May/June 1995): 46–49. Space tourism development and a look at the X-prize and aviation examples.

Commercial Space Transportation Study. Executive Summary. 1994. The executive summary of a detailed report about new space markets.

Fleeter, Rick. "The Smallsat Invasion," *Satellite Communications* 18 (November 1994): 27–30; Cosper, Amy. "Crank Up the Assembly Line," *Satellite Communications* 20 (February 1996): 24–30. Smallsats and new construction techniques.

Garcia, Jon C. "Heaven or Hell: The Future of the United States Launch Services Industry." *Har-vard Journal of Law & Technology.* Spring 1994. (7 Harv. J. Law & Tec. 333). Downloaded from Lexis; Kross, John. "Fields of Dreams." *Ad Astra* 8 (January/February 1996): 27–31; Saunders, Renee. "Rocket Industry Agenda for the Next Millennium." *Satellite Communications* 19 (July 1995): 22–24. The U.S. and international outlooks for spaceports and rockets.

Gedney, Frank, and Frank Gargione. "ACTS: New Services for Communications," *Satellite Communications* 18 (September 1994): 48–54. The ACTS satellite, a review of its first year of operation, and its influence on other satellite designs.

Hardin, R. Winn. "Solid-State Lasers Join the Space Race." *Photonics Spectra* 32 (June 1998): 114–118; Morgan, Walter L. "Pass It Along." *Satellite Communications* 22 (May 1998): 50–53. Laser applications in satellite communication.

Kerrod, Robin. *The Illustrated History of NASA: Anniversary Edition.* New York: Gallery Books, 1988. A comprehensive and richly illustrated history of NASA and the U.S. program.

Lewis Research Center, NASA. *ACTS Quarterly.* A Lewis Research Center newsletter that traces the development, launching, and testing of the ACTS satellite. As a collection, it is an interesting overview of the birth and life of a satellite, as well as the impact of external forces (for example, budgetary appropriations).

Military & Aerospace Electronics covers new launch vehicle/propulsion developments, among other topics. Three good articles include these: Keller, John. "Avionics Innovation Marks New Space Shuttle." (April 1997): 1, 7; Mchale, John. "Electrical Xenon Ion Engine to Power New Millennium Spacecraft." (June 1997): 1, 33; Rhea, John. "Avionics Key in Drive to Cut Space Launch Costs." (December 1998): 1, 8.

Sky and Telescope. The November 1993 issue had a series of articles/sidebars that focused on the Hubble Telescope and its repair. An example is Richard Tresch Fienberg, "Hubble's Road to Recovery," 16–22.

Space Law. Numerous articles and books cover space law issues (the last article covers space debris). These include Twibell, Ty S. " Legal Restraints on Commercialization and Development of Outer Space." *University of Missouri at Kansas City Law Review.* Spring 1997 (65 UMKC L. Rev. 589), downloaded from Lexis; Wasser, Alan. "A New Law Could Make Privately Funded Space Settlement Profitable." *Ad Astra* 9 (July/August, 1997): 32–35; Williams, Christopher D. "Space: The Cluttered

Frontier." *1995 Southern Methodist University Journal of Air Law and Commerce.* May/June 1995 (60 J. Air & Com. 1139), downloaded from Lexis.

U.S. General Accounting Office. "Aerospace Plane Technology: Research and Development Efforts in Japan and Australia." GAO/NSIAD-92-5, October 1991. One of a series of aerospace plane reports.

GLOSSARY

Advanced Communications Technology Satellite (ACTS): A NASA satellite that is a prototype for the future commercial fleet.

Arianespace: A commercial satellite launch agency.

Challenger: The space shuttle that was lost to equipment and systems failures.

Launch Vehicles: Both expendable (ELV) and piloted vehicles used to launch satellites and space probes. Two launch organizations have included NASA and Arianespace, and other countries have either developed, or are developing, their own capabilities.

NASA: The U.S. space agency given the mandate to oversee the civilian space and aeronautical programs. Also the successor to the NACA.

Reusable Launch Vehicle Technology Program: Program designed to create the next-generation U.S. launch vehicle.

Smallsat: A small, relatively inexpensive satellite.

Space Shuttle: The world's first refurbishable piloted spacecraft. A shuttle can carry a variety of payloads to a low Earth orbit.

X-Prize: Echoing back to an earlier era, a proposed monetary prize to advance reusable launch vehicle technology (for example, to develop space tourism).

9 Wireless Technology and Mobile Communication

Wireless technology comes in different flavors. As implied by the name, you are not physically connected with or tied to a communications line. One application supports data exchanges in a building while another extends this relay to another country. In essence, the wireless industry embraces an assortment of technologies and applications you can tap to solve your communications needs.

For our discussion, we also focus on systems that support mobile business and personal communication. These include cellular telephones, personal communications services (PCS), certain satellite relays, the virtual office, and wireless local area networks. But before we cover these topics, other wireless systems are quickly reviewed. It is important to examine a sample of these other operations to provide full coverage of the wireless universe.

formation through the air. It is cost effective and can support a wide and secure communications channel. An FCC license is not required, and operations could be set up in areas where microwave communication may not be feasible.[1] A laser relay is, however, line of sight. It could also be variably affected by smog and other atmospheric conditions.

Wireless technology has also been used by the cable industry. Wireless relays have existed for a number of years under the aegis of the Multichannel Multipoint Distribution Service (MMDS). New digital compression techniques could extend a system's capacity to well over 100 channels.[2]

Finally, the communication transactions supported by some of these wireless systems have been conducted as bypass operations. An organization bypasses, or goes around, the traditional public communications networks.[3]

Figure 9.1

An over-the-air (atmospheric) optical communications system. It can accommodate a range of information. (Courtesy of American Laser systems, Inc.)

WIRELESS SYSTEMS

Microwave, Laser, and Cable

Microwave systems have handled long, short, and intracity relays. A microwave system is cost effective and can accommodate a range of information with a wide channel capacity. Microwave applications may also be easier to implement than fiber-optic or copper-based ones. You do not, for instance, have the potential problem of obtaining clearances to lay the cable.

On a negative note, a microwave transmission could be affected by heavy rain. An FCC license has also been required and, as a line-of-sight medium, the transmitter and receiver must be in each other's line of view.

Another wireless system uses an infrared laser to relay voice, video, and computer in-

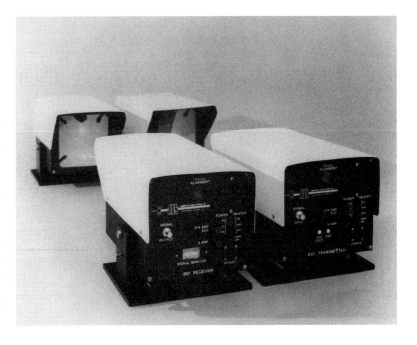

Thanks, in part, to the divestiture of AT&T in the early 1980s, a new dawn greeted the telecommunications world. Industry segments once dominated by AT&T became open markets for sales and leasing opportunities.

Companies also became more responsible for their own communications systems. Various intra- and intercity links were established that could be more responsive to an organization's unique communication needs. Thus, a company could more readily react to new communication demands since it used private or leased links.[4]

The latter concept is a key one. Wireless systems are analogous, in a sense, to ergonomic designs in the computer industry, In this case, communications systems will conform to our communication needs and not the other way around.

MOBILE WIRELESS SERVICES

In the context of our present discussion, mobile wireless systems could be viewed as personal bypass systems. Nevertheless, the important concept for us is mobility. Various technological and social changes have combined to make us a society on the go, a mobile society. Our information and communications tools are following suit.

- Computers are smaller and weigh less. Yet these smaller computers may be as powerful as desktop units.
- We conduct business from our cars and can use the same telephone to call home.
- Pagers are no longer relegated to only a few professions.
- Satellite-based relays can support everyday communication.

A goal of these operations is to free us from physical wires—constraints. As stated by one author when describing such a system, "PCS is a new wireless mobile technology for voice and data communication to and from *people*, not *locations* [my emphasis]."[5] Consequently, we may no longer be bound to a physical space to communicate or to exchange information. Communication would be centered on us rather than an office or other site.

Cellular Telephone
The cellular telephone industry, a key wireless player, is a communications fixture. A specified geographical region is divided into small physical areas called *cells*, each of which is equipped with a low-power transmission system. Because the cellular telephone is designed to support mobile activities, it allows you to use the system in a car and other locations.

In a typical operation, as you approach a cell's boundary while driving, the signal between the telephone and the transmitter becomes weaker. At this point, the telephone connection is basically picked up by the new cell the car is approaching. The telephone is then switched to a different frequency, to avoid potential interference with adjacent cells. This procedure is automatically completed by a sophisticated control network.

Cellular technology has also been integrated with portable PCs for remote data relays. In a related development, digital operations support a cleaner signal, for both data and voice, and will expand a network's capacity.[6]

The same digital technology will also give us a bonus, privacy. A shortcoming of conventional systems has been the ability to intercept conversations via widely available scanning devices.[7] But digital technology could provide some measure of security and help make a conversation unintelligible to the average scanner owner.

In sum, cellular technology has made it possible to relay information and to hold a conversation from the field. Cellular users have also benefited from call forwarding, other technological advancements, and an array of telephones that can match your specific needs.[8] But as with other communications systems, there are limiting factors, including interference created by terrain and human-made obstructions (e.g., buildings).

Personal Communications Services
As defined by the FCC, a PCS is "a family of mobile or portable radio communications

services" designed to meet the "communications requirements of people on the move."[9] The basic premise is to deliver mobile, wireless communications services. As of the mid-1990s, it was also a wide-open market.

Computer-to-computer and voice relays can be supported. Both local-area and broader geographical communications links can be established, and ties can be created between developing and existing networks. Communications tools can also combine the capabilities of two or more devices, as may be the case with a portable telephone that has a built-in paging function.[10] When viewed from a broad perspective, this emerging field can partly serve as a testbed for new personal bypass technologies and their applications.[11]

PCS companies also have an advantage over their cellular competitors. As outlined in a report, PCS companies "will be building networks from the ground up that take advantage of the newest network and digital technologies. They will design their systems from the start to be interoperable [potentially]. . . . Their corporate mission is to provide the American public with a lower cost, more user-friendly 'cellular-like' services."[12] Thus, PCS companies may have been better positioned to tap technological advances and their complementary applications. They are not bound by a preexisting infrastructure.

As described throughout the book, many communications developments are evolutionary since they build on existing systems. In some cases, this is a reflection of financial concerns—an overnight shift to a new technology base could be financially prohibitive and disruptive.

PCS companies, though, could take advantage of new technological opportunities and offer them to their subscribers. In fact, in one PCS promotional campaign, it was pointed out the service was based on digital technology, which would make for a more secure relay.[13]

The PCS industry also experienced another, somewhat new development. Traditionally, segments of the communications industry have used the airwaves for free. But in an attempt to help defray the budget deficit, auctions were initiated for specific U.S. PCS licenses. Companies paid for the right to use the spectrum allocations.

Satellite Communications

As discussed in Chapter 7, satellites are powerful long-distance and point-to-multipoint communications tools. A new type of satellite system may also support mobile and personal relays in the general business and consumer markets. This could be considered a space-based extension of cellular or mobile communications technology. It could also complement and supplement the services provided by geostationary communications satellites.

To actually create the system, a series of small satellites is launched and placed in low Earth orbits (LEOs).[14] In one configuration, a constellation of satellites could provide global coverage while their low altitudes would support portable transmitters and receivers on Earth. A potential application would be the relaying of short bursts of data, which may include messages from remote locations, over small and inexpensive terminals.[15]

Different companies have stepped forward in this field. Iridium, for instance, was set up as a multisatellite network that would function as an interconnected Earth- and space-based venture—you could talk with a friend half a world away through portable, mobile telephones.[16] The same system could also serve as a relay platform for nations without a developed communications infrastructure.

All in all, it is an interesting concept, especially in light of the history of our satellite system. As you may recall, LEO satellites were generally abandoned in favor of their geostationary counterparts for communications applications. Yet the new generation of LEO satellites, when properly configured and coordinated, could deliver certain satellite services right into our hands. It is almost like Dick Tracy and the wrist radio come to life.

But as of this writing, Dick Tracy may actually be using other, more established, wireless systems (e.g., PCS). In brief, this class of satellite operation ran into financial difficulties, despite its capabilities. The cost for a call

Figure 9.2.

A space and earth-based communication operation can support a sophisticated personal communications network. The various elements are highlighted in this diagram, including the satellites and personal communications device (subscriber unit-right bottom-hand corner). (Courtesy of Motorola Inc.; Iridium.)

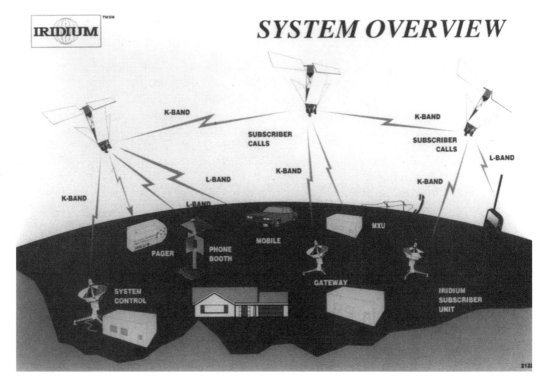

was, for instance, high when compared with other wireless tools. As one writer put it, ". . . both first and second generation terrestrial wireless have become so successful that most of the lucrative global markets have been mopped up."[17] As a consequence, there may not be a viable user base, other than for specialized applications. The systems, while conceptually sound and exciting, may have been launched too late.

Virtual Office and Wireless LANs

Various companies have either experimented with or have adopted the virtual office for elements of their workforce.[18] If you are a salesperson, you may no longer be confined to an office. You may be equipped with portable computers and other information and communications tools so you can work on the road. The virtual office has also been considered a "variant of telecommuting," discussed in a later chapter.[19] But because of some of the virtual office's wireless possibilities, it has been placed in this chapter.

Through the virtual office, a salesperson can work directly with clients in the field,

and depending on the application, can tap either wireless or wire-based communications systems. The former systems will, however, become more important as they are refined. You can be untethered, that is, free from constraints imposed by telephone lines or other such systems. In one sense, your communications tools become as mobile as you.

Besides serving the client base better, a virtual office can help cut real estate costs because fewer offices have to be set up and maintained.[20] If properly implemented, a company could enhance its productivity while cutting costs. The key to success, though, is proper implementation and addressing of key issues. These include securing a suitable communications channel, offering employee training, and providing adequate logistical support (for example, clerical help).

Companies have also established special work sites to provide "an office away from the office for transient workers."[21] For Xerox, this meant creating a large work space where employees could meet, exchange ideas, and plug in their PCs when they are not on the road.

A wireless local area network (LAN), for its part, is just what it sounds like. You can tap a LAN's resources without a direct, physical connection. In one application, doctors and nurses working in a hospital can readily gain access to information.[22] In another application, you could go to a meeting and retrieve pertinent data, or record notes, without physically connecting your PC or other device.

As of this writing, wireless components are more expensive than their wired counterparts. But you are no longer restricted to a desk or other physical location. This flexibility is especially valuable for an application that demands mobility and, in fact, for a facility that cannot be wired for a standard LAN.[23] It may also be used for locations where you can only install a limited number of data ports, yet must support numerous individuals.

CONCLUSION

Wireless systems may free you from having to plug your computer or telephone into an outlet for certain communications applications. They can also work with traditional communications systems, as may be the case with a virtual office. Even though you may be untethered, a conventional telephone line could be used for data relay. In many ways, the wireless and wired worlds are complementary.

Wireless technology may also allow you to receive e-mail while in the field and to relay data to your office. You could potentially tap into a network, dial a specific number, and reach another person, wherever he or she may be, via a portable telephone. It will be kind of like "super-call-forwarding," free of the constraints imposed by wires, distance, and location. Consequently, we may no longer be bound to a physical space to receive and relay information, whether it is a telephone call or a computer feed. Armed with a personal number, the information will be automatically routed to us.

New developments will also continue to refine our wireless systems. In one example, very small cells, called microcells, "are being

Figure 9.3
This photo reveals the compact size of a personal communicator device that could tap into a LEO satellite communications network. (Courtesy of Orbital Sciences.)

designed to take care of downtown users on city streets, and picocells, for inside office buildings."[24] Essentially, our communications world could be subdivided into smaller, physical regions (for example, a building or floor in a building) to handle our increasing wireless traffic.

On the flip side, wireless communication carries a price. Wireless components can be more expensive, standards issues have surfaced, connections can be sporadic, and data relays can be chancy.[25] Much like earlier computer monitors, concerns have also been raised about cellular telephones and potential health risks.

From a nontechnical standpoint, wireless technology has another implication. If a call can be routed to you, and if your home becomes a workplace, what impact will this have on your individual privacy? Will the "electronic clutter" become so pervasive that these tools become more of a distraction than an aid?[26] In brief, is it necessary to have access to information at every waking moment? Or do we also require quiet, reflective periods, free from distractions, even in a business setting? How do you strike a balance between the two?

Finally, the wireless field also illustrates the interdependence of technological developments. The launching of an Iridium-type network serves as an example. The technology now exists to build sophisticated small satellites that are matched by portable telephones and other communications devices. This entire system, in turn, is influenced by the satellite launch industry. Without cost-effective launch vehicles, the deployment of a constellation of satellites could be prohibitively expensive.

Thus, as introduced in Chapter 1, although you can examine individual applications in isolation, it may also be important to explore related areas. If you do not, you might miss key connections that could have a major impact on an industry's success. In this example, cost-effective launches may play a role in an Iridium-type system's future. So too does its acceptance by the targeted user group(s).

REFERENCES/NOTES

1. This has included high-density urban areas where physical obstacles and the lack of licenses have restricted microwave communication.
2. Mark Hallinger, "Wireless Cable Ready for the World," *TV Technology* 13 (July 1995): 1.
3. Dwight B. Davis, "Making Sense of the Telecommunications Circus," *High Technology* (September 1985): 20. *Note:* Companies have been using a bypass system for years. The Private Branch Exchange, or PBX, serves as a privately owned switchboard, of sorts, primarily for a company's telephone operation. Thus, it bypasses the local telephone system for this type of internal hookup, and it is the link to the outside world, the public telephone system. PBX systems, and all communications operations, were given a boost by the 1968 Carterphone decision. In essence, after this time, equipment manufactured by private companies could be connected with the public telephone system.
4. These also include fiber and copper lines.
5. Lou Manuta, "PCS's Promise Is in Satellites," *PCS* (September 1995): 14.
6. Anthony Ramirez, "Next for the Cellular Phone," *New York Times* (March 15, 1992): Section 3, 7.
7. Robert Corn-Revere, "Cellular Phones: Only the Illusion of Privacy," *Network World* 6 (August 28, 1989): 36. *Note:* Data relays could also be intercepted.
8. For example, you can buy telephones designed for cars or to fit in your pocket.
9. From an August 1992 FCC Notice of Proposed Rule Making and Tentative Decision; GEN Docket #90-314; downloaded from CompuServe, prepared by Scott Loftesness, August 25, 1992.
10. Jay Kitchen, "PCs: A Progress Report," *Telecommunications Wireless Report,* Special Supplement, 48.
11. Conversation with FCC engineering department, September 16, 1992.
12. Maxine Carter-Lome, "Cellular in the Lead, But Can't Stay Ahead," *PCS Today* (September 1995): 8.
13. Harry A. Jessell, "Sprint Unveils U.S. PCS Service," *Broadcasting & Cable* (November 20, 1995): 47.
14. There are also different categories of LEOs. Part of the distinction has to do with a satellite's size and relay capabilities (for example, signal strength).
15. Orbital Sciences Corporation, "Orbcomm Signs Marketing Agreement in Five More Countries," news release.
16. Jim Foley, "Iridium: Key to Worldwide Cellular Communications," *Telecommunications* 25 (October 1991): 23. *Note:* The ground-based sector would handle, in part, billing and the necessary authorization to use the system. This would be provided through gateways.
17. Alan Pearce, "Is Satellite Telephony Worth Saving," *Wireless Integration* (September/October, 1999): 19.
18. Michael Nadeua, "Not Lost in Space," *Byte* 20 (June 1995): 50.
19. Osman Eldib and Daniel Minoli, *Telecommuting* (Norwood, Mass.: Artech House, 1995), 1.
20. Patrick Flanagan, "Wireless Data: Closing the Gap Between Promise and Reality," *Telecommunications* 28 (March 1994): 28.
21. Nadeua, "Not Lost in Space," 52. *Note:* Much like telecommuting, interpersonal communication between employees was still important.
22. Elisabeth Horwitt, "What's Wrong with Wireless," *Network World* 12 (November 13, 1995): 65.
23. Flanagan, "Wireless Data," 30.
24. Robert G. Winch, *Telecommunication Transmission Systems* (New York: McGraw-Hill, 1998), 363.

25. Horwitt, "What's Wrong with Wireless," 64.
26. Patrick M. Reilly, "The Dark Side," *The Wall Street Journal* (November 16, 1992), technology supplement on "Going Portable," R12.

SUGGESTED READINGS

Bernard, Josef. "Inside Cellular Telephone." *Radio-Electronics* 11 (September 1987): 53–55, 93. An overview of this growing field.

Caldwell, Bruce, and Bruce Gambon. "The Virtual Office." *Information Week* (January 22, 1996): 33–36, 40. Telecommuting, requirements, and is it appropriate for you?

CIT Publications. "A Wireless Decade: A European Survey." *Telecommunications* 32 (September 1998): 70–79. Comprehensive examination of wireless systems in Europe. Individual countries are highlighted.

Eldib, Osman, and Daniel Minoli. *Telecommuting*. Norwood, Mass.: Artech House, 1995. Excellent and comprehensive examination of telecommuting.

Fink, Michael. "Lasers over Manhattan." *Communications News* (May 1994): 30; Guttendorf, Dick. "Flash Gordon Meets Ma Bell." Communications Industries Report, September 1995, 26. Over-the-air laser communications systems.

Flanagan, Patrick. "Personal Communications Services: The Long Road Ahead." *Telecommunications* 30 (February 1996): 23–28; Frieden, Rob. "Satellite-Based Personal Communication Services." *Telecommunication* 27 (December 1993): 25–28; O'Keefe, Susan. "The Wireless Boom." *Telecommunications* 32 (November 1998): 30–36. Terrestrial- and satellite-based personal communications services.

Grambs, Peter, and Patrick Zerbib. "Caring for Customers in a Global Marketplace." *Satellite Communications* (October 1998): 24–30; Corporate Information News Center. Iridium Press Release: "Iridium LLC Investors Commit Funding to Ensure Service During Restructuring Discussions." (December 9, 1999); "Iridium Constellation Finishes Launch Deployment." *Ad Astra* 10 (July/August 1998): 12. A look at the Iridium system; from coverage of its operation to its financial restructuring situation.

Horwitt, Elisabeth. "What's Wrong with Wireless," *Network World* 12 (November 13, 1995): 64–70;

Picker, Marisa. "It's a Wireless World." *Mobile Computing & Communications* 10 (May 1999): 95–101; Young, Jeffrey R. "Are Wireless Networks the Wave of the Future?" *Chronicle of Higher Education* XLV (February 5, 1999): A25–A26. Wireless technologies, applications, and implications.

Meeks, Brock N. "Spectrum Auctions Pull in $7.7 Billion." *Inter@ctive Week* 2 (March 1995): 7. PCS auctions.

Vitaliano, Franco. "How Work Becomes Remotely Possible." *BackOffice* (January 1996): 43–51. Portable technologies; includes a list of terms.

GLOSSARY

Bypass System: A private/leased communications system that bypasses standard commercial and public systems.

Cellular Telephone: A personal communications tool based on frequency reuse and a monitoring design.

Microwave and Laser: Two wireless, line-of-sight communications systems.

Personal Communications Services (PCSs): A family of mobile services designed to meet the communication requirements of people on the move. Computer-to-computer and voice relays can be supported.

Personal Satellite Communications: A new generation of satellite can deliver personal communications services—you are not restricted by location. In one configuration, satellites are placed in low earth orbits.

Virtual Office: Instead of working in a traditional office, you are equipped with a PC and mobile communications tools so you can work/communicate from the field. Sites have also been designed to serve this workforce when not in the field (for example, a place to plug in your PC).

Wireless LAN: As implied, you can tap a LAN's resources without a direct, physical connection.

10 Information Storage: The Optical Disk

The optical disk emerged as an important information storage tool in the 1980s. A popular application is the CD, a small, round disk that stores digital audio information in the form of microscopic pits. To retrieve this information, you place a CD in a player.[1] A laser subsequently reads back or recovers the information.

A laser's light scans the CD, and its beam is reflected to different degrees, in terms of its strength, when it passes over the pits and unpitted areas called *lands*. The reflected light, an optical representation of the stored information, is picked up by a light-sensitive detector. After processing, the final output for a CD is an analog signal.

The CD and other optical disks are also constructed like sandwiches: These include the information layer with the code of pits and a reflective metallic layer. The latter enables the read or playback operation.

OPTICAL DISK OVERVIEW

For our discussion, the growing optical disk family falls into two categories: nonrecordable and recordable media. CD and conventional compact disk read-only memory (CD-ROM) systems are nonrecordable. Write-once, read-many (WORM) and erasable systems are recordable.

Both categories of disks share some characteristics:

1. Information can be stored in the form of pits, or in erasable systems, through other techniques. This information is also digital, with the major exception of the videodisk.

2. Optical disks are fairly rugged since the stored information is physically protected from fingerprints and scratches. The laser is also focused beneath a disk's surface at the information layer, so dust and other minor surface obstructions may not adversely affect a playback.

3. Unlike a conventional vinyl LP, a disk is not subject to wear, because the playback mechanism is not a stylus but a beam of light. The same disk can be played multiple times with no discernible loss of quality.

4. Disks are not indestructible, however. For example, deep scratches can affect a playback. Some older disks may also have a manufacturing defect—corrosion of a disk's metallic layer.[2]

5. Optical disks are high-capacity storage media. This is partly a reflection of a laser's capability to distinguish between tightly recorded information tracks.

6. The different systems incorporate sophisticated error-detection and checking schemes to ensure the information's integrity. But the potential impact of errors has

Figure 10.1.
Some members of the optical disk family. They are a crucial element of our data (and media) storage system.

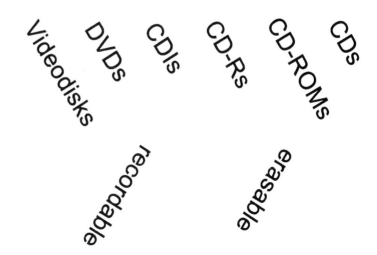

OPTICAL DISKS

Videodisks DVDs CDIs CD-Rs CD-ROMs CDs

recordable erasable

made this process more critical for a disk that stores data than for a CD.

7. Like floppy and hard disks, optical disks are random-access devices. They provide almost immediate access to the stored information.

8. The optical storage field is expanding. Even though some formats may disappear, the list continues to grow.

NONRECORDABLE MEDIA

Compact Disks

A CD is a long-playing, high-fidelity audio storage medium, and its excellent sound reproduction qualities are a reflection of its digital and optical heritage. Interfering noise is reduced, and a disk can store approximately an hour of music.

A CD player, equipped with a microprocessor, allows you to quickly access any of the disk's tracks and to select a predefined playback order. These functions, in addition to a disk's small size, durability, and capacity have contributed to its popularity with consumers and radio stations.

As described in Chapter 2, the CD industry is also governed by an established set of standards. This factor played an instrumental role in the CD's widespread acceptance.

The CD has, however, been faced with competition from digital audio tape (DAT) and other systems.[3] A DAT player can record as well as play back digital tapes, and the audio quality is equal to that of a CD.

CD-ROMs

A CD-ROM is a high-capacity data storage medium. A CD-ROM, which looks like a CD but can store 600 megabytes of data, is preloaded with information and/or programs. A CD-ROM is also interfaced with a computer via a CD-ROM drive and special driver software.

An early CD-ROM release was the electronic text version of Grolier's Academic American Encyclopedia. This disk highlighted the CD-ROM's storage properties: An entire encyclopedia of some 30,000 articles was recorded on a single disk, with room to spare. It was also integrated in a PC environment, and information from the encyclopedia could be retrieved by word processing programs. More recent encyclopedias have also incorporated sound, graphics, and animations.

Another interesting earlier disk, which spawned more recent releases, was the PC-SIG CD-ROM. The PC-SIG has been a source of public domain and user-supported software (shareware) written for IBM PCs. The programs have covered everything from computer languages to games, and the entire library could fill 1000 or more floppy disks.

This library was transferred to a CD-ROM. The application was and remains a valuable one for PC owners, since the CD-ROM is inexpensive when compared to an equivalent floppy disk library. This factor has made the CD-ROM an ideal distribution medium for computer software collections. Other applications include the following:

- Information pools can be compiled ranging from telephone number compilations to U.S. street maps.
- Companies have adopted CD-ROMs to complement their print lines. For software, a CD-ROM could hold tutorials and sample files referred to in the manual. Their size may have precluded their use when floppy disks were the primary distribution vehicle.
- Magazine collections can be compiled. You can browse through hundreds of articles, or years of back issues, all on one disk.
- Spacecraft images have been made available. Through the National Space Science Data Center and other sources, you can explore Venus and Mars from your armchair by viewing information generated from the Magellan and Viking Orbiter missions, respectively.[4]
- Complex software collections can be created. This may include a desktop publishing program and an extensive clip art collection. Clip art is a library of drawings and pictures you can legally use. At one point,

the Corel Systems Corporation tapped a CD-ROM's capabilities to bundle several thousand images with Corel Draw, its high-end illustration program. The floppy disk equivalent was approximately 500+ disks.[5]

- The new generation of interactive games has a large storage appetite, especially if the games incorporate digital audio, video, or computer animations. They can, however, be accommodated by a CD-ROM or multiple disks.

CD-ROMs are also valuable for libraries, which typically face storage and budgetary problems. CD-ROMs are cost effective, can be used with a PC to search for specific information, and can save space.

CD-Interactive

The compact disk-interactive (CD-I) was originally designed as a stand-alone unit equipped with an internal computer.[6] A goal was to make a CD-I system attractive to consumers since it was self-contained, simple to operate, and could be used with a television set. CD-I applications have cut across the entertainment, educational, and business domains. In one example, a Louis Armstrong disk could be used with a CD player, but when the same disk is played on a CD-I system, you could also view relevant graphics and retrieve information about Armstrong's work.[7]

Photo CD and DVD

Kodak entered this industry with the Photo CD system. As designed, up to 100 pictures could be stored on a disk. It could subsequently be used in a compatible CD-ROM drive, or for consumers, in a player for viewing on a television. The same system could also play audio CDs.

The capability to play different disks was an important one and an industry-wide development. Instead of buying multiple players and/or PC-based drives, you use one machine. This is particularly significant in view of the growing number of formats.[8]

This system also opened up new creative possibilities. By using the Photo CD infra-structure, you could create your own electronic, disk-based clip art collection from original pictures, and store the images on a disk and retrieve them when needed.

The Photo CD, when viewed in context with the overall CD family, is actually quite amazing. What started as a new audio delivery system, the audio CD, has blossomed into a series of disks that can accommodate an array of data and applications. Game manufacturers have also emerged as key players in this arena.

Finally, a new, high-capacity CD was developed in the 1990s. Designed for commercial and consumer applications, the product was initially headed for a format war, much like the earlier Beta and VHS scenario. But a compromise was reached in late 1995 by the companies and manufacturers that had a stake in this field.

Called the Digital Versatile or Video Disk (DVD), its storage capacity far exceeds a conventional CD-ROM or CD. This capability offers producers and consumers numerous advantages.

First, unlike CD-ROMs, it would now be possible to readily integrate high-quality audio-video cuts in a game, educational title, or other multimedia production.[9] Second, it would be a boon for computer data storage—a single disk could replace a collection of conventional CD-ROMs. Third, consumers would be able to view movies with support for wide-screen television sets, advanced television systems, and other enhancements. The disk could supplant the videocassette for distributing movies for home viewing. In one setup, you may also be using a disk for recording.

The DVD could similarly replace videodisks, which are described in the next section, in the educational and home markets. Its digital heritage, when combined with its information capabilities, could make a DVD-based product a superior educational tool.

Videodisks

The videodisk is the pioneer product of the optical disk family. The videodisk has been produced in different formats, including a discontinued nonoptical version manufactured by RCA.

Videodisks primarily supported two applications. In the first, the consumer category, high-quality movies have been distributed. Typically used by videophiles, individuals who demand a superior audio-video reproduction, a videodisk surpassed consumer-based VCRs in audio and video quality. It also supported a digital audio signal, even though the picture information is analog.

In the second category, videodisks have been interfaced with PCs to create a sophisticated interactive environment. A computer controls the player and retrieves the information from the videodisk. Commercial products have ranged from educational tutorials to manufacturing-related disks.

Videodisks have also been adopted by libraries, researchers, and schools. In one application, a disk could store thousands of still photographs. It could then be reproduced and distributed for a lower cost than the price for a comparable hardcopy collection.

This function, though, was also extended to CD-ROMs and, eventually, to DVD-based products. The same operations may also be fulfilled by high-speed data networks in conjunction with video servers, high-capacity and high-speed devices that are covered in later chapters.

RECORDABLE MEDIA

WORM Drives

Data can be written to and retrieved from a WORM disk, a permanent storage medium. Optical storage systems have been used for years, including a 14-inch optical disk configuration introduced by the Storage Technology Corporation in 1983. The company's high-capacity disk could hold 4GB of data, the equivalent of two million pages of double-spaced text.[10] Contemporary WORM systems, which can be interfaced with PCs, can hold well over 600 megabytes of data. A disk can also serve as an archival storage medium in view of its expected lifetime of 10 or more years. Ten years extends beyond the

mandatory time period certain types of records must be retained.

In one application, a financial institution could use a WORM drive to create a permanent record of transactions. This could be advantageous since the information cannot be altered, and the disk could facilitate a future audit.

A WORM configuration could also be used as an identification tool. A depositor's signature could be digitized and stored. When this individual conducts a transaction, the teller could recall the stored signature for verification.

CD-R, CD-RW, and DVD

A CD-recordable (CD-R) enables you to store your own information on a disk. The CD-R is a write-once, permanent medium. After the information is recorded, it cannot be erased.

The CD-R has two important advantages. It is a cost-effective and high-capacity storage medium. CD-Rs can also tap the enormous CD-ROM drive universe. CD-Rs you create can be used with conventional CD-ROM players. Thus, you can create a disk and give it to a friend for playback on his or her CD-ROM system. Similarly, you can create or "burn" an audio disk for playback on a CD player.

In contrast, the CD-rewritable (CD-RW) functions much like a high-capacity floppy disk. You can reuse a disk multiple times, storing new and erasing old data.

A disk can also be read and used with newer CD-ROM drives. But older CD-ROM drives may not be similarly compatible, due the physical nature of the disk (less reflective).

The DVD family, for its part, also supports a recordable option in various flavors. As of this writing, they include the DVD-RAM and DVD-RW.[11] Their primary advantage? A storage capacity of 2 or more gigabytes. As the technology matures, and more computers are equipped with compatible DVD readers, this opens up new capabilities for producers. For example, you may be able to use a high-quality video sequence in a program that would have been too large for a CD-ROM.

This enhanced storage capacity had already been tapped in DVDs used for movie distribution. It became possible to include additional, related information on a disk.

The information can range from multiple language tracks to still shots from the movie. While the traditional videodisk supported some of the same options, the videodisk never achieved a substantial market penetration.

This capability has another implication. Instead of just using a DVD to distribute a movie, the process can now be one of making a multimedia production. You have to pick and choose the material as well as organize this information for easy retrieval.

Finally, magneto-optical (MO) configurations, which employ optical and magnetic principles to store and retrieve data, have been manufactured. Like the CD-RW, it is a reusable medium.

Figure 10.2
A CD-I disk. The Art of the Czars: St. Petersburg and the Treasures of the Hermitage. *An example of the rich pool of information electronic publishers have tapped. (Courtesy of the Philips Corp., photo credit Richard Foertsh.)*

Summary

Regardless of the optical drive or media, current and future systems provide us with the means to store large quantities of information. This is important: Data storage demands are increasing as the nature of information becomes more complex. In desktop video and multimedia applications, for instance, 24-bit graphics as well as digital audio and video clips can be used. While data compression can be applied, this information is still data storage intensive.

The different classes of optical systems will also continue to coexist. At first glance, a conventional CD-ROM may appear obsolete in the face of some other configurations. But CD-ROMs are cost effective, have a large storage capacity, and as of this writing, have the edge in market penetration.

Magnetic media, hard and floppy drives, will also continue to be used. Hard drives are faster than their recordable optical counterparts and are fairly compact, a critical design asset for portable PCs.

Fixed storage hard drives can also be less expensive than recordable optical drives, although this economic advantage decreases as the data storage requirement increases. Basi-

cally, when dealing with mass storage needs, it is cheaper to buy another optical disk than another hard drive.

But as introduced in Chapter 3, removable magnetic storage systems have added a new twist to this scenario. Iomega's widely adopted Zip and Jazz drives, for instance, have data storage capacities ranging between 100 megabytes to over a gigabyte. Jazz drives have also been used for audio-video applications. So when you fill a disk with video clips, you simply buy a new disk rather than a new drive.

However, in the late 1990s, the price for CD-Rs and CD-RWs fell below the $250 mark. Blank CD-R disks often cost less than $1 each, in contrast with $80 for a Jazz disk. But a Jazz drive/disk combination has a higher storage capacity and greater storage/access rate, which as described, can make it compatible with certain audio-video operations.

In essence, each medium has its benefits and appropriate application areas. You might choose to use a CD-R to distribute audio cuts, but you would choose a Jazz or fixed hard drive to store a video sequence you are digitizing. The media are complementary and, in the end, have resulted in a cost-effective and flexible storage bonanza.

OTHER ISSUES

Convergence
The rapid growth of the optical disk field highlights the growing convergence between different technologies and their respective applications. Products such as the CD-I married computer applications with standard television technology. A television became part of a computer-based entertainment and educational system.[12]

Products such as the Photo CD, for their part, have cut across the traditional and silverless photographic fields. Pictures produced as standard prints or slides could also be stored on a disk and viewed on a television. The same pictures could subsequently be manipulated with a computer.

The convergence factor has even bridged different application areas. For example, the mass storage capabilities of CD-ROMs were married to the instant update capability of online services.

A CD-ROM stored data that would normally require an extended download time, that is, the period of time to relay the information from the company to you. The online service, for its part, would supply new and updated information. Consequently, you could retrieve recent stock price information from the online service and tap the CD-ROM's data pools.[13] Until high-speed communications lines become commonplace, which would speed up data relays, this hybrid system could prove to be a workable compromise.

Privacy
The storage capabilities of optical media have raised privacy concerns. In one case, a potential CD-ROM produced by the Lotus Development Corporation triggered a public outcry. Lotus planned to take advantage of a CD-ROM's storage capabilities and sell a disk loaded with demographic data about Ameri-

can consumers. The company received so many complaints, though, that the product was dropped from its line.

This example highlights a key issue of the information age: an individual's right to privacy. Although the new technologies can enhance our communications capabilities, they can also be invasive.[14] The available options to protect yourself include consumer pressure, employing other emerging technologies, and adopting privacy regulations, as was the case with the European Community (EC).

The EC had proposed a series of strict regulations governing the collection and dissemination of personal information. The rules recalled the World War II era when information from telephone records was used for political purposes. The proposed regulations were designed to help protect an individual's privacy.[15]

Although privacy is important, some people believed the regulations were too strict and would impede the flow of information between countries. A similar situation has prevailed in the United States. Some individuals, as well as government agencies, believe there has to be a balance between protecting an individual's privacy and the government's ability to retrieve specific types of information. Questions were also raised about privacy with regard to e-mail in a business environment.[16] This topic is covered in a later chapter.

Information Retrieval
The development of optical media has enabled us to gain access to entire libraries of information through a series of small disks. Yet to tap this information effectively, suitable search methods must be devised.

In one example, a CD-ROM that serves as a database is equipped with software that functions as the information-retrieval mechanism. Depending on the package, it may also allow us to search through the information in different ways. Some systems employ keyword searches while others support more sophisticated mechanisms.

This searching capability highlights the power of the PC when combined with a mass storage device. Instead of looking through different books and magazines, you can let the PC do the work for you. If the data are stored on a disk, the retrieval process can be simplified and enhanced.

CONCLUSION

Optical disk technology has played an important role in the communication revolution. As we generate more information, these disks serve as effective storage and distribution media. They are also cost effective and can accommodate a range of applications.

REFERENCES/NOTES

1. Lasers initially store this information on a master disk. This disk plays a key role in producing copies.
2. Paul Freiberger, "CD Rot," *MPC World* (June/July 1992): 34.
3. See Chapter 14 for further information.
4. As of early 1992, the price was $20 for the first disk and only $6 per additional disk in a series.
5. Advertisement, "CorelDraw," *NewMedia,* April 1992, back cover.
6. A more sophisticated business model was later released.
7. CD-Interactive Information Bureau, "CD-I Launches Titles," *CD-Interactive News* (January 1992): 4.
8. These include the CD-ROM XA and other formats.
9. Frank Beacham, "Consortium Proposes New CD Format," *Computer Video* 2 (March/April 1995): 10.
10. Storage Technology Corporation, "7600 Optical Storage Subsystem," product description brochure.
11. Please see "Writable DVD: A Guide for the Perplexed," by Dana J. Parker, *EMedia Professional,* downloaded January 1999, for an excellent overview of different writable systems.
12. As discussed in a later chapter, the establishment of an enhanced and advanced television standard will only accelerate this trend.
13. Domenic Stansberry, "Going Hybrid: The Online/CD-ROM Connection," *NewMedia* 5 (June 1995): 37.
14. Caller ID triggered a similar response. See Bob Wallace, "Mich. Bell Offers Caller ID Service with Call Blocking," *Network World* 9 (March 2, 1992): 19, for a description of why blocking options were adopted.
15. Mary Martin, "Expectations on Ice," *Network World* 9 (September 7, 1992): 44.
16. See Chapter 20 for specific information.

SUGGESTED READINGS

Bell, Alan. "Next-Generation Compact Discs." *Scientific American* 275 (July 1996): 42–46; Dodds, Philip V. W. "Comparing Oranges and Mangoes: Another View of the Emerging Digital Videodisk." *SMPTE Journal* 105 (January 1996): 46–47; Sugiyama, Kenji and Neil Neubert. "Elements of a New Authoring System for Digital Video Disc (DVD)." *SMPTE Journal* 106 (November 1997): 762–767. In-depth coverage of the DVD.

Gunshore, Robert L., and Arto V. Nurmikko. "Blue-Laser Technology." *Scientific American* 275 (July 1996): 48–51; Liebman, Sheldon. "Optical Storage and Recording Systems: Natural Home of the Blues?" *Advanced Imaging* 13 (February 1998): 60–62. Lasers and optical disk systems.

Lambert, Steve, and Suzanne Ropiequet, eds. *CD ROM: The New Papyrus.* Redmond, Wash.: Microsoft Press, 1986; Ropiequet, Suzanne, ed., with John Einberger and Bill Zoellick. *CD ROM Optical Publishing: A Practical Approach to Developing CD ROM Applications,* Vol. 2. Redmond, Wash.: Microsoft Press, 1987. Microsoft Press's two-volume CD-ROM book series. Although published in the 1980s, they remain excellent CD-ROM and electronic/optical publishing resources.

Pountain, Dick. "The Fine Art of CD-ROM Publishing." *Byte* 19 (June 1994): 47–54. The creation of a CD-ROM, including palette optimization.

Pozo, Leo F. "Glossary of CD and DVD Technologies." Online download. Excellent and comprehensive listing of terms and systems.

Stansberry, Domenic. "Going Hybrid: The Online/CD-ROM Connection." *NewMedia* 5 (June 1995): 34–40. The convergence of CD-ROMs and online services.

Waring, Becky, and Alexander Rosenberg. "New CD-ROM Hardware Swells the Consumer Market." *NewMedia* 2 (May 1992): 12–15. A look at CD-ROM systems with sidebars on Kodak's Photo CD system and the role of game manufacturers in this market.

GLOSSARY

Compact Disk (CD): A prerecorded optical disk that stores music. The CD player uses a laser to read the information.

Compact Disk Read-Only Memory (CD-ROM): A prerecorded optical disk that stores data. CD-ROM applications range from the distribution of computer software to electronic publishing. Recordable systems (the CD-R) also exist.

Compact Disk-Recordable (CD-R): A permanent, recordable CD system.

Compact Disk-Rewritable (CD-RW): An erasable, CD recordable system.

Digital Video or Versatile Disk (DVD): A high-capacity optical disk developed in the 1990s. Designed for commercial and consumer applications, it could replace the home VCR.

Erasable Optical Disks: A class of optical disk where data can be stored and erased.

Optical Disk: The umbrella term for optical storage systems.

Photo CD: An optical disk that can store pictures. Professional and consumer applications have been supported.

Videodisk: The pioneer of prerecorded optical disks. Videodisks have been used for interactive training applications and for movie distribution (to consumers; pre-DVD).

Write-Once, Read-Many (WORM): A nonerasable but user-recordable optical disk.

11 Information Retrieval: Hypertext and Hypermedia Systems

The megabytes and gigabytes of information that can be stored with contemporary optical and magnetic systems can pose a problem. How do you organize and communicate this information?

As described in the CD-ROM chapter, software can be used as a retrieval tool. At this time, an important information retrieval system is *hypertext*, a word coined by Theodor Nelson, a pioneer in this field.

OPERATION

Hypertext

Hypertext is a sophisticated information management and retrieval mechanism that works in a nonlinear fashion. It cuts across magnetic and optical storage domains, the World Wide Web, and other information systems.[1] The concept for such a mechanism was originally conceived in the 1940s by Vannevar Bush, President Roosevelt's science advisor.

Hypertext operates in much the same way that humans think. You may, for example, use hypertext to conduct research about the new communications technologies. While exploring this topic, you see the term *laser diode*. Because it is important, the words are highlighted or underlined in our hypertext environment, indicating they are linked to other information. You move the on-screen cursor to a highlighted word and click the mouse button. The linked information is subsequently retrieved and displayed, either on a separate page or in a window. When you are finished reading, you can return to the original page with another click of a button.

As you continue, the term *semiconductor* is similarly marked. You could then retrieve information, through a series of links, about semiconductors, the semiconductor industry, and related economic implications.

By following this pattern, you can retrieve information in a natural way, that is, the way people think. You follow a train of thought that enables you to make associations between diverse topics. In this context, hypertext is no longer just a search mechanism. It provides a new way to organize, link, and communicate bodies of information and knowledge. You can also reveal previously hidden connections between topics.

A hypertext system performs these tasks effortlessly, at least for the user. The various links can also be retraced so you can return to a given source.

Hypermedia

We were introduced to hypertext programs on a wide-scale basis by HyperCard, a Mac product. Much like a series of index cards, HyperCard consists of blank, user-defined cards, which make up a stack.[2] HyperCard supports text, graphics, and other information, and you can tap a special programming language to create more sophisticated applications.

Besides HyperCard, IBM PCs and other platforms have their own hypertext software. HyperWriter, a program developed and distributed by Ntergaid, Inc., is used to explore hypertext and hypermedia concepts and applications.

The term *hypermedia* has been associated with multimedia, a subject discussed in a later chapter. Briefly, "multimedia is the integration of different media types into a single document. Multimedia productions can be composed of text, graphics, digitized sound. . . ," video, and other information. Hypermedia software, in turn, allows you to "form logical connections among the different media composing the document."[3] When you click on a highlighted term, you may see a picture or hear a sound, instead of simply a page of text.

With a hypermedia program, the links between the information, whether textual or graphic, can be represented in different ways. These include highlighted text, text placed between different symbols, and buttons. Buttons are visual markers, typically labeled, that can help create a more effective user interface.

To activate a link, simply move the cursor to the appropriate point in the document and click a mouse button or use a keystroke. The links join the different topics, essentially the document's information elements.[4]

HyperWriter supports various classes of links: those to and from text, graphics, and action. The text can be full pages of information or pop-up windows. Pop-up windows are useful for short pieces of supporting data, such as defining a term mentioned in the text. Graphics links, as implied, tie information to a graphic while action links can execute external programs and complete other tasks.

HyperWriter also supports animations and digital audio and video. This function can enhance a document and, depending on the application, deliver appropriate information. The latter may include incorporating a digital video clip in a manufacturing project to demonstrate equipment maintenance procedures.

Another essential feature is a series of navigational aids that help you find your way through a document. These range from color coding of specific link types to the use of icons that can quickly move you to a specific page (for example, the opening or home page).

Regardless of its form, a navigational aid, when properly designed, enhances the user interface. This is a vital consideration for any hypermedia author. The information is more accessible to the reader and the communication process is improved.

Hypertext and hypermedia programs also feature text editors, text and graphics import capabilities, and other functions. Text styles and colors can also be altered, and with respect to HyperWriter, different visual backgrounds can be created for aesthetic and user-interface purposes.

APPLICATIONS AND IMPLICATIONS

Hypermedia and hypertext systems are used in many applications. Software companies have created help systems to support their programs. By clicking on a term, you can retrieve linked information and read about an operation without opening a manual.

In another application, a hypermedia program can serve as an information kiosk engine. You can click on a specific area of a map to retrieve information about the location you chose, possibly including bus schedules and street names. This type of application is particularly effective when used with a touch-screen interface.

Hypermedia and hypertext software can also be used for electronic publishing. In one example, REGIS was created for the United Nations. It featured information about aquaculture in Africa.[5]

Figure 11.1

An example of a pop-up window. Note the link markers on either side of the word asteroid in the background text. (Software courtesy of Ntergaid, Inc.; HyperWriter!)

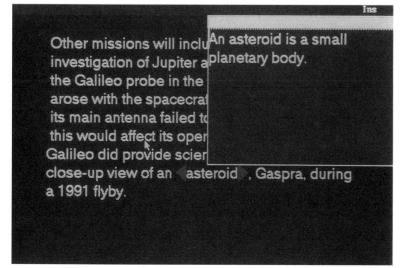

In another example, hypertext and hypermedia programs have served as CD-ROM catalog front-ends. Click on a picture, and information about that product appears on the screen. As described in Chapter 18, the hypertext concept has also been extended to the Internet through the World Wide Web.

Finally, hypermedia systems are conducive to browsing. When you choose the links you want to follow, learning becomes more dynamic. You control what you read and the trails you may follow.

Depending on the system, it may also be possible to guide the reader via a more predefined path through the document. In this way, important information is not bypassed.

But despite its advantages, a hypermedia system, like other organizational and communications tools, can be used ineffectively. When creating a document, you should keep these guidelines in mind:

1.   Much like a video or film production, a hypermedia project should be planned. Each link, like each frame, should serve a purpose.

2.   A reader can become lost in a document if there are too many consecutive links. A prototype of the final project, like a storyboard for a television production, can help point out potential problems.

3.   Know your audience. For example, the interface for a commercial catalog may be very different than one geared for the general public. The former may be designed to retrieve information as quickly as possible. The latter may have more visual effects and graphics to hold the audience's attention.

4.   Remember the old adage "form follows function." This concept should be applied when you design your product.

5.   Examine your distribution venues. Are you going to use a floppy, CD-ROM, or the Internet? The distribution medium may, for instance, have an impact on the project's graphics. For the Internet, you may use compressed and/or smaller sized graphics to speed up the information relay through the typical telephone line connection. Similarly, if you design a CD-ROM product, can it be

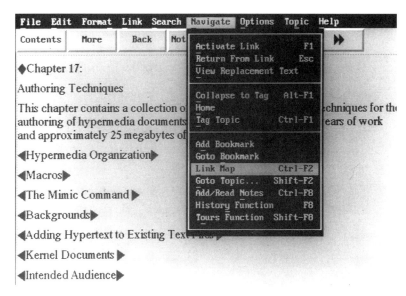

readily adapted for the Internet? Does your software support this option?

6.   Pay attention to other technical and nontechnical needs. Can your product be distributed on multiple computer platforms? Do you have to pay a fee for distribution rights? Most software is bundled with a separate player program that is distributed with your product. This program has a read-only capability and is used to navigate through your document. Some companies charge a fee for a reader; others do not.

7.   Can the product be set up in modular fashion? Instead of creating a single, large file, can it be subdivided into smaller and more manageable files that are linked together?

CONCLUSION

Through the use of hypertext and hypermedia programs, we can now create electronic documents that are only limited by hardware and software constraints. For example, hypermedia programs can be potent educational tools in which a static book can be transformed into a fluid, living document. Instead of simply reading about Beethoven, you can read about Beethoven while seeing his picture and hearing his music. In the art world, instead of viewing just one side of a

Figure 11.2

Some of the navigational tools supported by hypermedia packages. (Software courtesy of Ntergaid, Inc.; HyperWriter!)

statue, you may be able to view all sides through an animation or video clip. The various hypermedia links would enable you to explore these and other subjects in more depth and variety.

REFERENCES/NOTES

1. Please see the Internet chapter (Chapter 18) for specific information.
2. Greg Harvey, *Understanding HyperCard* (San Francisco: Sybex, 1989), 3.
3. Ntergaid, Inc., "HyperWriter," flyer.
4. Ntergaid, Inc., *HyperWriter! User's Guide and Reference Manual* (Fairfield, Conn.: Ntergaid, Inc., 1991), 5-5. *Note:* Topics are also referred to as cards or nodes.
5. Knowledge Garden, "New from Knowledge Inc.," information release.

SUGGESTED READINGS

Foskett, William H. "Reg In Λ Box: A Hypertext Solution." *AI Expert* 5 (February 1990): 38–45. Traces the development of a hypertext system concerned with regulations about underground storage tanks.

Nelson, Theodor. "Managing Immense Storage." *Byte* 13 (January 1988): 225–238. A description of the storage engine of the Xanadu project, a model for a new information storage and management system.

GLOSSARY

HyperCard: The first widely released hypertext program. Designed for Macintoshes.

Hypertext: A nonlinear system for information storage, management, and retrieval. Links between associative information can be created and activated. This concept has been extended to pictures and sounds.

Navigational Aids: Tools to help you navigate through a document. They range from color coding of links to the use of icons to quickly move to a specific page.

12 Desktop Publishing

This chapter examines desktop publishing (DTP), an application used to produce newsletters, brochures, and other documents. DTP tools include PCs, software, printers, and scanners. When combined, they create a near-typeset-quality publishing system that can literally fit on a desktop. DTP also provides us with an electronic composition tool. Formerly inaccessible publishing capabilities are now practical. For example, instead of following the cut-and-paste method to create a layout, a design is electronically composed.[1] A monitor's screen serves as a window in this process.

Various factors led to the proliferation of PC-based DTP systems in the late 1980s. More sophisticated PCs and complementary software flooded the market and the laser printer became an affordable option.

HARDWARE

The Computer

The hardware end of the typical DTP system consists of four major components: a monitor, printer, scanner, and this section's focus, the computer.

The Macintosh helped launch the PC-based DTP industry. It was more graphically oriented and easier to set up than comparable IBMs. A stock Mac, with its software library, was also better equipped to handle desktop publishing and graphics applications. When combined, these factors helped individuals who were not graphic artists to design their own projects. Graphic artists, on the other hand, could now experiment with different concepts.

IBM PCs, for their part, despite some initial hardware and software disadvantages, have also emerged as a major force in this field.

Their dominant position in the overall computer market, and the introduction of new equipment and programs, contributed to this development.

Other computers also support this application. But as described in Chapter 3, our focus is on Macs and IBM PCs.

The Monitor

Many people own 15-inch monitors. While suitable for different tasks, they cannot display a full page of readable text. Consequently, different viewing modes are used during the design process.

In the full-page mode, a page can be displayed in its entirety. This mode can reveal the placement and spatial relationship between the page's graphic and textual elements. But the page outline is noticeably reduced in size. Most, if not all, the text may be replaced by small lines or bars (a process called *greeking*) since the characters are essentially too small to be reproduced on the screen.

Other modes provide magnified or enlarged views of specific page sections. The text can be read, and fine details of the document's style can be checked. Both full-page and magnified views are regularly used.

Special monitors have also been manufactured that generate enhanced document views, especially in the full-page mode. Larger, conventional monitors, which are becoming the norm, are also helpful in this regard.

Printers

A laser printer can produce near-typeset-quality documents. Most printers support a 300 or 600 dots per inch (dpi) resolution. In the context of this discussion, the term

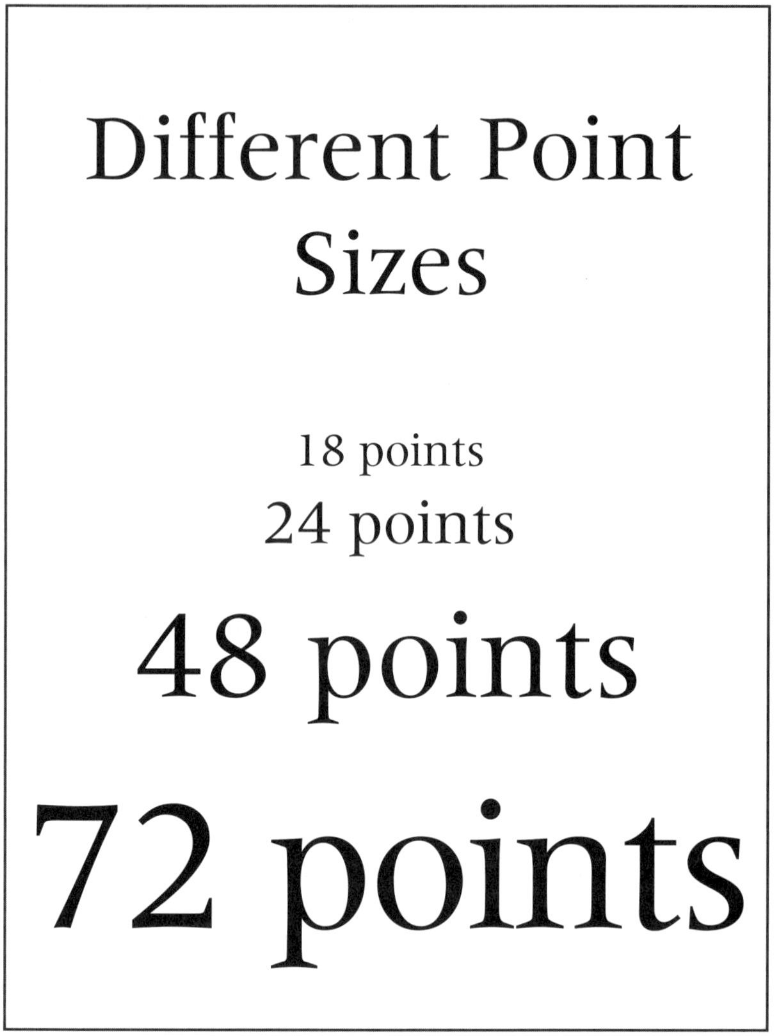

Different Point Sizes

18 points

24 points

48 points

72 points

Figure 12.1

Different point sizes. (Software courtesy of Adobe Systems, Inc.; PageMaker.)

As of this writing, most if not all printers designed for the general business and consumer markets share several broad characteristics.

First, a printer should be equipped with enough memory to take full advantage of its printing capabilities. The memory requirement increases for a color printer.

Second, various typefaces and fonts are available for the printers. The upshot? You can design a document that fits your publishing needs. A typeface is a unique print style. The different characters of a given typeface conform to a style, a set of physical attributes shared by all the characters. Two examples of common typefaces are Helvetica and Times Roman.[2]

A font is a typeface in a specific size. The size is measured in points, and as the point size increases, so too does the character's size. As a frame of reference, 72 points equals approximately 1 inch.

Third, although the typical printer cannot generate typeset-quality documents, it is satisfactory for creating newsletters, an organization's in-house magazine, and even a book on a tight production schedule and budget. High-quality line drawings, a series of black lines on a white background, can also be printed. These may range from a building to an interior view of an engine that's slated for a technical document.

Fourth, the typical laser printer cannot support full-color output. Its graphics capabilities may also be limited for reproducing black-and-white photographs. Details may not be sharply defined or too few gray levels may be reproduced.

Color Printing.   Color printers became increasingly popular in the mid- to late 1990s. They range from thermal wax to laser to dye sublimation units.[3] The latter two systems can support high-quality output, but have been comparatively expensive.[4]

Cost-effective color ink-jet units, another printer type, are suitable for numerous applications. By incorporating various technological improvements, printing quality improved, particularly when a special paper

resolution refers to the apparent visual sharpness or clarity of the printed characters and graphics. This working definition is used throughout the chapter. There's a relationship between the dpi number and a document's perceived quality. In general, a higher dpi figure could result in a higher quality document.

When the first reasonably priced laser printer appeared in the early 1980s, it created a stir in the computer industry. The printer could handle some of the printing jobs that had been reserved for traditional typesetting equipment. This trend, started by the Hewlett-Packard LaserJet and the Apple LaserWriter printers, has continued unabated.

designed for these units is used. The paper, which looks much like glossy photographic paper, helps prevent the ink from smearing. The end result? An image that can look like a conventional photograph. Specialized printers have also been designed to reproduce large-format prints.

Depending on your needs, you may even opt for commercial printing. While new hardware and software releases may make it easier to prepare color images for this task, it is still a complicated process.[5] Consequently, it is a good idea to discuss a project with a commercial printer to obtain the best results. You may also decide to leave the work to individuals who are well versed in this field.[6]

Finally, color can be an important DTP element. Color can catch a reader's eye and can help convey information more effectively. Think of a bar chart showing a radio station's ratings versus its competitors. Different colors may make it easier to differentiate the information. In another example, color may produce a more visually appealing ad.

PostScript.   PostScript is a page description language (PDL). Developed by Adobe Systems and popularized by Apple, it emerged as a DTP standard.

In essence, PostScript is one of the software mechanisms that makes it possible to output information to printers and other devices. For our present discussion, PostScript is also a programming language

whose sole purpose is to precisely describe the placement and appearance of text and graphics on a page or pages. . . . The language must establish conventions for the printing device. These include telling it . . . what type fonts are in use. . . . PostScript is an interpreted language . . . the computer must "interpret" the statements into a machine-executable form. A PostScript printer must have its own internal computer to do this.[7]

Thus, PostScript defines graphic and alphanumeric elements. A PostScript-based system is also hardware independent, unlike some other standards. This is a valuable feature since a document can be manipulated by a wide range of equipment. You can also create a proof copy with a standard printer and then generate the final output with a commercial unit.

Scanners

A DTP project may include photographs, line drawings, and other artwork originally produced in hardcopy form. This operation is made possible by using a scanner, a piece of equipment interfaced with the computer. For example, a black-and-white photograph can be placed on the scanner, much like a piece of paper on a copy machine. The image is read or scanned, and the picture information is digitized and fed to the computer.

At this point, the image can be manipulated with graphics software. It can be edited, the contrast and brightness levels can be changed, and special filters can be applied. The now altered image can then be saved and imported by the DTP software.

If a hardcopy is produced with our laser printer, a digital halftone method is employed. Because the printer cannot produce true shades of gray and only prints black dots on a page, the halftone method creates gray-level representations or simulated gray shades. The image is divided into small areas or cells. The picture's various gray shade representations are subsequently generated by turning dots in these areas either on or off. This varying density of black dots, and ultimately the cells, creates the various apparent shades of gray throughout the picture.

There is also a balance, especially when using a typical PC-based configuration, between the number of gray levels and the picture's resolution. As the number of levels increases, the apparent resolution drops.[8]

Finally, for desktop color work, the final print quality is affected by proper color registration, the quality of the printer, and other factors.

Gray-Level and Color Scanners.   When introduced in the general market, scanners did not capture color or gray-level information. This has changed, and the typical

scanner now supports 256 levels of gray and 24+ bit color. Contemporary systems also capture an image in a single pass or scan; previously, three passes had been the norm.[9]

Scanners also come in different flavors. In one example, when desktop units were still expensive, manufacturers developed less expensive handheld scanners. You physically moved the scanner down the page to capture the information. While cost effective and suitable for various tasks, there were limitations.[10] Specialized scanners have also been created, including very high resolution units geared for commercial printing.

The second component of a scanning system is the software—the mechanism by which the computer controls the scanning process. Pioneered by Ofoto, the latest generation of software can automatically generate higher quality scans or images. This capability also makes the technology more accessible. You do not have to be an expert for general work. You supply the aesthetic framework, and the program can help you produce a better product. Manual adjustments and various image editing functions are also supported.

In a related area, film scanners are becoming more popular. Newspapers have used such systems to send photographs between local and international locations. In a typical situation, a roll of film is developed, and select film images are digitized by a scanner. They are then compressed, relayed over a telephone line, and received and reproduced at the home office.[11] This system saves time and has provided for a more error-free relay.

Optical Character Recognition.   As introduced in Chapter 4, a scanner can also be used with optical character recognition (OCR) software. In a typical application, a printed document is scanned and the software recognizes the text. The information can then be saved and used with a word processing program.

An OCR system can save time and labor but may have limitations. Only a certain number of typefaces can be recognized, and the characters must be legible and fairly dark. Some characters will also be incorrectly read and must be replaced during an editing session.

To overcome some of these problems, the program may support a learning mode. You teach the software/computer to recognize incompatible type. More recent packages also recognize a wider range of type and have reduced the number of read errors.

SOFTWARE

The heart of any DTP system is the software. Our discussion focuses on two PC program categories, word processing and page composition programs.[12]

Word Processing Software
Newer word processing programs can complete some of the jobs once reserved for DTP software. These include generating articles, business forms, and newsletters. As described in Chapter 3, programs incorporate different functions to assist the writer. These include the following:

- Spelling and grammar checkers.
- A macro capability. You complete a command with one or two rather than multiple keystrokes.
- A what-you-see-is-what-you-get (WYSI-WYG) display on your monitor. As you compose a page, you visually see its margins, graphics and other elements. You

Figure 12.2.

An OCR operation. The text is reproduced in the top of the screen (top window). The window in the left, bottom corner, shows the original, scanned text. The learning mode is also activated so the system can be "taught" to identify letters, for example, that were previously unrecognized. (Software courtesy of Image-In, Inc.; Image-In-Read.)

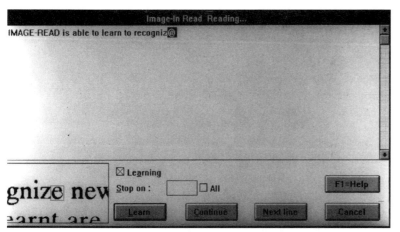

don't have to wait for the page to be printed to view it.[13]

- An ability to create tables, charts, and indexes.

A sophisticated word processing program can handle a range of applications, including those that do not demand the full power of a DTP system. A word processing program can also be easier to use, faster, and can produce high-quality output.

Page Composition Software

Page composition (DTP) programs also support a WYSIWYG display and an interactive interface. Like many word processing programs, as you move a graphic or column of text, the changes take place in real time. This visual feedback helps you determine if a page design is satisfactory, and it allows you to quickly experiment with different layouts.

DTP software also sports numerous enhancements. For example, text and graphics can be accurately placed via alignment aids, and extensive text and graphics manipulation modes are supplied. The discussion that follows provides a general overview of select operations.[14]

Text Manipulation. When you create a document, the program displays an outline of a blank page. Next, you can open up one or more columns for text placement. Text and graphics can also be placed in resizable, movable blocks or frames. For example, you can create a newsletter that's formatted like a newspaper. The first page consists of columns of text, assorted illustrations, and the newsletter's masthead.

For the masthead, a mouse is used to create a long and narrow frame across the top of the page. An appropriate typeface can then be selected, as can stylistic elements. These include printing the text with a shadow effect.

A DTP program also has a word processing module, but you may prefer to use your favorite software to type the text. It may be faster, and you may be more comfortable with its functions. The DTP program can subsequently import this file.[15] When retrieved,

it can be placed in designated columns and spaces.

If the file is large and one column fills up, the text can be routed to another column. The routing can be automatic, or you can manually designate the next column the text should fill.

A program can also compensate for editing. As words are added or deleted, the text flows or snakes from column to column until the proper space adjustments are made.

This control over the text also extends to the physical spacing between individual characters and sentences. In kerning and leading, respectively, the space between specific pairs of letters and between individual lines can be altered. This capability can enhance a document's appearance and readability.

DTP programs can also import different file types. These include data from graphics, word processing, and spreadsheet programs. It may also be possible to rotate a line of text and to produce other special effects.

The trend appears to be toward the creation of more self-contained programs. As modules are added and refined, the program may be able to handle more tasks without tapping other software.[16]

Graphics Manipulation. Many DTP programs can create simple graphics. But like the

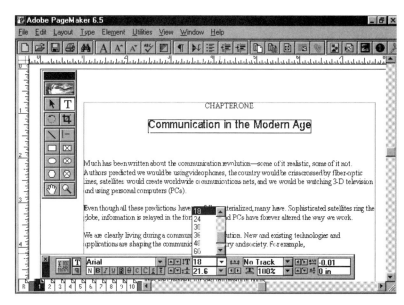

Figure 12.3

The PageMaker's DTP program. Its Toolbox is used for drawing, rotating text, and other options (upper left corner). The control panel, visible at the bottom, can be used to quickly manipulate text/graphics. (Software courtesy of Adobe Systems, Inc; PageMaker.)

word processing function, you will probably continue to use a dedicated graphics package.

Once a graphic is created—an artist's rendition of a mountain to illustrate, for instance, an article in a newsletter—it is imported. At this point, the graphic can be resized. It can also be cropped, so only a portion of the entire image appears. You can then move the graphic to other positions, and if supported, wrap text around the image. Thus, instead of seeing separate and distinct blocks of text and graphics, they can be more integrated.

Templates. A DTP program may be sold with templates. A template specifies the design of newsletters, books, and other documents.[17] Instead of spending hours designing a publication, you can use a premade template, which delineates the document's physical appearance. Custom templates can also be created, stored, and recalled when necessary.

Besides helping novices, a template can be useful to people who are in a hurry or who cannot create an effective design. For an organization, it can also bring a sense of order and uniformity to reports. A large company, for instance, can house several or more divisions that produce their own reports. If a format is not adopted, a report's structure could vary from division to division. This could hinder communication, especially if a document does not present the material in a clear and logical fashion.[18] It may also detract from the company's look or corporate identity, recognizable physical attributes that readily identify the company to internal and external parties.

DTP GUIDELINES

Now that we've covered DTP basics, it is appropriate to discuss some general guidelines:

1. Institute a DTP training program. The software may be complicated and have a steep learning curve.

2. A DTP system will not turn everyone into an artist. It is simply a tool to present ideas and information more effectively. Paraphrasing Clint Eastwood in one of his *Dirty Harry* roles, you've got to know your own limita-

tions (and strengths). Do what you do best. But if you need an artist or other professional, hire one. You will save time and money and are likely to get a superior final product.

3. Plan and effectively use a page's white space. It can provide visual relief for the reader and serves as a design tool. White space can highlight and focus attention on specific page elements. This concept also extends beyond the DTP industry. It is an established technical writing axiom.

4. Pay attention to the basics: grammar, typos, and spelling mistakes. Proofread the document after you are finished. Do not try to catch every kind of mistake in one reading.

5. When designing a document, keep it simple, if appropriate. Although a DTP program may support many typefaces and special printing effects, do not use them all on the same page. The document may be difficult to read, and the information may be lost in a maze of fonts and double-underlined text.

6. Read. Hundreds of DTP books and magazine articles have been written that cover everything from aesthetics to scanning to tips from professional DTP users. If you are learning how to use a DTP system, it is also important to practice your craft. Experiment with what you've learned and develop your own style.

7. Explore your software's other capabilities. For example, it may also support document generation for the Internet and/or other electronic publishing venues. Instead of printing on paper, you print to an electronic medium, thus extending your publishing options via the same program.

8. Use your imagination. If you have a tight budget, do not have color equipment, and the project requires a color picture, a color copier may work. Design the page, position the picture, and press the Start button. Depending on the job, the output may be satisfactory.

APPLICATIONS

Personal Publishing
Prior to the DTP revolution, if you wanted to publish a book, you generally had two

options. You signed a contract with either an established publishing house or a vanity press. For the latter, you paid a publisher for printing and possibly distributing your work. DTP systems provide authors with a third choice—to act as your own publisher.

As the author/publisher, you are in control of the entire process. You can make last-minute changes and updates, and a standard laser printer may suffice as the printing press. Or a higher dpi commercial unit can be used for the final copy.

In a variation of this theme, you can initiate what Don Lancaster has called book-on-demand publishing. You do all the work yourself, including the printing and binding. But instead of producing an initial run of 500 or more copies, you print a book only when someone orders it. This reduces the up-front costs for materials, and each copy could literally be an updated version of the original book.[19]

New authors/publishers do, however, face some constraints. These vary from individual to individual and may include your budget, experience, and DTP system's level of sophistication, all of which can have an impact on your final product.

You must also face the dual problem of promoting and distributing your work. Although the structure for this market is evolving, it may still have to mature to provide a more established support mechanism. You also do not have access to the editorial and technical expertise afforded by a traditional publishing house.

Other Applications

DTP systems have been employed in other applications. These include the publication of technical manuals, year-end reports, ads, information flyers, posters, and newsletters. They have also been adopted by traditional media organizations because DTP systems can save time and money.

The New Yorker slowly integrated DTP technology in its operation. The move was initiated to speed up certain tasks and to save money.[20] In the newspaper industry, PCs equipped with the appropriate software have been used for photographic preparation and editing.[21]

Publishing companies have also used DTP technology. Since DTP systems are cost effective, an organization can potentially publish more books and take a chance with a manuscript geared toward a narrow audience. DTP may also be applicable for books that are regularly updated.

Individuals have also benefited. Besides book-on-demand publishing and other ventures, customers can use DTP stations in printing and copy shops. You can produce a resume or other document with this system, typically at a lower cost than the traditional typesetting route.

Finally, in a related application, magazine publishers have created electronic versions of their products and have placed them on the Internet. Although the core of the information may remain unchanged, a conversion process is typically used to prep the information for this environment.

The Internet and related electronic media can offer a publisher additional capabilities through the use of interactive links, animations, and digital audio and video cuts. A publisher can also update information in a more timely fashion and provide in-depth coverage. If space limitations force an article(s) to be cut from the print version, it can possibly be housed online. Back issues can also be made available, and this service could be used to attract new, potential subscribers. Publications that have taken this route have included *Advertising Age*, *NewMedia*, and *HotWired*, the Web version of *Wired*.[22]

CONCLUSION

The DTP industry is still maturing. Besides the developments outlined in this chapter, the industry will also benefit from the convergence of different technologies and applications. Entire font and clip art collections can, for instance, be stored on CD-ROMs. There is also an overlap in the area of graphics software. A CAD program can be used to design a building. In a DTP project, the same program can create an illustration.

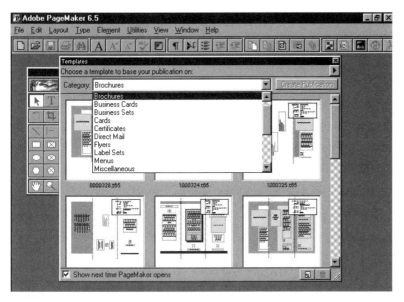

Figure 12.4.

Desktop publishing programs typically include templates or premade publication designs. In this shot, some of Page-Maker's brochure options are highlighted.
(Software courtesy of Adobe Systems, Inc; PageMaker.)

Advancements in one field can also have an impact on another field in the new technology universe. Desktop publishing is no exception. Newer, powerful PCs speed up various tasks, and for DTP, a project can completed more rapidly. These same machines can also support sophisticated software that was once the domain of larger and more expensive computer systems. In one application, DTP users can correct and prepare images for publication.

In two other examples, progress in the overall laser market will have an impact on DTP systems (for example, laser printer development), while the proliferation of networks could promote DTP-based operations. An electronic publishing environment would make it possible for more than one person to retrieve, review, and edit a document.

The concept of networking also cuts across national and international boundaries. High-speed digital lines can tie offices around the world in a global communications net. For DTP, you can gain access to the data stored on other networks, information can be rapidly exchanged, and ultimately important resources can be shared.[23]

The DTP field has also helped promote the growth of a *personal communications tool*. With a DTP system, an information consumer can now become an information producer. Information can also be tailored for a narrow rather than a mass audience, as may be the case with newsletters, pamphlets, and even book-on-demand publishing projects.

REFERENCES/NOTES

1. *Cut and paste* refers to the manual positioning of columns of text and/or graphics on a page.
2. You can buy font libraries on CD-ROMs.
3. Please see Tom Thompson, "Color at a Reasonable Cost," *Byte* 17 (January 1992): 320, for a discussion of thermal wax units.
4. Dye sublimation units also produce a continuous-tone output, in contrast to halftone-based systems, described in a later section of this chapter. Continuous-tone images look more like conventional photographs. For specific information, see Tom Thompson, "The Phaser II SD Prints Dazzling Dyes," *Byte* 17 (December 1992): 217.
5. John Gantz, "DTP Is Inching Toward Color, But Don't Hold Your Breath," *InfoWorld* 13 (June 10, 1991): 51. See also Janet Anderson, Philip A. Borgnes, Carol Brown, et al., *Aldus PageMaker Reference Manual* (Seattle, Wash.: Aldus Corporation, 1991), 76–83, for an excellent overview of color printing via a DTP system.
6. If you decide to go the commercial route, you can use process- or spot-color printing. For more information, see Eda Warren, "See Spot Color," *Aldus Magazine* 3 (January/February 1992): 45.
7. Daniel J. Makuta and William F. Lawrence, *The Complete Desktop Publisher* (Greensboro, N.C.: Compute! Publications, 1986), 86–87.
8. Image-In, Inc., *Image-In* (Minneapolis: 1991), 178. *Note:* Line drawings are not affected by this factor.
9. By capturing color and gray-level information, you can take full advantage of image editing software; scanning the same image multiple times, to produce a color output, has also been used in NASA's outer space probes and missions.
10. Problems could include uneven scans.
11. Barbara Bourassa, "Mac Systems Speed Photo Transmissions," *PC Week* 9 (February 17, 1992): 25.
12. Besides word processing and page composition software, another program category has

supported page formatting/design options. A series of codes implements a page design. This type of program has been written for professional computer typesetting systems and PCs. A print preview mode may have also been supported, and this type of program could typically handle difficult formatting jobs.

13. Depending on the system, WYSIWYG may more aptly be called "what-you-see-is-probably-what-you'll-get" in the output. There may not be an exact one-to-one correspondence.

14. Note that the terms used can also vary from program to program even though the basic concepts hold true.

15. Most DTP programs can import from a broad range of word processing programs.

16. As described, word processing programs have also been enhanced. One price, though, for this increased sophistication, may be software that is almost too powerful for simple tasks. A program may also place a higher processing and data storage demand on the host PC.

17. Virginia Rose, *Templates Guide* (Seattle, Wash.: Aldus Corporation, 1990) 3.

18. When working with text, it is also possible to use a style sheet. In essence, a style sheet defines the attributes of a document's different elements, such as a headline and body text. A headline may be centered and set in a specific typeface. To use this style, highlight the appropriate word(s) during editing, select the headline style, and the text will be automatically reformatted.

19. Don Lancaster, "Ask the Guru," *Computer Shopper* 9 (September 1989): 242.

20. James A. Martin, "There at *The New Yorker*," *Publish* 6 (November 1991): 53.

21. Jane Hundertmark, "Picture Success," *Publish* 7 (July 1992): 52.

22. Pat Soberanis, "Born Again Publications," *Publish* (November 1995): 82. *Note:* CD-ROMs have also played a role in this electronic publishing environment.

23. Lon Poole, "Digital Data on Demand," *MacWorld* (February 1992): 227.

SUGGESTED READINGS

Aldus Magazine 3 (January/February 1992). The following articles cover selecting a paper type, the history of paper, and a history of offset li-

thography: Beach, Mark. "Paper in the Short Run." 33–36; Stratton, Dirk J. "Down the Paper Trail." 80; Vick, Nichole J. "Oil and Water." 19–22.

Alford, Roger C. "How Scanners Work." *Byte* 17 (June 1992): 347–350; Larkin, James. "Scanning 101." *Aldus Magazine* 4 (June 1993): 21–26. Scanning operations and basics.

Bishop, Philip. "Crimes of the Art." *Personal Publishing* (May 1990): 19–25; Parker, Roger C. "Desktop Publishing Common Sense." *PC/Computing* (March 1989): 151–156; "Desktop Quality Circa 1992." *Business Publishing* 8 (January 1992): 23–29; "Publish Special Section; 101 Hot Tips." *Publish* 7 (July 1992): 63–88; Warren, Eda. "See Spot Color." *Aldus Magazine* 3 (January/February 1992): 45–48. While the articles may be older, they offer an array of tips, guidelines, aesthetics, and effective design.

Gass, Linda, John Deubert, et al. *PostScript Language Tutorial and Cookbook.* Reading, Mass.: Addison-Wesley Publishing Company, 1985. Tutorial on PostScript.

Hitchcock, Nancy A. "How New Digital Papers Will Impact Designers." *Electronic Publishing* 22 (December 1998): 32–40; Vaughn, Bill. "Are We Running Out of Trees." *Aldus Magazine* 4 (September/October 1994): 32–40. Two looks at paper: design and production issues (for example, alternative sources/shortages).

Pennycook, Bruce. "Towards Advanced Optical Music Recognition." *Advanced Imaging* 5 (April 1990): 54–57; Ward, Noel. "Digital Printing Goes Mainstream." *Electronic Publishing* 24 (February 2000): 32–36. Two desktop publishing applications.

GLOSSARY

Desktop Publishing (DTP): A term that describes both the field and process whereby near-typeset-quality documents can be produced with a PC, a laser printer, and software. DTP also implies that you have access to enhanced layout and printing options.

Font: A font is a typeface in a specific size.

Optical Character Recognition (OCR): Either a stand-alone unit or a software option for a scanner that makes it possible to directly input alphanumeric information from a printed page to a computer.

Personal Publishing: Desktop publishing systems make it possible for individuals to produce and potentially market their own work.

Scanner: An optical/mechanical device that is interfaced with and subsequently inputs graphics or text to a computer.

Typeface: A specific and unique print style.

13 Desktop Video and Multimedia Productions

The term *multimedia* can describe the integration of graphics, audio, and other media in a presentation or production. This chapter explores multimedia authoring software used to create such a production. Other topics include hardware, applications, and aesthetic considerations.

The chapter also covers desktop video. Analogous to desktop publishing, we can now create our own video productions. PCs are an integral element in this process and are used for applications ranging from editing to creating graphics.

The desktop video and multimedia fields are also complementary. Desktop video tools can contribute to a multimedia production, and desktop video can be categorized as a component of the broader multimedia market.[1]

MULTIMEDIA

Multimedia presentations are not new. Videodisks and accompanying software have served as a multimedia platform for years. But other products and factors, discussed in the sections that follow, have primarily spurred the field's growth. Other multimedia resources that are not discussed in this chapter include hypermedia and presentation programs, which may have certain multimedia authoring capabilities. A conventional programming language may also support this field.

Software Overview

In the context of our present discussion, the focus is on the dedicated multimedia authoring program. Many are user-friendly in that you do not have to be a programmer to create a finished product. One software category has used a visual metaphor for this task. A series of icons, representing different functions, is linked to create the presentation. The icons are used as audiovisual and program control building blocks—one icon may play an animation while another may create a loop. When reached, a loop causes a series of events to repeat.

Other software categories support menu- and text-based scripting interfaces. With a menu-based program, commands that perform various functions are selected from pull-down menus. In one example, you can assign an on-screen button an action, such as playing a sound, by selecting the command from a menu. Depending on the software, it may also have a complementary programming language for additional functions.

A text-based interface uses a series of written commands. The program, as has been the case with Asymetrix's ToolBook line, may also be equipped with control panels and other

Figure 13.1
This shot highlights two features of After Effects, a powerful and popular digital postproduction tool. (Software courtesy of Adobe Systems, Inc; After Effects.)

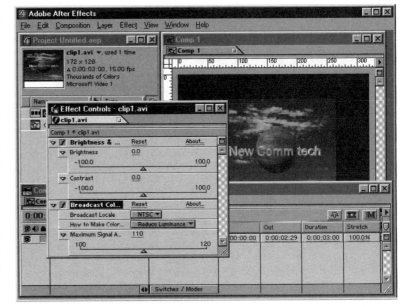

prescripted multimedia objects. They help simplify and speed up the authoring process.

As you create your presentation, you may also be able to use transitional effects. These could include a fade-to-black and a supporting audio effect.

Impact.   Authoring software has opened up multimedia production to a broader user base. If you are visually oriented, you can use visual tools to create a project, much like a desktop publishing document. You can import text and graphics and create on-screen buttons. These elements can subsequently be moved, resized, and linked to other information, or as described, to actions. From another standpoint, it is almost like using a PC-based Colorform set.

Nevertheless, although authoring software may be easier to use than conventional languages, you must still abide by a programming convention. To sustain an effective multimedia environment, the presentation must flow logically from event to event.

You should initially draft a set of criteria that drive the program's design. These should include the program's purpose and the best way to satisfy these goals with your software's tools. You must also adhere to technical and

aesthetic guidelines. The former may include the number of on-screen colors a program can support.

Other Capabilities.   An authoring program can also create a highly interactive production. If you click on a specific area of a graphic, an event can be triggered. A digitized voice can be heard or a video clip played.

Authoring systems may also be more flexible than conventional programming tools for specific data handling tasks. The seamless integration of audiovisual elements in a multimedia production and hardware control are two examples. The key word in the last sentence is *seamless*. Unlike a conventional language that may require add-on software modules or special programming hooks, the capability is built-in and fully integrated in multimedia authoring packages.

A program may also extend your production capabilities to the Internet. Macromedia's Director, one of the leading multimedia production tools, has made it possible to play optimized versions of its projects in this environment. The same may hold true for presentation programs. In one example, a computer presentation slated for a group of investors could similarly be exported and reviewed via the Internet.

Hardware Overview

Optical media have emerged as key multimedia distribution tools. Consequently, CD-ROM and DVD drives, the latter of which can also read CD-ROMs, are essential multimedia components.

Another important multimedia feature is a PC's audio capability. Typical applications are playing music and audio cuts, as well as editing your own pieces. Unlike a traditional setup where you may physically cut and splice tape, you work with a computer. You can digitize your voice, view it as a waveform, and subsequently edit and alter it. You can also select an audio effect, such as an echo or reverb, mix your voice with music, and save different variations of the same piece. In essence, the software provides you with a computer-based audio console.

Figure 13.2

PCs, when combined with software, can create an audio editing system. You can view an audio sequence and edit/manipulate a specific section (which appears highlighted). (Software courtesy of Syntrillium Software Corp.; Cool Edit Pro.)

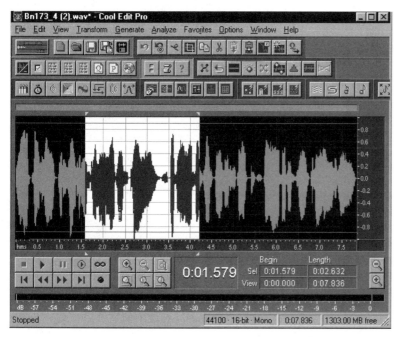

Sound effects are also common and can serve as an audio cue. For one project, a bell, chime, or other sound can provide a user with feedback confirming the selection of an on-screen button.

These effects, as well as music clips, can also be purchased. This is an important consideration because audio can make a presentation more effective.

The video display is another consideration. The latest video cards can support lifelike (photorealistic) images, and in conjunction with the PC and software drivers, may offer accelerated operation.[2]

Beyond these devices, the primary hardware consideration is the computer itself. Multimedia authoring is generally hardware intensive, regardless of the platform. The PC should also be equipped with a fast, high-capacity hard drive and as much RAM as you can afford.

Finally, Apple and Amiga computers were well ahead of the IBM PC world with respect to hardware compatibility. Stock systems were also able to support a multimedia operation more effectively. But the scale started to balance with later Microsoft Windows releases and, in fact, they helped fuel this market's growth. Windows 95 also supported a plug-and-play capability, much like that of Macintoshes.[3]

DESKTOP VIDEO

Desktop video is basically what it sounds like. You can set up a video production system on a desktop. PCs can control video equipment and are used to create video productions.

Some equipment may be geared for consumers, others target professionals, while a third category crosses both brackets (prosumer). The equipment's durability, speed, and sophistication can also vary, based on the intended application, budget, and buyer. Nevertheless, the new generation of hardware and software has blurred some of the distinctions between the professional and consumer worlds.

As discussed in a later section, this development also empowers people. Individuals and smaller organizations can now tap into the power of production tools that were once the domain of established media groups.

The following subsections cover the basic equipment and software used in desktop video applications. Although it is important to examine the capabilities of individual elements, it is also important to view them as a system.

A component may also have multiple applications: A PC may be used for creating graphics and for editing.

Hardware Overview

Frame Grabbers and Motion Video Capture Cards. A digitizing system usually consists of a camera and an interface device. Much like a scanner, the system converts images into a computer-compatible format. Yet, unlike typical scanners, a camera can work with three-dimensional objects.

A frame grabber can capture and digitize an image in real time, in 1/30th of a second, the standard video frame rate. In a typical application, you feed a video camera's signal to the frame grabber. A specific frame, essentially a still image from the video, is captured. The image can be manipulated by a graphics program.

Other cards allow you to capture motion video, moving images. This capability plays an important role in PC-based digital video production and editing systems.

A card's capabilities can vary. It may be capable of capturing high-quality, full-screen motion video. As covered in the next chapter, these type of cards are used in corporate and professional video production environments.

Other cards may only capture information at a reduced size and frame rate. The technical quality of this information can also vary.

Edit Controllers. PC-based edit controllers are used to control editing VCRs. When you edit a production, a story is electronically created by selecting, organizing, and joining individual shots.

Desktop video systems can range from cuts-only to A/B roll to nonlinear configurations.

In the first setup, shots are linked. In the second, you can make more sophisticated transitions, such as incorporating a dissolve, where one image is gradually replaced with the next. These transitions are more complex and require additional VCRs and other equipment.

In the third environment, you can edit a production using a visual metaphor—a timeline where video clips are arranged in sequential order. This topic is discussed in depth in the next chapter.

All in all, an editing system has also become a realistic production option for desktop video users. A fairly sophisticated system costs thousands of dollars less than earlier PC-based and stand-alone systems.[4]

Musical Instrument Digital Interface. PCs can also take advantage of the Musical Instrument Digital Interface (MIDI). This standard made it possible for electronic musical instruments developed by different manufacturers to communicate with each other, and ultimately, with computers.

As part of a larger system, MIDI can also help create a powerful composing tool. In a typical application, a computer is linked with a synthesizer, an electronic musical instrument. This connection is made through the PC and synthesizer's MIDI ports, and a program can subsequently turn the computer into a sequencer. A sequencer essentially is a multitrack machine that can record and play back multitrack compositions.

You can manipulate the synthesizer's notes with the PC, since these notes, the events, are viewed by the computer as another form of data. The actual sounds are not recorded, but rather, information detailing the performance. The "speed at which the key was pressed" is one such piece of information.[5]

You can also create and edit musical compositions. A section of a track can be copied, the key can be transposed, and the tempo can be altered. The final piece can then be played back under the control of the PC.

The computer–MIDI marriage offers other advantages. A synthesizer can create a range of sounds that can vary from a harpsichord to a pulsating tone, a special effect suitable for a science fiction movie. The parameters that constitute a sound can also be saved on a disk, and it is possible to store, manipulate, and recall entire sound libraries. Collections of premade sounds are also available.

The MIDI revolution was brought about by the adoption of the standard in the early 1980s. Electronic equipment manufacturers agreed to follow this common standard to enable musicians to link different synthesizers, helping to fuel the growth of the electronic music industry. Previously incompatible and expensive instruments could now be interfaced, and musicians were handed a set of creative tools.

Other Considerations

Compression. The term *compression*, for our purposes, refers to data compression. Digitized video, audio, and certain image files have enormous storage appetites. By using software and/or hardware schemes, data can be compressed and stored more efficiently.

Compression can be lossy or lossless.[6] *Lossy* means some of the data are lost through the compression scheme. For desktop publishing and certain other applications, this may not be a problem since the loss can be negligible.[7] *Lossless*, on the other hand, implies there is no loss of data. Lossless techniques are used in the medical field and other areas where data loss may be unacceptable.[8]

Popular compression options have included the Joint Photographic Experts Group (JPEG), the Moving Picture Experts Group (MPEG), and by extension, MPEG-2. JPEG and MPEG were originally and nominally geared for still and moving images, respectively. MPEG-2 has emerged as an important tool for numerous video applications.[9]

Compression's importance also extends beyond the multimedia and desktop video fields. The development of efficient standards is critical for enhanced data relays through communications networks. Both developments, advanced communications systems and compression techniques, go hand in hand. Compression will be discussed again in chapters covering high-definition television, the Internet, and teleconferencing.

QuickTime. The QuickTime architecture or standard gave the multimedia and desktop video fields an enormous boost. Released by Apple in the early 1990s, it brought, in part, compressed digital audio and video to the Mac at a reasonable cost. Designed to run on most models, QuickTime made it possible for Apple users to readily tap these resources. For example, playback did not require additional hardware, and digital video and synchronized audio became an integral component of the overall Mac system. Multimedia and desktop video support was now a built-in function, not an afterthought or a hardware and software kludge.

Since that time, enhanced QuickTime versions have been released. More important to the overall market, QuickTime supports Macs, IBMs, and Internet distribution. QuickTime serves as a representative example of how this type of standard, which includes Microsoft's Video for Windows, can be used.

In a typical application, you can capture an audio-video clip via your Mac and capture board. After it is saved, the clip and accompanying audio can be manipulated by an editing program.

One package, which illustrates this software's capabilities, is Premiere. The program can join clips using wipes, dissolves, and other transitions. Special creative and image correction filters can also be applied, and animations, still images, and multiple audio tracks, are supported.

Premiere's visual interface makes it an intuitive program. Clip sequences can be quickly rearranged, transitions can be added or deleted, and audio tracks, graphically represented as waveforms, can be manipulated.

Once you complete the project, it can be previewed. If it looks and sounds good, it can be made into a QuickTime movie, where its size and other output characteristics can be controlled.

The movie can be used by itself or with other QuickTime-aware or -compatible software. Instead of importing a still image, you import a movie. Place the movie in a document, click on it with a mouse, and the movie plays. It can turn a static document into a moving one. These characteristics contribute to QuickTime's value as a communications tool. It can be used with numerous software packages, and as indicated, does not require special hardware for playback.

Hardware components do, however, have an impact on the system's performance. These range from the computer model to the use of a hardware-based versus a software-based compression system.[10]

It is also important to examine QuickTime and other standards from a systems approach. When first introduced, a QuickTime movie may have looked great to a computer user. It was cost effective and extended a PC's production capabilities. But a person working with video systems, on the other hand, may have looked at the same movie and wondered what the fuss was all about.[11] Images were typically small, and as the movie's size was scaled up, clarity decreased. The frame acquisition and playback rates could also be low, which could make motions look jerky (not smooth).

The truth probably lies somewhere between both extremes. When first introduced, PC-based digital video generally did not

Figure 13.3

Premiere's work environment. The construction window shows two clips that will be edited together. (Software courtesy of Adobe Systems, Inc.; Premiere.)

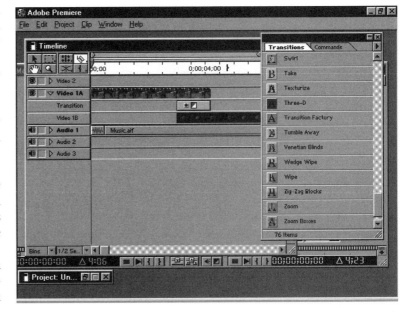

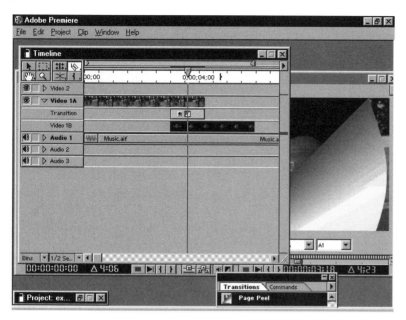

Figure 13.4

Premiere also supports a wide range of transitions, one of which is a Page Peel, as highlighted in this shot. (Software courtesy of Adobe Systems, Inc.; Premiere.)

match the quality of conventional high-end systems. But they were relatively inexpensive to implement and could be accommodated on a network.

Typical Application

You may have to design a computer-based presentation on short notice. To open the project, you may want to incorporate a short digital movie to grab the audience's attention. To complete this work, follow these steps:

1. Use an audio program to digitize the narration, your voice. If applicable, add a special effect (for example, an echo).
2. Mix it with a music clip.
3. Digitize the different video clips.
4. With a program like Premiere, create an opening title and edit the video clips together.
5. Save the final combination as a digital movie, and use it as part of the presentation's opening. By using a PC-based configuration, you can quickly produce and seamlessly integrate the movie in the presentation.[12]

It is also important to note that QuickTime-type products have improved with age. But limitations do still exist, depending on your hardware and software's capabilities. For example, there is a range of video capture cards on the market. Some PCs even sport internal video inputs. They may not, however, be able to support a broadcast-quality output. The digital video you make may be the perfect complement for your presentation, but it may not be suitable for ABC.

There is another caveat in this discussion. FireWire-based editing systems proliferated in the late 1990s. This standard supports high-quality audio-video input and output. The upshot of this development? Higher quality video editing, including the production of clips for multimedia projects, became available to a broader user base. But to take full advantage of the standard, you still had to have a powerful computer fitted with the appropriate components (e.g., A/V-ready hard drives). This criterion applies to PC-based video editing in general: You must have a sophisticated system to produce broadcast-quality productions.

Convergence

What it boils down to is a sense of balance or perspective. Each production environment has its own relative merits and primary application areas. They should all be examined in the context of how they may fit in the overall communications system.

The convergence factor should also be considered. As indicated in different sections of the book, we continue to see a convergence of technologies and applications, including those taking place in the video editing market.

The next chapter describes professional video editing equipment. A system may allow you to edit compressed digital video. When you are finished, the final product may either be assembled on or printed to (recorded) videotape. This type of system can provide you with the best of both worlds: the convenience of nonlinear editing, which can speed up the editing session, and the quality afforded by a standard video configuration.

This capability may also be incorporated in preexisting products. Premiere, for instance, has supported an edit decision list (EDL) export feature. An EDL is essentially a list of editing instructions that a high-end editing system can use for automatic assembly, in which the master tape, the final program, is assembled from the source material. Important EDL elements include selected transitions and relevant time code numbers.

Briefly, *time code* refers to an identification process in which video frames are assigned specific reference numbers. A computer-assisted editing system can use this information to identify and select the designated scenes as the final tape is assembled. Consequently, Premiere, originally a QuickTime-editing program, is now capable of interfacing with more advanced systems.

Videotape: It's Still Here

In this new world of digital video, A/V-ready hard drives, and optical storage, videotape is still here. It is a highly efficient and effective storage medium. For example, fast hard drives have supplemented and replaced tape in different broadcast applications. But these systems can still be expensive, and for the foreseeable future, cannot match tape's high storage capacity/low cost ratio. Thus, both tape and nontape systems will coexist. Even when hard drives or other data storage systems become more cost effective, tape may still be used as an archival medium. This topic is covered in more detail in the next chapter.

Graphics Software

Graphics programs contribute to desktop video and multimedia applications. Besides the typical software, other products are used. CAD programs, for example, can serve as illustration tools. American Small Business Computers, creators of the DesignCAD software series, discovered that their programs were also used for business graphics, technical illustration, and desktop publishing.[13]

Other software may even be more specialized, including these types:

- Programs that create realistic human figures;
- Programs that generate landscapes ranging from Mt. Saint Helens to the planet Mars

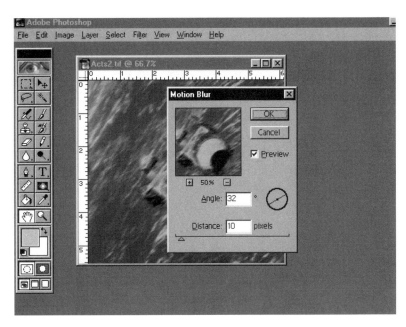

Figure 13.5

Image editing and graphics programs play key roles in desktop video and multimedia production. In this example, a Photoshop filter is used to create the effect of motion blur. Some of Photoshop's tools are also visible on the left side of the screen. (Software courtesy of Adobe Systems, Inc; Photoshop.)

to surrealistic scenes created in your own mind; or

- Character generator programs that create titles that can be overlaid on your video.[14]

Video Toaster.   NewTek's Video Toaster is a product that spanned the professional and nonprofessional markets. Introduced for the Amiga and more recently made available for IBM PCs, the Toaster raised the level of desktop video production by a notch.

The Toaster brought professional video production capabilities to the desktop at a low cost. The system functions, in part, as a character generator, frame grabber, and a switcher. Switchers are used for image transitions. Video Toaster also supports digital video effects, and LightWave 3-D, its 3-D graphics program, actually emerged as a cross-platform, industry standard.

Depending on the production situation, you may also need a time base corrector (TBC) to take full advantage of its capabilities. A TBC "takes the unstable video from a VTR and acts as a shock absorber, outputting rock stable video that can be integrated with other video sources in a system to maintain good picture quality."[15]

TBCs were once out of the financial and technical reach of most desktop video users.

The situation changed, though, when the Toaster's popularity and the growing desktop video market prompted the introduction of inexpensive TBCs, including internal models that fit inside a PC.

APPLICATIONS AND IMPLICATIONS

Multimedia and desktop video systems have emerged as powerful information tools. In the business world a multimedia production can make a speech more interesting and informative through the use of video and other media. This concept has been extended to teleconferencing and other fields.

Education

The educational market is served by multimedia and desktop video products. As indicated in the hypermedia chapter, a student can use a multimedia book that incorporates sounds and video clips. When combined with hypermedia links, the student can explore and experience this new world in a nonlinear fashion. The document's interactive nature, whether it is created with a hypermedia or multimedia program, can also help make learning active.

Figure 13.6

Computer graphics, animations, and the law— from a case involving an aircraft accident. (Courtesy of FTI Corporation, Annapolis, MD.)

In the medical field, surgeons could use QuickTime or another digital video product to inexpensively document new techniques. Besides a written description, a surgical procedure could be annotated with video and voice. Digital video additionally offers a rapid turnaround time, a reasonable learning curve through programs such as Premiere, and the potential to exchange these data over communications networks.[16]

Legal and Broader Implications

Multimedia and desktop video productions have emerged as legal tools. In a patent infringement lawsuit over hip prostheses, a firm created an interactive presentation that depicted how the devices worked. The goal was to show the devices in action and to make the information accessible and interesting to the jury.[17] The actual production process included using scanned images, image editing software, and animations.

Computer graphics and animations have been used to recreate real-life events. In one case, an animation of a roller coaster depicted the G-forces an individual would be subjected to during a ride. The data were gathered by instruments and incorporated in the animation to show how these forces caused a rider to suffer a stroke through a ruptured blood vessel.[18] Other cases have ranged from pinpointing the causes of fires to depicting how automobile and aircraft accidents could have taken place.

Exploring the World

As you know, the world is an intricate and rapidly changing place. Volumes of information are generated with each passing hour, and we must cope with this continuous information stream. To make matters worse, much of this information is difficult to comprehend.

Graphics, animations, and audio-video clips can help solve this problem by distilling a mountain of information into a more accessible form. For a report that examines the growth rate of urban centers, a series of graphics could potentially replace pages of census material.

Multimedia and desktop video productions can also be used to explore complex events. In one example, a series of graphics and animations has depicted the Strategic Defense Initiative (SDI), also known as "Star Wars." This satellite-based plan, proposed during the Reagan administration, was the core of a multilayered defense system. The missiles, satellites, and other components, which all played a role in the SDI plan, were brought to life via the computer.

Animations made it possible for viewers to at least grasp how the system would theoretically function. The interactions between the different SDI elements were also depicted in a dynamic fashion. With only a text-based description, this information would not have been conveyed as effectively. The subject may have been too difficult to comprehend.

But despite this positive attribute, there is an inherent danger in converting a large information base, especially about controversial and intricate subjects, into a series of images. The real-world situation may be portrayed in too simplistic a manner. Or the graphics and animations could be manipulated to promote a particular point of view—possibly a distorted view of the facts.

Because we tend to believe what we see, a production that effectively uses a mixture of media could be misleading. We also generally do not have an opportunity to refute the information on a point-by-point basis.

The opposite is true of a courtroom situation when computer-based information is used. In court, expert witnesses could be called and the information could possibly be challenged and rebutted.

Finally, the potential to present an altered view of the real world becomes an even more pressing concern in light of a virtual reality system's capabilities. A computer-generated world, designed and controlled by a human operator, could serve as an analog for real-world situations. Closer to home, images, including news pictures, can be manipulated and altered with a computer. Both topics are explored in later chapters.

Training, Sales, and Advertising

Training applications are well suited for desktop video and multimedia. A production could cover tasks ranging from car engine repairs to basic PC operations. Video clips of a real engine could be used, and the production could incorporate an interactive interface.

Store owners are also served. An electronic sales catalog can either replace or supplement a print version, and interactive kiosks where customers can get information about products have popped up in supermarkets and other outlets.

The advertising and public relations industries have also benefited. Video clips and animations can be rapidly generated and used in a presentation. In a related area, a sophisticated multimedia system was developed to showcase the city of Atlanta. The presentation was used to promote Atlanta as the site for the 1996 Summer Olympics.[19]

Other Applications

Video artists have also adopted desktop video tools. Inexpensive video systems have been used by artists for years. Earlier projects included personal documentaries as well as video feedback, where different visual patterns could be created, controlled, and displayed on a television screen.

Today, artists can tap sophisticated PCs and video equipment. In essence, video can be digitized for either still or motion displays, images can be colored, and animations can be produced.

As covered in a later chapter, desktop video and multimedia systems also extend the concept of the free flow of information. Like desktop publishing, production tools are now accessible to more people.[20] These tools have also contributed to development of a personal media. Instead of everyone receiving the same information, more personalized information can be created and received. For example, by using a video camera and a computer, we literally become producers and editors.

Another implication is that we can now tap more individualized information and entertainment pools. This topic was briefly discussed in Chapter 6 and resurfaces in Chapter 18. Various organizations, including MIT's Media Lab, are pushing both concepts

Figure 13.7

PCs, in combination with video equipment, have provided artists with a powerful set of tools. Shown here is a still from "Godzilla Hey," by Megan Roberts and Raymond Ghirardo (1988). The work combines digital video imagery and sound with analog video synthesis. Produced at the Experimental Television Center with Amiga PCs and the FB-01 video frame buffer with proprietary software, designed by David Jones. (Courtesy of Megan Roberts and Raymond Ghirardo.)

beyond current boundaries. The Media Lab, as directed by Nicholas Negroponte, has been one of the world's premier research institutes. Besides exploring the convergence of different media, personalized interactive media have been investigated. Two examples are personal electronic newspapers and television.[21]

In this new world, information could be retrieved from different sources and subsequently delivered to you through the assistance of intelligent systems and human–machine interfaces. In one setting, a computer could scan a night's worth of programming and then summarize and possibly replay the portions that would be of personal interest to you.[22]

Other developments, including an interesting look at this institution, can be found in Stewart Brand's book *The Media Lab*. Negroponte has also discussed this topic, and his view of our digital future, in his book *Being Digital*.

PRODUCTION CONSIDERATIONS

To wrap up this chapter, we should examine some basic production issues along with broader, aesthetic issues. For the latter, the growth of the multimedia and desktop video markets may have outstripped the development of a sound aesthetic base. An examination of traditional film and television frameworks can be valuable.

Production Elements

Like desktop publishing, there are some basic conventions you should follow when creating a desktop video or multimedia project, including the following:

1.   Check all your work on a conventional television or monitor while you are working. Quality decreases during the conversion process. Colors may not be as rich and reso-

lution can be lost. Some programs also have an option to help ensure the colors you use will technically conform to a television environment (NTSC-safe).

2.   Do not place titles and other visual elements at the edge of the screen. They might be cut off when displayed on a television. Similarly, make sure the text is not too small to read. You should also avoid a fancy script typeface. Although a fancy typeface might be fine on the computer's display, it might be illegible on a standard television.

3.   When recording, do not use old tape. Videotape that has been reused a number of times may not give you a clean recording.

4.   *Be paranoid.* Always save your work and back it up—back it up—back it up.

Aesthetic Elements

Element Integration.   All the media elements that compose a production should be fully integrated, much like a film or television show. There should be motivation, that is, a reason, for using any given element. Do not use an animation, for example, simply because you own animation software. Use one for a specific goal. It can range from demonstrating a new piece of equipment to serving as an attention grabbing device.

Preproduction.   This issue of what elements to use in a presentation can be worked out during the preproduction phase, the time before the program is created. The idea is to establish different criteria to help guide the program's design. The preproduction stage also serves a more practical purpose. It is less expensive to make changes at this stage than after the program's final assembly.

During the preproduction phase, different questions, including the following, should be asked:

- What is the presentation's goal?
- Who is the potential audience?
- What are the budgetary limitations?
- What is the best way to satisfy the goal? For example, should video be used?

The last question is particularly important. Should you use conventional video or would a QuickTime clip suffice? This flexibility is one of the hallmarks of multimedia production. The key again, though, is ensuring the presentation's integrity. Use a tool only if you have a reason to use it.

You must also make sure that you have the proper clearances for sounds and images. Clip art and music are two options, as are original and public domain materials.

Prototyping.   Besides a preproduction plan, you may want to prototype a small-scale version of the program. You can try out your ideas and plan any necessary changes.

A storyboard would also be helpful. A storyboard depicts, in sequential order, the major events in a production. It can be made of still pictures and may include audio. If the storyboard is computer based, you may be able to use some of its components in the final product. These can include graphics, animations, and QuickTime movies.

Enhancement.   Once you create your production, seek ways to improve your craft. Continue producing, watch related media products, including movies and television programs, and read.

If you are planning a desktop video project, you can learn how music can be an effective component by examining its role in certain movies. Music can heighten the tension in a scene or can serve as a counterpoint to what we're watching. This principle, observing for analytical purposes, also applies to lighting, scriptwriting, shooting, editing, and other production techniques.[23]

Finally, the goal of this process should be the development of your own style. Your production, whether it is video art, an electronic catalog, or a multimedia presentation, can have your own personal signature. With all the tools at your disposal, this should be an enjoyable prospect.

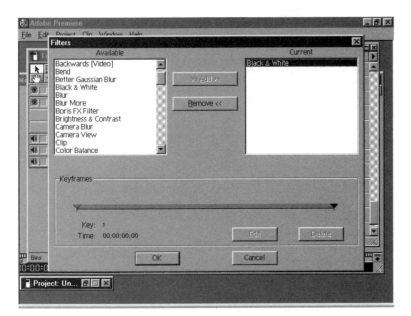

Figure 13.8

Premiere's options allow you to manipulate audio and video clips in different ways. In this example, a filter could be applied to a clip to achieve a specific effect (for example, to change a color clip into a black-and-white clip). (Software courtesy of Adobe Systems, Inc.; Premiere.)

REFERENCES/NOTES

1. Tom Yager, "Practical Desktop Video," *Byte* 15 (April 1990): 108.

2. David A. Harvey, "Local Bus Video," *Computer Shopper* (July 1992): 181. *Note:* This concept can be extended to other computer peripherals to similarly speed up their performance.

3. A plug-and-play capability simplifies the installation of peripherals because the software handles, whenever possible, this procedure.

4. Product prices and capabilities range from consumer to professional levels.

5. David Miles Huber, *The MIDI Manual* (Carmel, Ind.: SAMS, 1991), 20. See Jeff Burger, "Getting Started with MIDI: A Guide for Beginners," *NewMedia* 1 (November/December 1991): 61, for additional information and for a chart that highlights MIDI's storage efficiency when compared with conventional digitized audio.

6. Bruce Fraser, "Scan Handlers," *Publish* (April 1992): 56.

7. The same principle applies to video production operations, and in this case, the generally low resolution of the television screen can make a loss somewhat negligible.

8. Chris Cavigioli, "Image Compression: Spelling Out the Options," *Advanced Imaging* 5 (October 1990): 64.

9. Please see the *Media Cleaner Pro 4* (1999) software manual, by Terran Interactive, for an excellent discussion of compression schemes

and their advantages/disadvantages (for example, pp. 31–35).

10. See Denise Salles and Judith Walthers von Alten, "Making, Playing, and Sequencing a Movie," Chapter 7 in *Adobe Premiere User Guide* (Mountain View, Calif.: Adobe Systems, 1992), for a comprehensive overview of the latter topic.

11. Ben Calica, "The Clash of the Video and Computer Worlds," *NewMedia* 1 (September/October 1991): 58.

12. Depending on the project, a digital environment could offer other advantages. With a conventional audio-video setup, an individual may have to learn how to operate a range of equipment versus a few software/hardware packages in a PC-based operation. More important, since you can now see the audio and video information you are manipulating, the learning curve, versus that for conventional equipment, may be reduced.

13. American Small Business Computers, personal communication.

14. Character generators range from dedicated broadcast to PC-based units.

15. Tedd Jacoby, "Old Problems, New Answers," *Video Systems* (April 1988): 80. *Note:* TBCs are also used in high-end, conventional editing systems.

16. Steve Blank, "Video Image Manipulation with QuickTime and VideoSpigot," *Advanced Imaging* 7 (February 1992): 54.

17. Charles Rubin, "Multimedia on Trial," *NewMedia* 1 (April 1991): 27. *Note:* This case was settled out of court. Opposing lawyers would also have objected to the use of the presentation.

18. Carrie McLean, "Houston Lawyer Makes a Case for Computer Animation in the Courtroom," *Presentation Products* 5 (May 1991): 18.

19. Mike Sinclair, "Interactive Multimedia Pitches Atlanta Olympic Bid," *Advanced Imaging* 5 (March 1990): 38.

20. The produced electronic documents may present even more powerful messages than their paper counterparts.

21. Nicholas Negroponte, speech delivered at Ithaca College, Ithaca, New York, May 29, 1992.

22. Ibid.

23. Relevant television programs include *The Twilight Zone* and PBS's documentary *The Civil War*. In cinema, the list runs the gamut from older classics to more modern films: *Battleship*

Potemkin and *Alexander Nevsky, Grand Illusion, The Adventures of Robin Hood, Citizen Kane, Casablanca, The Third Man, Psycho, The Godfather, Raging Bull, Glory, Braveheart,* and *Apocalypse Now.* As a group, the films serve as examples for production and aesthetic principles ranging from music to lighting to shot composition.

SUGGESTED READINGS

Ashdown, Ian. "Lighting for Architects." *Computer Graphics World* (August 1996): 38–46. Architectural rendering, graphics, and the visual importance/impact of lighting (in an image).

Barnett, Peter. "Implementing Digital Compression: Picture Quality Issues for Television." *Advanced Imaging* 11 (April 1996): 30–33, 88. Compression and technical concerns for broadcasters (for example, analog versus digital signal quality control).

Bernard, Robert. *Practical Videography; Field Systems and Troubleshooting.* Boston: Focal Press, 1990; Millerson, Gerald. *The Technique of Television Production.* Boston: Focal Press, 1990; Zettl, Herbert. *Television Production Handbook.* Belmont, Calif.: Wadsworth Publishing Company, 1999. These are video production texts. The topics include, depending on the book, field production, editing, shot composition, and aesthetics. Zettl's book has an excellent CD-ROM that complements the text (for example, interactive production examples).

Brand, Stewart. *The Media Lab.* New York: Penguin Books, 1988. As stated in the text, the book provides an interesting look at MIT's Media Lab.

Corbitt, Pat. "CD-ROM Production: Getting It All Wrong." *Computer Video Production* 3 (May/June 1996): 25–26. If you are a new multimedia producer and are going to create a CD-ROM, read this article.

De Leeuw, Ben. "Moving in Real-Time." *3D Design* (September 1998): 31–35; Weinman, Lynda, and Jack Biello. "Animation A to Z." *NewMedia* 3 (August 1993): 57–62. Animation techniques.

Digital Magic. August 1996. This issue covered different types of character animation and the creative tools, including a story on the techniques used in the movie *Dragonheart.* An example is Barbara Robertson, "Best Behaviors," S12–S19.

Ganzel, Rebecca. "Digital Justice." *Presentations* 13 (November 1999): 36–48; Jackson, Jerry W. "Case-Making Courtroom Car Crash Visualization." *Advanced Imaging* 7 (July 1992): 48–51; Lichtman, Andrew. "Forensic Animation." *Amazing Computing* 6 (January 1991): 42–44, 46–47; Schroeder, Erica. "3D Studio Gives Crime-Solving a New Twist." *PC Week* 9 (March 9, 1992): 51, 58. Courtroom imagery, types of, other imaging applications, and animation use (in a courtroom).

Hall, Brandon. "Lessons in Corporate Training." *NewMedia* 6 (March 1996): 40–45. Corporate training and different tools and techniques.

Huber, David Miles. *The MIDI Manual.* Carmel, Ind.: SAMS, 1991. A detailed guide to MIDI systems and operations.

Kelsey, Logan, and Jim Feeley. "Shooting Video for the Web." *DV* (February 2000): 54–62. DTV techniques for Internet-based projects.

Luther, Arch C. *Digital Video in the PC Environment.* New York: McGraw-Hill Book Company, 1989. Provides an excellent perspective for the development of the DVI system. The author was one of its developers.

Lyn, Craig. "Master Series, Part 9: Mapping." *DV* (January 1999): 68–69; Oken, Eni. "Color. Color Everywhere." *3D Design* (September 1998): 53–63; O'Rourke, Michael. *Principles of Three-Dimensional Computer Animation.* New York: W. W. Norton & Co., 1995. Practical 3-D graphics/animation theory and techniques; Oken's article is an excellent color primer.

McGarvey, Joe. "Interactive Indeo Takes the MPEG Challenge." *Inter@ctive Week* 2 (October 1995): 28; Nelson, Lee J. "Video Compression." *Broadcast Engineering* 37 (October 1995): 42–46. Compression and standards.

Negroponte, Nicholas. *Being Digital.* New York: Vintage Books. 1995. A fascinating look at the possibilities brought about by the communication and information revolution.

Wallace, Lou. "Amiga Video: Done to a T." *AmigaWorld* (October 1990): 21–26. An initial look at the Video Toaster.

GLOSSARY

Authoring Software: Software that can simplify and enhance the creation of a multimedia presentation.

Compression: Compression refers to reducing the amount of space required to store information (for example, digitized video). Compression can also speed up information relays.

Desktop Video: Advancements in video technology have made it possible to assemble cost-effective yet powerful video production configurations. PCs typically play a major role in this environment.

Edit Decision List (EDL): Essentially a list of editing instructions used by a high-end editing system to assemble the final tape from the original source material.

Musical Instrument Digital Interface (MIDI): A MIDI interface makes it possible to link a variety of electronic musical instruments and computers. The MIDI standard also enables musicians to tap a computer's processing capabilities.

QuickTime: Apple Computer's PC-based digital media system.

Video for Windows (VFW): Microsoft's first PC-based digital media system.

14 The Production Environment: PCs, Digital Technology, and Audio-Video Systems

Besides the desktop publishing and video revolutions, another revolution is sweeping the broadcast and nonbroadcast production industries. It is a revolution based on the adoption of computer and digital technologies. This chapter covers these developments as well as complementary topics. The latter range from convergence issues to digital recording.

PRODUCTION EQUIPMENT AND APPLICATIONS

Switchers and Cameras

The switcher has benefited from the integration of computer technology. In brief, a switcher is used to select the pictures produced by a video facility's multiple cameras. It also creates visual transitions, can be used in postproduction work, and serves other functions.

A computer-assisted switcher can help an operator in these tasks. In one application, a complex visual effect may be required. The actions to create the effect can be preprogrammed, stored, and recalled at the press of a button. This capability, to immediately execute a command in the middle of a production when time is always critical, is an important one.

Switcher configurations can also be stored and later retrieved. This option enables an operator to quickly reconfigure the switcher for different production situations (for example, a news show versus a commercial).

The influence of computer technology also extends to video cameras and robotic camera configurations. In the former, various parameters can be set up with a computer-based control system, to help free an engineer's valuable time. In the latter, individual camera operators are replaced by a robotic system. Television news and other production situations generally call for set shots. A robotic system may prove acceptable in this environment.

A human operator can monitor and control a system's multiple cameras with various interfaces, including graphics tablets and joysticks. Prestored camera shots and movements can also be recalled, which lends itself to repeatability.[1]

A robotic system may have some limitations. For example, it may not equal the

Figure 14.1
One of the fallouts of the communication revolution: the ability to create sophisticated productions with PC-based products. This is a shot from Todd Rundgren's "Change Myself," produced by Rundgren using the Video Toaster. (Courtesy of NewTek, Inc.)

145

speed or capabilities of individual camera operators. This is an important consideration for sporting events and other dynamic shoots. There may also be a cost factor, depending on a system's sophistication.

Digital Special Effects and Graphics Systems

A digital special effects system, also called a digital video effects generator (DVE), manipulates a digitized picture or video sequence to create a special effect.[2] The final product may be recorded on videotape or used in real time.

A DVE can support 2-D and 3-D effects; an example of a manipulation is a compressed television picture. After the video signal is digitized and processed, the original picture can be reduced in size and repositioned on the screen. The original picture is still visible, but now it is physically smaller. Similarly, you can initiate an on-air zoom—a title "zooms in" to fill the screen. Other, more advanced effects include manipulation of an image so it appears to flip over, much like a page turning in a book.

You could use dedicated hardware or modify a PC for this operation. Depending on the configuration, the PC could be less expensive, but you may pay a price in speed and flexibility.

The same scenario applies to graphics systems. A PC can create graphics suitable for a production. This setup could also be cost effective, accommodate numerous applica-

tions, and allow you to tap the freelance market, which is the pool of people with PC graphics experience.[3]

Dedicated systems, however, have their own advantages. They can handle intricate 3-D images, digitized pictures, and large files. This differential is diminishing, though, as faster PCs and complementary hardware and software are developed.

Video Editing

Computer and digital technologies have also influenced video editing. As described in Chapter 13, editing can encompass the simple joining of scenes or more complex transitions.

Computer-assisted editing systems provide editors with a fine degree of control over the finished product. Two pertinent terms are offline and online editing.

During offline editing, a program's sequencing is laid out and can be used to create a rough cut of the final program or an edit decision list (EDL). The final master tape is then automatically assembled during a more expensive online session with a computer-assisted editing console and other support equipment. PCs have contributed to this operation. In an offline session, they have generated EDLs that are subsequently input to the online system.[4]

Newer editing configurations have extended this role. One application is PC-based nonlinear editing. Briefly, nonlinear editing allows you to retrieve stored video sequences in random access fashion, much like conventional computer data. In a PC environment, the scenes are digitized and stored on A/V-ready hard drives or erasable optical media, for later use.

A linear system, in contrast, is in keeping with the more traditional editing method. You must search through a videotape to find the specified scenes. This process ultimately eats up more time.

A PC-based system can also marry a nonlinear capability and a visual interface. The different audio and visual elements are displayed on a monitor as you assemble the production. In one configuration, digitized material can be

Figure 14.2
The editor's workspace—the Canopus editing system. Note the audio and video clips stored in the electronic bin (top right); the display window to view your work in progress (top left); the timeline where your clips are placed in sequential order (bottom). (Software courtesy of the Canopus Corp.; Rex Edit.)

represented by video frames, which serve as visual references for the scenes.

Some of the characteristics of working in this environment are as follows:

1.   Video is digitized and, based on the system, you can select the video's quality level (for example, VHS to broadcast).

2.   The different scenes are selected from an electronic pool or bin and are sequentially placed on a time line. The scenes can be quickly rearranged and/or deleted. For many individuals, this visual metaphor has made the editing process more accessible. It can also lead to more creative freedom. You can rapidly experiment with edit points, different shot arrangements, and other audio-video elements.

3.   Transitions between the scenes are selected from an options list. At some point, if you want to delete a transition, it is a simple operation, much like deleting a phrase with a word processing program.

4.   High-end systems can complete page turns and other complex transitions in real-time—almost instantly. Other systems may require several or more seconds to complete even a simple transition. Hardware acceleration, versus software-only systems, can be an important factor in this process.

5.   You can generate titles and tap a PC's graphics capabilities. A graphic or animation, which you create, can be imported and used. You can also manipulate the audio tracks and add music.

6.   In many cases, you can immediately play back an edited sequence and subsequently make changes.

7.   You may be able to directly print (record) the production to videotape; the quality is affected by various technical considerations.[5] This capability has somewhat blurred the distinctions between online and offline systems. If so equipped, you could also use an EDL to assemble the final production from the original source material.

8.   Nonlinear systems flooded the market by the mid-to-late1990s. The final output improved while prices dropped. Both factors helped open the field to more users. Newer

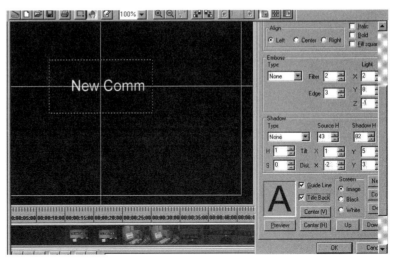

systems could also accommodate multiple video formats.

9.   These developments were fueled by faster PCs, new software/compression capabilities, cheaper memory and hard drives, and the adoption of FireWire and other standards.

10.   FireWire facilitates the connection of different equipment that supports this standard to "personal computers for desktop video post-production."[6] FireWire also supports a fast data transfer rate and enables you to create this link (for example, camera-video deck) with a single cable. Newer video cameras, including those based on the Mini-DV standard, are also equipped with a FireWire port. If you don't own a deck, you can use the camera to export recorded sequences to the PC. Once edited, you can use the camera to record your project.[7]

Although PC-based nonlinear systems are valuable, some factors should be taken into consideration:

• Install as much memory as you can afford.
• Be prepared for occasional lock-ups. System freezes can occur.
• Digital video has a high storage overhead, and to work efficiently and effectively, you may have to buy more drive space than you anticipated. Follow this simple rule of thumb: When you think you have enough storage, you'll probably need more.

Figure 14.3
This shot highlights the Canopus system's title creation tool. This visual environment has enhanced the editing process for many users. (Software courtesy of the Canopus Corp.; Rex Edit.)

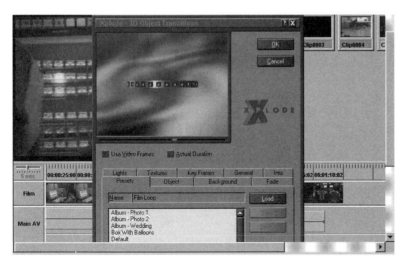

Figure 14.4
Editing systems may also support 3-D-like transitions. In this example, a spinning film reel functions as the transition. Video from the edited clips can be viewed in the individual frame cells. (Software courtesy of the Canopus Corp.; Rex Edit.)

Figure 14.5
One of the effects that may be available from an audio editing program. Note the presets in the window (top right). They allow you to quickly manipulate your audio piece. (Software courtesy of Syntrillium Software Corp.; Cool Edit Pro.)

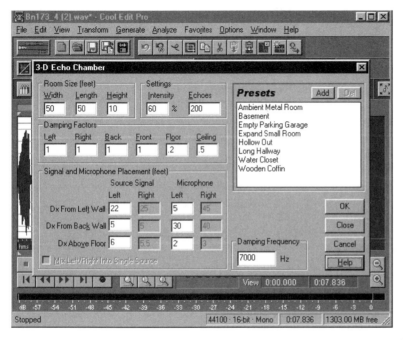

- Although PC-based nonlinear systems can speed up certain tasks, such as rearranging scenes, other tasks may be slower. Depending on the system, the latter include generating certain transitions and effects.
- Even though the visual interface can simplify the overall editing process, the software may have a high learning curve.
- Can the system can handle the quality of video as well as the format required for your applications?
- Does the company have a good technical support policy?
- Pay attention to aesthetics. Editing is still a craft that must be learned and practiced.

Audio Consoles and Editing

Audio equipment has also been influenced by computer and digital technologies. A computer-assisted audio console, for example, can help an operator to manipulate sound elements. As covered in the previous chapter, your voice or other audio piece can be digitized, edited, and manipulated. In this production environment, you can take advantage of random access editing and the visual representation of the audio sequence as a waveform on a monitor. This setup provides you with aural and visual clues to help you quickly identify edit points.

For facilities with limited budgets, this application is supported by professional quality audio cards that work with PCs. Complementary software enables you to produce multitrack projects that can be recorded on tape, a hard drive, or other media. You can also create a sequence and use it in a video editing project.

But regardless of the system, you pay a price for this performance and convenience. With editing, for instance, you can edit tape with a razor blade, grease pencil, and other common tools. Digital systems cannot match this setup on a strictly per-cost basis.

DIGITAL RECORDING

Why Digital?

Digital audio and video systems have certain advantages over their analog counterparts.

Chapter 2 covered some of these characteristics, which include a more robust signal. A signal's quality is also preserved after multiple generations, unlike a typical analog system where the output suffers as you progressively "go down" a generation.

Audio Recording and Playback

A digital audio tape (DAT) machine can record and play back digital tapes. The audio quality is equal to that of a CD, and the information can be stored on a compact cassette. DAT's recording and operational characteristics have made it particularly attractive to professionals. Studio and field machines have been designed, time code can be supported, and the output is excellent.

The consumer market, though, has been a different story. On face value, DAT systems should have been a success. Yet various factors combined to create a flat U.S. consumer response. For example, the recording industry claimed consumers would make CD-to-DAT copies, that is, digital-to-digital recordings, thus potentially reducing CD sales. Protection schemes were subsequently proposed, including the Copycode system. In practice, a DAT machine would shut down if it detected a special "notch" in prerecorded media. But this system was dropped because it did not work all the time, and some individuals indicated it affected a playback's quality.[8] When this factor was combined with the threat of litigation against manufacturers who sold unmodified machines, the result was the flat market.

But this picture has somewhat improved. DAT equipment has been embraced by the audio industry and audiophiles.[9] A royalty and new protection scheme have also defused the contentious atmosphere that delayed DAT's introduction in the overall U.S. market. Nevertheless, as of this writing, DAT has failed to achieve a broad-based U.S. consumer acceptance.[10]

In the 1990s, Sony introduced a new digital audio format, the MiniDisc (MD). The equipment line has included a compact recording system that supports a random access capability, digital quality sound, and can compensate for physical jarring. Other mod-

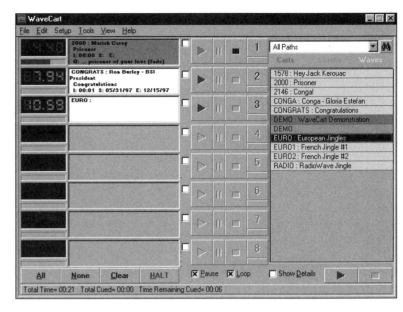

els were later released for the professional market, including a unit that could replace the traditional cart machine, a common fixture in most professional facilities.[11]

Digital cart systems have also appeared on the market. Audio can be stored, much like computer data, for the immediate access to and the playback of different audio cuts. You can also tap your existing PC, if equipped with a high-quality audio card, for this application.[12] Install the software, and you are greeted by virtual carts machines—on screen representations of this equipment. You can use your keyboard and mouse to subsequently control and playback the audio cuts.

Video Recording

As indicated at the beginning of this section, one of the advantages of a digital VTR is its high-quality, multigenerational capability. Two earlier formats include D-1 and D-2. As described by Herbert Zettl,

the D-1 system . . . is especially useful if you plan on extensive post-production that involves many tape generations . . . and extensive special effects. By keeping the R,G,B components separate, the image quality remains basically unaffected by even the most complex manipulation. . . . The D-2 system, on the other hand, does not require any modification or replacement of the

Figure 14.6

A PC-based virtual cart system can replace multiple, traditional cart machines. The virtual carts are displayed on the left; the different music cuts, which are subsequently loaded in the carts, are displayed on the right. A cart can be started at the press of a key and can be played sequentially, automatically. (Software courtesy of Broadcast Software International; WaveCart.)

existing equipment. Because the D-2 VTR processes the video signal in its composite NTSC configuration, you can use it instead of regular . . . VTRs and in tandem with regular switchers and monitors.[13]

Other digital formats, and complementary equipment, have subsequently been produced. These include DVCAM and DVCPRO, two competing formats from Sony and Panasonic, respectively. The Mini-DV, another format, emerged as a prosumer standard supported by different manufacturers. Mini-DV systems also became popular professional tools. Cameras, for instance, are compact, produce a high-quality output, and sport FireWire ports and multiple shooting modes (e.g., wide screen).

Tapeless Systems

Tapeless systems have also become popular. For our purpose, this term is an umbrella phrase. It describes equipment that does not use tape as its primary recording and/or playback medium. We have already met some of these configurations: video servers and non-linear editing stations equipped with fast hard drives and optical media.

In one application, a server can play back a television station's ads. Traditionally, a setup was expensive to maintain, prone to mechanical failure, and could be a complex mix of VTRs and a robotic control system. A server, which can require less maintenance and operator supervision, can replace this mechanical configuration. These devices can be more maintenance free and can run with less operator supervision.[14] The end result would be a more cost-efficient station.

A tapeless environment has also been extended to other areas. In one application, Avid helped introduce a disk-based camera for fieldwork. Instead of recording video on tape, it is recorded on a removable drive that can be married to a digital editing station. Thus, you can quickly go from a shoot to editing to air.

When viewed from a broad perspective, tapeless systems have certain advantages:

1.   Tapeless systems are less prone to physical breakdowns and usually have fewer mechanical components and parts to wear out (for example, a VTR's heads).

2.   Tape is prone to wear and tear. When reused, defects that can adversely affect playback may surface. A tapeless system's playback quality should remain unchanged.

3.   The quantity of consumables, including replacements VTRs and tape, will be reduced.[15]

4.   A tapeless system can simultaneously serve multiple users. In a typical scenario, the same video clip could be accessed by two editors to compose different stories.[16]

Yet despite these advantages, there are some limiting factors as we migrate toward a tapeless environment. First, the broadcast industry is still dominated by a tape-based standard. Because an overnight switch would be too expensive, the changes will most likely be evolutionary. The technology base is also too new. It has not met the test of time in the eyes of some individuals.

Second, technical limitations, including potential software problems, must be ironed out.[17] Basically, if software failures occur, whole systems could be rendered inoperable.

Third, system redundancy must be considered. If a single VTR breaks in a traditional setup, it could easily be replaced by another unit. In contrast, how effective are the server's backup and replacement capabilities?

Fourth, media costs are still comparably high for tapeless systems. Videotape and other tape formats are cost effective, especially in view of their storage capacities. Hard drives, for instance, cannot as yet match this capacity-to-cost ratio.

Consequently, as we head toward a tapeless future, tape will still be used for an undetermined time. This is especially true for smaller stations with limited budgets.

We may also see tape/tapeless hybrids in the interim. In one example, servers used for video-on-demand applications may be supplemented by tape systems. The server could store frequently requested and timely information while other programming could be archived on tape. When requested, it would be transferred to the server and delivered.

Data tape, rather than videotape, may also be used for archival purposes.[18] Ultimately, in this type of environment, we could draw on the relative strengths offered by both technologies and media.[19]

INFORMATION MANAGEMENT AND OPERATIONS

Computers have been adopted for information management and control operations. These include scriptwriting, newsroom automation, and computerizing a station's traffic and sales departments. Even though these tasks may not be as visible as the production end of a facility, they are nonetheless vital.

Scriptwriting, Budgets, and News

A critical job in any production facility is scriptwriting. The script is the heart of a program, and scriptwriting software has been released to support this function. The software can help free you from numbering scenes, formatting dialogue, and other time-consuming mechanical chores. The software completes these functions, and you can concentrate on writing.

A program typically supports a standard two-column television script and other formats. It may also link specific column sections. If changes are made, the corresponding audio or video section tags along.

The scriptwriting process can be further enhanced through supporting programs. In one instance, software can be used to create a storyboard, essentially a road map of a production. A storyboard can help you visualize the final product, and it may even be possible to use short QuickTime movies with other media.

Other specialized and general release programs are also widely used. For budgets, spreadsheets can track business expenses, crew fees, travel and postproduction costs, and equipment rentals. Another option includes the use of CAD programs for facility design.

Figure 14.7
Multiple videotape formats have coexisted for years. This includes the S-VHS (right) and the newer, high-quality Mini-DV formats (left). Note the size difference.

For news, computer technology has helped transform the traditional news department into an electronic newsroom. By integrating hardware and software, different tasks are enhanced. For example, wire service stories can be fed directly to a computer. This information can be retrieved, printed, and saved.

A news department can also establish an electronic news morgue.[20] As part of a networked operation, reporters and editors can gain access to current and past stories. Digital video clips and other information could likewise be accommodated, particularly as compression techniques improve. News footage from outside agencies could also be delivered over high-speed communications channels for subsequent retrieval.[21]

Figure 14.8

Computers play a key role in the broadcast industry. In this example, they are used to produce shots for weather forecasting. (Courtesy of AccuWeather Inc.)

This digital vision, where information is networked and is readily accessible, could facilitate the overall production and news processes. Digital video workstations, for instance, could enable journalists, who may have traditionally worked in a text-based environment, to tap into video and related resources. For facilities with limited personnel, the same individual could write and edit the story, potentially all on the same desktop.[22]

Depending on the manufacturer and package, other applications could be supported. These include an interface to the station's production facilities where the newscast's text could be fed to a prompter and a closed captioning setup. A prompter is a device used by on-air talent to maintain good eye contact with a camera while reading news copy. Closed captions are the normally invisible subtitles for programming that can be displayed on a television set by means of a special decoder. The captions support the hearing impaired audience. You can "read" the dialogue, a talent's spoken lines, for instance, via the printed captions.

Consequently, a product can be used alone, as may be the case with a newswire service feed to a single PC. Or an integrated system that links the newsroom with the production end can be created. In fact, this principle has been extended to tie character generators, video cart machines, and robotic camera systems in a central control network. It could potentially encompass the entire facility.[23]

Operations and Information Management

Various media organizations have adopted computers for marketing research. Arbitron has served the broadcast industry for years. The company has tapped computer technology to speed up the delivery of information to client stations. Arbitron has also employed different audience measuring techniques. In one setup, viewers have responded to on-screen prompts as part of one interactive television measuring system. This has provided the company with more accurate information about audience viewing preferences.

Arbitron even investigated an AI-based passive system that could identify different viewers without prompting.[24]

Besides Arbitron, companies have supported telephone surveys with software. In brief, questionnaires could be designed and the data collected and analyzed to reveal the audience's characteristics. In one typical application, these new insights could be used to attract a specific demographic group.

Programming, Traffic, and Sales

Software can also be used as an organizational and managerial tool. A package can generate a detailed list of a day's programming events and can handle other jobs. A radio station serves as an example.

In one application, a program can support a music library, essentially a sophisticated database. Depending on the system, songs can be cross-referenced, coded according to their tempo and intensity, and linked to age group and demographic appeal codes. These data are useful in attracting specific audiences. Similarly, the software may help a program direc-

tor to devise the music rotation schedule—the list of songs played during the day.

Programs have also accommodated traffic duties. Log maintenance is an essential task, and computers have sped up this process.

A program could also automate the scheduling of commercials, and sales force performance could be evaluated in different categories. These include the dollar amount of sales, the number of sold spots, and each salesperson's commissions.

An accounting package may also be used to complete a comprehensive system that could cover everything from music rotation to commissions. Accounting software performs a multitude of billing, payroll, and projection functions. In sum, the accounting software could handle most, if not all, of a station's financial and bookkeeping needs.

CONCLUSION

The broadcast and nonbroadcast worlds have been influenced by computer and digital

Figure 14.9
Another example of a virtual set—a computer-generated set that could be used for a news program. Powerful computers and software have made it possible to create customized environments that otherwise might be impossible to build because of, for example, money and time constraints. (Courtesy of Evans & Sutherland. Copyright 2000, Evans and Sutherland Computer Corporation.)

technologies. Some of these effects have included the following

- The introduction of new audio systems
- The widespread adoption of digital recording systems
- The birth of the electronic newsroom
- The ability to create amazing visual effects
- The provision of the impetus for an all-digital production facility

The last item can be considered an evolutionary process. Digital equipment initially had to operate in what was an analog sea, and an all-digital plant would reverse the situation—for example, an analog video camera's signal would be immediately digitized. This process would streamline certain operations and would help preserve a signal's integrity.

In view of its superiority, engineers are designing such digital facilities. This development also complements the creation of a fully integrated facility that can tie different equipment and systems into a centralized communications and control network. Although stand-alone systems will continue to be used, the trend is to unite these elements, as may be the case with an automated station.

But even though these tools are powerful, they still require human input. For a graphics system, this may be the creative ability to visualize a graphic and the skill and aesthetic judgment to execute the final product.

All of these developments have "pushed the envelope." Only this time, it is a creative envelope, and one whose potential may be unlimited.

REFERENCES/NOTES

1. Robert Saltarelli, "Robotic Camera Control: A News Director's Tool," *SMPTE Journal* 100 (January 1991): 23.
2. DVE is a registered trademark of NEC America, Inc.
3. Linda Jacobson, "Mac Looks Good in Video Graphics," sidebar in "Macs Aid Corporate Video Production," *Macweek* (December 3, 1991): 40.
4. PC-generated lists have been saved on floppy disks and paper tape. The editing instructions could also be typed in for an online system. See Lon McQuillin, "The PC and Editing: One Approach," *Video Systems* 11 (February 1985): 30–34, for an interesting look at how the PC has functioned in this production environment.
5. These include the system's capabilities, the quality of the video, and the compression ratio (if relevant) when the video is digitized and subsequently recorded.
6. Frank Beacham, "Camcorders Take Another Great Leap," *TV Technology* 13 (November 1995): 1.
7. Using the camera for these tasks can, however, have an impact on its operational lifetime—the extra wear and tear.
8. Brian C. Fenton, "Digital Audio Tape," *Radio-Electronics* 58 (October 1987): 78.
9. An audiophile, like a videophile, demands the best performance from equipment—in this case, audio equipment.
10. Part of the problem is the lack of software, prerecorded music tapes. For details, see Jenna Dela Cruz, "Digital Audiotape," in August E. Grant, ed., *Communication Technology Update* (Boston: Focal Press, 1994), 259.
11. Ken C. Pohlmann, "Digital I/Os Added to Sony's Latest MD," *Radio World* 19 (March 22, 1995): 4.
12. There are also dedicated audio systems designed for this application.
13. Herbert Zettl, *Television Production Handbook* (Belmont, Calif.: Wadsworth Publishing Company, 1992), 302.
14. Claire Tristram, "Stream On: Video Servers in the Real World," *NewMedia* 5 (April 1995): 49.
15. Chris McConnell, "Curt Rawley: Avid Advocate for a Disk-Based Future," *Broadcasting & Cable* (April 3, 1995): 64.
16. Avid, "Server Control Production," *Across Avid* 2 (Issue 1): 6.
17. Peter Adamiak et al., "Digital Servers," *Broadcasting & Cable* (April 3, 1995): insert, S-3.
18. Mark Ostlund, "Multichannel Video Server Applications in TV Broadcasting and Post-Production," *SMPTE Journal* 105 (January 1996): 11.
19. Claire Tristram, "Bottleneck Busters," *NewMedia* 5 (April 1995): 53.
20. The idea of security is an important one since a newsroom computer, like any other

computer, may be vulnerable. The electronic newsroom should also be equipped with backup systems to maintain at least a basic level of operation in times of emergency, when the automated system may be rendered inoperable. For information, see William A. Owens, "Newsroom Computers . . . Another View," sidebar in James McBride, "Newsroom Computers," *Television Engineering* (May 1990): 31.

21. Adamiak, "Digital Severs," 4.
22. McConnell, "Curt Rawley: Avid Advocate for a Disk-Based Future," 65.
23. Conversation with Basys Automation Systems, October 1992. *Note:* A still store "stores" images for later retrieval and display.
24. The Arbitron Company, "At Arbitron, These Technologies Aren't Just a Vision, They're Reality," brochure.

SUGGESTED READINGS

Beacham, Frank. "Camcorders Take Another Giant Leap." *TV Technology* 13 (November 1995): 1, 16; Haigney, Karen R. "Looking to Go Digital?" *Television Broadcast* (June 1996): 1, 84; Livingston, Phil. "The Case for Digital Video Recording." *Advanced Imaging* 11 (June 1996): 16–18, 21. The new digital video formats and equipment.

Boucher, Roland J. "Disk-Based Spot Playback Systems." *Broadcast Engineering* 36 (April 1994): 49–52. Using disk-based systems for commercial playback.

Broadcast Engineering is an important publication for the broadcast industry. Three articles that covered various issues include Epstein, Steve. "The Advantages of Tape." 39 (February 1997): 100–102; Hopkins, David. "Digital Effects Systems: Moving to Software-Based Open Systems." 39 (February 1997): 88–92; Tucker, Tom. "Audio-to-Video Delay Systems for DTV." 40 (November 1998): 82–85.

Chan, Curtis. "Advances in DAT Recorders." *Broadcast Engineering* 36 (August 1994): 34–40; Dick, Brad. "Moving into R-DAT." *Broadcast Engineering* 30 (August 1988): 63–70, 74. A look at DAT—early to more recent systems.

Davidoff, Frank. "The All Digital Studio." *SMPTE Journal* 89 (June 1980): 445–449; Nasse, D., J. L. Grimaldi, and A. Cayet. "An Experimental All-Digital Television Center." *SMPTE Journal* 95 (January 1986): 13–19; an early look at digital broadcast systems.

Douglas, Peter. "Automating Master Control for Multichannel." *Broadcast Engineering* 40 (April 1998): 106–111; Hannig, Kelly. "Program Automation Systems." *Broadcast Engineering* 35 (July 1993): 60–62. Broadcast automation systems.

Hamit, Francis. "Imaging and Effects at *Star Trek: New Universes to Conquer.*" *Advanced Imaging* 10 (April 1995): 22–24; Robertson, Barbara. "The Grand Illusion." *Computer Graphics World* 21 (January 1998): 23–34. Creating special effects for *Star Trek* and *Titanic,* respectively.

Harris, Dave. "Selecting a DAW." *Broadcast Engineering* 36 (August 1994): 20–24. Digital audio workstations and their capabilities.

Houston, Brant. *Computer-Assisted Reporting.* New York: St. Martin's Press, 1996; Vigneaux, Stevan. "Digital News Gathering on a Desktop." *Broadcast Engineering* 35 (September 1993): 50–56. Practical guide for learning computer-based tools for journalists and nonlinear editing, production, and news operations.

McConnell, Chris. "Curt Rawley: Avid Advocate for a Disk-Based Future." *Broadcasting & Cable* (April 3, 1995): 63–65. An interview with Avid's president and the broadcast industry's tapeless future.

Ostlund, Mark. "Multichannel Video Server Applications in TV Broadcasting and Post-Production." *SMPTE Journal* 105 (January 1996): 8–12; Paulsen, Karl. "Servers Found on Video Menu." *TV Technology* 14 (February 9, 1996): 44–45, 65; Roth, Todd. "Video Servers: 'Shared Storage' for Cost-Effective Realtime Access." *SMPTE Journal* 107 (January 1998): 54–57; Tristram, Claire. "Stream On: Video Servers in the Real World." *NewMedia* 5 (April 1995): 46–51. Video server applications and operations.

Pohlmann, Ken C., ed. *Advanced Digital Audio.* Carmel, Ind.: SAMS, 1991; Rumsey, Francis. *Digital Audio Operations.* Boston: Focal Press, 1991. Examinations of the digital audio field.

Rafter, Patrick. "The Line Between On-line and Off-line." *Video System* 21 (April 1995): 28-30; Turner, Robert R. "1,001 Questions to Ask Before Deciding On a Nonlinear Video Editing System." *SMPTE Journal* 103 (March 1994): 160–173. Nonlinear editing and related issues.

Reed, Kim. "FireWire's Future." *DV* (January 1999): 35–44. An excellent look at FireWire.

Yamanaka, Noritada, Yoshikazu Yamamura, and Kazuyuki Mitsuzuka. "An Intelligent Robotic Camera System." *SMPTE Journal* 104 (January

1995): 23–25. Intelligent robotic camera systems enhance operations.

GLOSSARY

Audio Console: The component that controls microphones, CD players, and other audio equipment.

Computer-Assisted Editing: The process of using a computer to help streamline and enhance the editing process.

Digital Audio Tape (DAT): A recordable digital tape system that can match a CD's audio quality. DATs have also been used for computer data backups.

Digital Effects Generator: A production component that digitizes and manipulates images to produce visual special effects.

Electronic Newsroom: A newsroom equipped with computers for newswriting, creating databases, and other functions.

Graphics Generator: The component that is used to create computer graphics. The graphics can be produced either with special, dedicated stations or properly configured PCs.

MiniDisc (MD): A newer digital audio system from Sony.

Nonlinear Video Editing: Allows you to retrieve stored video sequences in random-access fashion, much like conventional computer data. In a PC environment, the scenes are digitized, and a visual interface is used for editing.

Switcher: The video component that is used to select a production's camera and videotape sources. It can also be used in post-production work and may incorporate computer technology.

Tapeless Systems: An umbrella phrase describing equipment that doesn't use tape as the primary recording and/or playback medium.

15 High-Definition Television and Digital Audio Broadcasting

The last chapter covered computer and digital developments that are shaping the television, radio, and nonbroadcast media. This chapter focuses on high-definition television (HDTV), an application designed to produce an enhanced television display. An overview of an analogous system geared for radio, digital audio broadcasting (DAB), is also presented.

HIGH-DEFINITION TELEVISION

As promoted for a number of years, HDTV was touted as the new television standard. Its display would visually approach the quality of film, and the system would support an enhanced audio signal.[1]

An HDTV set would also have a 16:9 aspect ratio, that is, the ratio between a screen's width and height. In contrast to a conventional 4:3 configuration, an HDTV screen would present a more powerful image to viewers.

The push for HDTV, which was included under the Advanced Television (ATV) designation, stemmed from a natural process: the improvement of technology and the growth that takes place in any industry. The trend toward larger television screens accelerated this research. In general, as the screen size increases, the picture quality decreases. But

Figure 15.1
Engineers set up the encoding room at the foot of the transmission tower at WRC-TV/Washington in preparation for the first over-the-air digital HDTV simulcast, in September 1992. (Courtesy of the Advanced Television Research Consortium; AD-HDTV.)

HDTV and other digital systems do provide enhanced displays.

Background

Japan and CBS.   The Japan Broadcasting Corporation, NHK, has been one of the world leaders in the HDTV field, and its personnel conducted breakthrough research. This included experiments to test a range of technical and nontechnical parameters, such as the relationship between the number of lines in a picture, the optimum viewing distance from a screen, and the screen's size. For example, viewers judged the quality of television pictures composed of different lines of resolution from various distances.[2] These considerations were then weighed against bandwidth limitations and other factors.

NHK also developed HDTV equipment and held trial HDTV transmissions in its own country and the United States.

In the United States, CBS was an early HDTV supporter. The company promoted a two-channel, 1050-line component scheme, in contrast to NHK's 1125-line configuration. The CBS operation was also designed for a DBS relay. A 525-line signal and enhanced picture information would have been carried in two separate channels. A receiver would have combined both to produce a 5:3 HDTV display.[3] For viewers without HDTV receivers, the single 525-line channel would have been converted for viewing on conventional sets.

The work conducted by NHK, CBS, and others helped spur the HDTV field's growth. This development has also been an evolutionary one, with a minor revolution thrown in to boot. The evolutionary phase encompassed an emerging infrastructure built to support HDTV productions and relays. The revolution was the birth of an all-digital system.

Global Issues.

"He who controls the spice, controls the universe."
—From the movie version of *Dune*

During the 1980s, attempts were made to create an international standard. Although some progress was seen, technical, political, and economic issues stood in the way. One problem was a concern that Japan would gain an edge if its HDTV system dominated the industry. By extension, the situation could have affected consumer electronics sales.

For Europe, manufacturers believed their home markets could have been swamped by Japanese-manufactured HDTV goods.[4] Other issues stemmed from the European Community's overall goal to promote European technologies and programming. In the United States, official support was initially given for a standard based on Japan's system. But this support faded as the prospect for a global standard faded.[5]

Another factor was the American semiconductor industry. To some, HDTV represented this industry's future, since HDTV systems would be heavily tied to semiconductor technology.[6] The semiconductor industry claimed that if the United States was not a leading HDTV manufacturer, the country's overall semiconductor industry could suffer and, by extension, PC and dependent markets.

Thus, an international consensus could not be reached. It was feared that the country that dominated this industry would have a hammerlock on a multibillion-dollar business and related industries.

This concern and perception is best summed up by the line that opened this section, from *Dune*, Frank Herbert's classic science fiction work: "He who controls the spice, controls the universe." Only in this case, the spice was and is HDTV.[7]

United States

History.   Although standardization attempts were made, HDTV developments kept pace. In Japan, work continued on its MUSE system, while the Europeans launched a new initiative in the mid-1980s, the Eureka Project.[8] Both systems were geared for satellite delivery.

In contrast, much of the focus and regulatory maneuvering in the United States has been on the terrestrial system, over-the-air broadcasts. This is a reflection of the system's unique status. Unlike cable or another optional service, terrestrial broadcasts are free. Even though the industry is supported by commercials, we do not pay a set fee to view the programming.

The FCC accelerated the development of HDTV operations in the late 1980s and early 1990s. In a 1988 decision, for example, the agency supported, at least for terrestrial transmissions, a configuration that would conform to existing channel allocations and would be backward compatible.

Another issue was and still is the channel space requirement. An HDTV relay packs in more information than a standard relay. For the United States, it would exceed, under normal circumstances, current 6-MHz channel allotments. Thus, various organizations attempted to develop a system that operated within these constraints.

In one example, the David Sarnoff Research Center supported its upgradable configuration, the Advanced Compatible Television (ACTV) system. Under ACTV-I, an enhanced picture, but not one equal to the 1125-line standard, would have been relayed over existing channels. At some future date, an augmentation channel, a second channel with enhanced information, would have been integrated in the system (ACTV-II) to produce a higher quality picture.

An ACTV-I configuration, also known as extended definition television, would have functioned as a bridge between conventional television and HDTV because it

- Was backward compatible.
- Accommodated an enhanced picture or display with an appropriate set.
- Did not disrupt current spectrum allocations.
- Could handle an HDTV relay through the augmentation approach.[9]

Related issues have included the differences between the broadcast, cable, and satellite industries. Limited spectrum allocations translated into tight channel requirements for broadcasters. Cable and satellite operators

had more flexibility and could launch, as another option, HDTV pay services.

Evolution and Revolution.   The FCC refined its decision in the early 1990s by announcing its support for simulcasting. In this operation, a station would continue to relay a standard signal. The station would then be assigned a second channel for an HDTV relay. This approach was viewed by some as having an important advantage over augmentation. With simulcasting, the HDTV transmission was not tied to the NTSC standard. Thus, a superior, noncompatible system could be developed.[10]

The final piece to the puzzle was added when the FCC, led by Chairman Alfred Sikes, encouraged the development of a digital system. If successful, this configuration would have advantages over analog operations.

Consideration was also given to a flexible standard that could handle future growth. Three terms were associated with this philosophy: scalability, extensibility, and interoperability.

In a scalable video system, the aspect ratio, number of frames per second and the number of scan lines can be adjusted . . . to the requirements of the individual picture . . . or viewer's choice. Extensibility means that a new television system must be able to operate on diverse display technologies . . . and be adaptable for use with new, higher resolution displays developed in the future. Interoperability means a television system can function at any frame rate on a variety of display devices.[11]

Basically, part of the HDTV deliberation process was the consideration of a standard that could accommodate different configurations and enhancements.

In preparation for this standard, several proposed terrestrial systems, which included an analog system, were scheduled for testing by the Advanced Television Test Center. Following the tests, the FCC was slated to select a standard in early 1993 (it was subsequently moved to a later date). The FCC was also working on a terrestrial HDTV broadcast timetable. This covered items such as the length of time NTSC signals would be supported before a full switchover was made to HDTV.

But many broadcasters were not happy with these decisions. One factor was the financial cost.[12] Because HDTV was new to everybody, both producing organizations and consumers would have to buy new equipment. The cost would be particularly painful for small market stations, even with eventual price reductions as equipment became more widely available.

Other broadcasters, at this time, were dissatisfied with the FCC's timetable and HDTV mandate. It was proposed that the second allocation could be used for digital multichannel television relays or, possibly, for interactive and/or data services. The goal was to provide broadcasters with additional programming options so they could better compete with cable and satellite services. Eventually, HDTV relays could be phased in.

The Mid-1990s: A New Name and Standard.

After this initial work, a standard was finally hammered out for HDTV, which now fell under the Digital Television (DTV) designation. It was recommended to the FCC as the U.S. digital television standard.[13] As summed up in an FCC document,

The Commission proposed adopting, as the technology for terrestrial broadcast in the United States, the Advanced Television Committee (ATSC) DTV standard. . . . The proposed standard is the culmination of over eight years of work by the federal Advisory Committee on Advanced Television Services (ACATS), the ATSC, the Advanced Television Test Center (ATTC), and the members of the Grand Alliance [an "alliance" of initial HDTV competitors that joined forces]. . . . The technology provides a variety of formats that will allow broadcasters to select the one appropriate for their program material, from very high resolution providing the best possible picture quality to multiple programs of lower resolutions, which could result in more choices for viewers. Even at the lower resolutions, the recommended system represents a clear improvement over the current NTSC standard. The recommended system also permits transmission of text and data.[14]

The proposed standard's video compression was based on MPEG-2. A relay would also sport an enhanced audio output.

For the broadcast industry, a station would not be locked into a single format. There was an option for "multiple simultaneous" standard television (SDTV) relays. While not true HDTV, a station could accommodate multiple feeds.[15]

A Funny Thing Happened on the Way to HDTV.

Zero Mostel and a supporting cast helped make *A Funny Thing Happened on the Way to the Forum* a hit. In 1996, the computer and film industries, among others, helped bring the standards adoption process to a crashing halt. The proposed standard ran into a technical roadblock.

As stated by FCC Chairman Reed Hundt, who also expressed his own concerns

the dedicated and hard-working members of the Advisory Committee tried in good faith to produce a consensus standard. Unfortunately, they did not succeed. Important players in two huge American industries—Silicon Valley and Hollywood—object strongly to some elements of the standard. These groups support much of the Grand Alliance's work as endorsed by the Advisory Committee, especially the creation of a dynamic digital broadcast standard. But the failure to reach consensus over the *interlaced format* and the *aspect ratio* [my emphasis] has led to a time-consuming and important debate in which all advocates are making serious points.[16]

It was a contentious atmosphere. Opponents indicated, for instance, that scanning and aspect ratio incompatibilities would hamper the marriage between computer and television products and would have an impact on the film industry. Supporters countered that other industries were not left out of the standards deliberation process. More pointedly, this was essentially a television broadcast standard—the broadcast industry should take the lead.[17] Other contributing elements included the *Dune* factor, convergence, and the government.

As for *Dune*, the question was this: Which industry was going to control the flow of "digital media coming into the home," basically, who was going to control the spice?[18] This had broad economic implications for the different players (for example, broadcasters and computer/consumer electronics manufacturers).

The second factor was convergence. What were once separate entities, such as the broadcast and computer industries, were now competing as well as collaborating with each other. Because multiple industries had an interest in HDTV, it became the game ball to toss about.

The government, for its part, was seemingly unwilling to grapple with this "hot potato." The FCC was divided, and since it was an election year, it was suggested the Clinton administration did not want to anger any particular group by taking sides.[19]

In late 1996, however, a consensus was reached. As described in a *Broadcasting & Cable* article, the "deal called for the adoption of the Grand Alliance standard minus the controversial picture formats that had divided the industries."[20] Basically, the core of the original proposal was retained. But now, the broadcast and computer industries, among others, could support different display formats (for example, types of scanning). The deal helped clear the road for digital television.

Other Considerations

Other key issues have shaped the DTV landscape. Examples include the following:

1.   Even though DTV is not yet an established global broadcasting tool, it has been and is being used in the United States and other countries. The systems vary, however, much like the current analog standards.[21] In the United States, regular HDTV broadcasts began in April 1999 on *The Tonight Show*.[22] Other programs, including coverage of John Glenn's launch on a space shuttle mission, were held prior to this time.[23]

2.   As envisioned, a television station would have until the year 2006 to make the change to a digital operation. At this time, a station would surrender its second channel, and the released spectrum space could be

Figure 15.2
An HDTV display. Note the size and shape of the screen compared with the typical TV in current use. (Courtesy of the Advanced Television Research Consortium; AD-HDTV.)

auctioned. But various factors could delay the time line, and potentially, the date a station would surrender the channel.[24]

3.   Although the U.S. DTV focus has been on terrestrial broadcasting, other media can deliver programming, potentially on an optional basis and on a more flexible timetable. DirecTV, for one, helped pioneer the delivery of digital signals to consumers.

4.   A DTV system could have numerous nontelevision applications. A display could be attractive for videoconferencing activities, the electronic meetings described in Chapter 19. Special educational programming highlighting delicate surgical procedures could also be produced, and the Defense Department has similarly expressed an interest in this field.

HDTV transmissions could help promote another entertainment form. A satellite could deliver signals to special movie theaters equipped with large, high-resolution screens. The programming could include movies, concerts, and sporting events.

Stations could also support datacasting— using their channels to handle "everything from broadcasting stock quotes to downloading an electronic catalog."[25] As demonstrated in early 2000, stations could "create web sites with multiple streams of video and broadcast them to PCs over their DTV channels."[26] Stations could also support, as described, multiple SDTV relays of more conventional programming.

5. As of early 2000, different factors slowed down the broad acceptance of DTV systems. Some standards issues were still unresolved, sets were expensive, and there was a dearth of programming.[27]

6. A central question, and one that is still unanswered, is the consumer's role. Does the average consumer want DTV? Is the average consumer willing, for example, to pay a substantially higher price for HDTV sets? These are crucial issues, especially in a volatile consumer electronics market where different products are competing for our dollars. Other factors include the economy and the ability of producers and distributors to sustain this new market until a critical viewing mass is reached.

Japan also serves as an example. Only a limited number of true (and expensive) HDTV sets were initially sold. In contrast, an enhanced definition wide-screen system was more successful. Consequently, is HDTV that critical?[28] Or is an improved picture, on a wider screen, a more important factor for consumer acceptance?

This scenario has been supported in some U.S. studies. In one case, it is believed that most consumers will initially opt for digital set-top converter boxes rather than true HDTV sets. You could watch digital programming and potentially gain access to other services, but would not receive "true" HTDV programs.[29] Another option would be to buy an SDTV set. Both solutions would be considerably less expensive than buying an HTDV unit.

7. In a related area, will PCs be used for entertainment and information programming? Will the PC, or television set for that matter, emerge as the core of an advanced home entertainment and information center?

Summary
As of this writing, the DTV field is still in a fluid state. It also appears it will remain unsettled until various issues, including limited programming options and the high cost for HDTV sets, are resolved.

DIGITAL AUDIO BROADCASTING

History
Pioneered in Canada and Europe, digital audio broadcasting (DAB) can deliver a CD-quality relay either through terrestrial means or by satellite. This service could be extended to car owners through a small roof-mounted dish.[30]

A frequency allocation (L-band), granted during a 1992 World Administrative Radio Conference, was initially supported. But the United States backed another plan since this allocation was already used for other services.

The National Association of Broadcasters (NAB), for its part, was cautious in its approach toward the technology. It "recommended that any domestic inauguration of DAB be on a terrestrial only basis, with existing broadcasters given first opportunity to employ the technology."[31] Broadcasters should be given the first crack at this technology, not potentially competing services.

This recommendation stemmed from two related concerns. The first was that a satellite-delivered service could adversely affect the broadcast industry. The second concern was localism.[32] Basically, it was stated that a satellite-based DAB operation could not match a conventional radio station's public interest commitment to the local community. The public would not be as well served.

In contrast, satellite-delivered companies indicated their services could have numerous benefits. National programming could supplement local radio broadcasts. Listeners with limited radio choices could also have a

Figure 15.3
Side-by-side comparisons of standard and HDTV sets. Note the aspect ratio and size differences. (Courtesy of the Advanced Television Research Consortium; AD-HDTV.)

broader selection, while pay services and narrowcasting could be supported. A complementary terrestrial and satellite system could likewise develop.

Other factors also played a role in this process. This included work on in-band on-channel (IBOC) systems. As envisioned, special techniques could make it possible to relay programming on existing allocations. This configuration could accommodate, at least in terms of spectrum space, the current radio broadcast industry. Much like DTV, it could also provide stations with an upgrade path to digital relays.[33]

By the mid-1990s, little had changed on the domestic front. U.S. efforts were still unfocused even though the international community had, with some reservations, embraced the Eureka-147 DAB standard.[34]

In fact, while pilot programs were launched in Europe, certain testing procedures to select a U.S. terrestrial standard were brought into question. The debate between radio broadcasters and potential Digital Audio Radio Service (DARS) operators also continued, especially in light of the FCC's decision to "allocate spectrum in the (S) band for . . . DARS," thus, laying the foundation for a satellite-based industry.[35] At the same time, however, the FCC repeated its support for local broadcasters.

Update
In late 1999, the FCC actually began a rule-making procedure to explore the possible ways to initiate digital audio broadcasting. The goal? To create a digital broadcast path that terrestrial station owners could follow. The criteria would include the capability to launch enhanced audio services without disrupting the current infrastructure. According to the FCC,

In 1990, the Commission opened a proceeding to consider the authorization of digital radio services. The proceeding initially addressed both a satellite DARS and a terrestrial DAB service. As the record developed, however, it became evident that the IBOC DAB systems then under consideration for a terrestrial service were not

technically feasible, and the proceeding ultimately focused on satellite DARS spectrum allocation, service and licensing issues. Nevertheless, the Commission advanced several principles relevant here. Most importantly to the instant proceeding, the Commission emphasized its conviction that "existing radio broadcasters can and should have the opportunity to take advantage of new digital radio technologies."

We anticipate that technical advances will soon permit both AM and FM broadcasters to offer improved digital sound. Some of the systems being tested are designed specifically to permit digital broadcasting within the existing AM and FM bands. We fully support these developments, and we see great promise in these innovations for providing improved services to consumers. These innovations will also help promote the future viability of our terrestrial broadcasting system, which provides local news and public affairs programming.[36]

But despite this step forward, broadcasters still face competition from satellite radio operators. The Internet is another factor. Numerous individuals now listen to Internet-based radio stations. They can also select *the* songs they

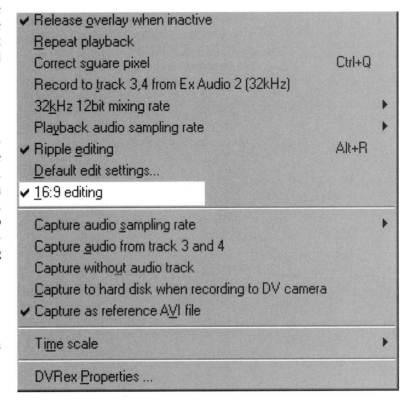

Figure 15.4
Many contemporary production tools support a wide screen format, as is the case with the Canopus nonlinear editing system. (Software courtesy of Canopus Corp.)

want to hear, *when* they want to hear them from a vast online music pool (for example, MP3 sites).[37]

So, much like HDTV, the DAB scene is still somewhat hazy as of this writing. Nevertheless, it could have a bright future. The promise of a high-quality audio relay, regardless of its source, may prove enticing to consumers. Only time will tell.

REFERENCES/NOTES

1. Technically speaking, it may not equal film's resolution.

2. Tetsuo Mitsuhashi, "Scanning Specifications and Picture Quality," *NHK Technical Monograph* 32 (June 1982): 24.

3. CBS/Broadcast Group, "CBS Announces a Two Channel Compatible Broadcast System for High-Definition Television," press release, September 22, 1983.

4. Elizabeth Corcoran, *Scientific American* 266 (February 1992): 96.

5. Even though there may not have been an official agreement, elements of the international production community did embrace the production standard.

6. "EIA Sets Itself Apart on HDTV," *Broadcasting* (January 16, 1989): 100.

7. Spice, also known as melange, was a rare and valuable substance.

8. U.S. Congress, Office of Technology Assessment, The Big Picture: HDTV and High Resolution Systems, OTA-BP-CIT-64 (Washington, D.C.: U.S. Government Printing Office, June 1990), 33.

9. Sheau-Bao Ng, "A Digital Augmentation Approach to HDTV," *SMPTE Journal* 99 (July 1990): 559.

10. "FCC to Take Simulcast Route to HDTV," *Broadcasting* (March 26 1990): 39.

11. Frank Beacham, "Sikes: A New Ballgame for HDTV," *TV Technology* 10 (May 1992): 3.

12. "HDTV: A Game of Take and Give," *Broadcasting* (April 20, 1992): 6.

13. Chris McConnell, "Broadcasters Arm for ATV Fight," *Broadcasting & Cable* (October 21, 1996): 6.

14. FCC, News Report No. DC 96-42; "Commission Proposes Adoption of Digital Television Broadcast Standard," May 9, 1996. (MM Docket No. 87-268.)

15. Richard Ducey, "An Overview of the American Digital Television Service Program," Paper, downloaded from www.nab.org, June 1999.

16. Reed Hundt, Chairman, FCC, "A New Paradigm for Digital Television," downloaded from www.fcc.gov, September 30, 1996. *Note:* TV uses an interlaced format as described in footnote 1, Chapter 17; computers use a progressive scanning process.

17. Frank Beacham, "Pressure Builds for DTV Compromise," *TV Technology* 14 (October 25, 1996): 27.

18. Ibid.

19. McConnell, "Broadcasters Arm for ATV Fight," 7.

20. DTV Standard: It's Official," *Broadcasting & Cable* (December 30, 1996): 4.

21. Different flavors of DTV are described in Ken Freed, "The World Adapts to Digital," *TV Technology* 16 (November 2, 1998): 1, 34.

22. Kristine Garcia, "Heeere's HDTV," *Digital Television* 2 (May 1999): 1.

23. Paige Albiniak, "HDTV: Launched and Counting," *Broadcasting & Cable* (November 2, 1998): 6.

24. Howard Fields, "Broadcasters May Keep Spectrum . . . Forever," *Television Broadcast* 22 (October 1999): 28.

25. Glen Dickson, "Getting Together Over Data," *Broadcasting & Cable* (March 27, 2000): 9.

26. Harry A. Jessell, "Broadcasting's Killer App?," *Broadcasting & Cable* (March 27, 2000): 10.

27. Naina Narayana, "Broadcasters Admit 2006 Deadline is Far-Fetched," *TV Technology* (January 12, 2000): 16.

28. Mario Orazio, "HDTV: Hot Dang—Total Video," *TV Technology* 13 (December 1995): 25.

29. Marcia L. De Sonne, "HDTVs—It's Where the Buyers Are—Consumer Interest, Set Availability, Set-Top Converters," downloaded from www.nab.org, June 1999. The National Association of Broadcasters web site is a rich resource for broadcasting topics, including DTV.

30. Carolyn Horwitz, "DAB: Coming to a Car Near You?," *Satellite Communications* (October 1994): 38.

31. NAB. "Digital Audio Broadcasting," Broadcast Regulation 1992; A Mid-Year Report, 140.

32. Ibid, 142.
33. Conversation with FCC, October 1992.
34. "U.S. DAB on the Slow Track," *Radio World* 19 (December 27, 1995): 37.
35. Alan Huber, "FCC Lays Groundwork for Satellite Radio," *Radio World* 19 (February 8, 1995): 1.
36. Before the Federal Communications Commission, Washington, D.C. 20554, In the Matter of Digital Audio Broadcasting Systems and Their Impact on the Terrestrial Radio Broadcast Service, MM Docket No. 99-325, downloaded from www.fcc.gov, November 1999.
37. Chuck Taylor, "Digital Radio Promises Too Little, Too Late In Face of Competition," *Billboard* (January 29, 2000), downloaded from Nexis.

SUGGESTED READINGS

ATSC. Like the FCC, this is a rich resource for the digital television standard. Documents have been available from the ATSC web site. Examples include A/53 ATSC Digital Television Standard (September 16, 1995); see also A/54 Guide to the Use of the ATSC Digital Television Standard (October 4, 1995); A/52 Digital Audio Compression (AC-3) Standard (December 20, 1995).

Beacham, Frank. "Digital TV Airs 'Grand' Soap Opera." *TV Technology* 14 (August 23, 1996): 1, 8; Beacham, Frank. "Pressure Builds for DTV Compromise." *TV Technology* 14 (October 25, 1996): 1, 27; McConnell, Chris. "Broadcasters Arm for ATV Fight." *Broadcasting & Cable* (October 21, 1996): 6–7, 12. The digital television standard controversy.

Boyer, William H. "Don't Touch That Dial." *Satellite Communication* 22 (May 1998): 44–47; Reynolds, Melanie. "DAB Firms Tune in for Product Launch." *Electronic Weekly,* downloaded from Nexis, February 2000; Lucent Digital Radio. "Submission to the National Radio Systems Committee." Downloaded from www.nab.org, January 2000. DAB issues.

Brown, Peter E. "PC OEMs Look to DTV as Panacea for Home Market Ills." *Digital Television* 2 (November 1999): 16; Rice, John. "What Flavor Is Your DTV?" *TV Technology* (November 30, 1998): 28. PCs and DTV.

Ducey, Richard V. "An Overview of the American Digital Television Service Program." A paper written by an NAB Senior Vice President. An excellent look at the development of the U.S. DTV system.

FCC. The FCC is a rich resource for HDTV, DAB, and related documents. A number are also available online from the FCC web site. Examples include the items cited in the Reference section as well as these: Commission Begins Final Step in the Implementation of Digital Television (DTV). July 15, 1996 (MM Docket No. 87-268); Advanced Television Systems and Their Impact Upon the Existing Television Broadcast Service. August 14, 1996 (MM Docket No. 87-268); FCC Commences Rulemaking to Consider Terrestrial Digital Audio Broadcasting." November 1, 1999.

Freeman, John. "A Cross-Referenced, Comprehensive Bibliography on High-Definition and Advanced Television Systems, 1971–1988." *SMPTE Journal* 101 (November 1990): 909–933. A comprehensive bibliography and a rich resource for the HDTV field's earlier years.

Kienzle, Claudia. "Internet Poised to Steal HDTV's Thunder." *TV Technology* 17 (October 6, 1999): 1, 12. Potential Internet developments and the capability to relay digital information.

Meadows, Lynn. "AT&T Pulls IBOC from DAB Tests." *Radio World* 20 (October 16, 1996): 1, 11. Overview of in-band DAB problems.

McConnell, Bill. "Deciding on Digital Public Interest." *Broadcasting & Cable* (October 12, 1998): 19; Advisory Committee on Public Interest Obligations of Digital Broadcasters. "PIAC Legacy Project: Executive Summary." Downloaded from www.benton.org/PIAC, June 1999. Public interest and digital television.

Strachan, David. "HDTV in North America." *SMPTE Journal* 105 (March 1996): 125–129. Comprehensive examination of U.S. HDTV developments.

TV Technology. Mario Orazio writes an ongoing column that covers a wide range of topics, including those related to HDTV.

"TV Transitions from a Global Perspective." *Broadcasting* (October 15, 1990): 50–52. HDTV and digital television from the perspective of the world community.

U.S. Congress, Office of Technology Assessment. The Big Picture: HDTV and High Resolution Systems. OTA-BP-CIT-64 Washington, D.C.: U.S. Government Printing Office, June 1990. A comprehensive background paper/report about domestic and international HDTV systems and related topics, such as consumer acceptance.

Zou, William Y. "SDTV and Multiple Service in a DTV Environment." *SMPTE Journal* 107 (October 1998): 870–878. Review of SDTV elements.

GLOSSARY

Advanced Television (ATV) Systems: A generic name for higher definition television configurations.

Digital Audio Broadcasting (DAB): A CD-quality audio signal that could be delivered to subscribers by satellite or terrestrial means.

Digital Television (DTV): A generic designation for digital television systems.

High-Definition Television (HDTV): The field concerned with the development of a new and improved television standard. Superior pictures will appear on a wider and larger screen; both digital and analog configurations have been developed.

16 The Production Environment: Colorization and Other Technology Issues

The previous section of the book explored production technologies. Besides influencing their respective industries, they've also had a rippling effect and have affected society. This chapter explores five such topics and their impact. These are colorization, image manipulation, electronic music, copyright, and the democratization of information.

COLORIZATION

Colorization is the process in which a black-and-white movie is manipulated by computer to produce a colorized version of the original film. Black-and-white television programs have also been affected.

In brief, when NASA started the exploration of the Solar System, thousands of photographs of other worlds were transmitted to Earth. They were fed to computers and subsequently underwent image processing. This operation had two primary goals: image correction and enhancement. If a picture was marred by noise and other defects, they could be eliminated or minimized. Enhancement techniques helped an individual to better interpret the photographic data. In one setup, an image's contrast could be altered to highlight a characteristic of a region of Mars.

Cost-effective and powerful computers extended image processing to a broader user base. Physicians processed X-rays to reveal previously invisible details. Astronomers adopted a technique called *pseudocoloring*.

In a black-and-white image, different regions of a galaxy or other celestial object may be reproduced as almost identical gray shades. This may make it impossible to visually discern its physical characteristics.[1] This is where computer processing steps in. We assign colors to the different gray shades, which are then reproduced in the appropriate colors. This image will highlight the object's physical features because they are now represented by different and contrasting colors.[2]

Image processing has also influenced the communications industry. The product may change, but the intent is essentially the same: to manipulate a picture for a specific purpose. In this case, for a desktop publishing project or for colorization.

Yankee Doodle Dandy

Starting in the 1980s, the colorization technique was applied to black-and-white films. Movies such as *Yankee Doodle Dandy* and *It's a Wonderful Life* were colorized to the cheers and jeers of supporters and opponents alike.[3]

Colorization is a multifaceted task. A videotape copy of the film is created, and in one step, colors are assigned to appropriate objects in a frame. If possible, the colors correspond to those used when the movie was shot. A computer eventually processes the film, frame by frame, and a colorized version of the original is generated.[4] A former gray hat and jacket may emerge as yellow and blue.

Colorization proponents argued that the process would introduce older, and possibly classic movies, to new generations of viewers. If you grew up with color films and television, would you be willing to watch black-and-white products? Colorization also did not harm the original film, and you could still watch the black-and-white version.

Opponents countered that colorization should be halted for aesthetic and ethical reasons. Although a company may have the legal right to colorize a film, a distorted version of the original may be created. Colors may bleed, the lighting and makeup, which were originally designed for a black-and-white medium, are altered, and physical details may be lost.[5]

But a more salient point is the ethical issue of altering another person's work. According to many artists, including Frank Capra who directed *It's a Wonderful Life*, no one has the right to change a film or another piece of art. The artist's original vision and the work's integrity are destroyed.

Finally, ask yourself this question—even if colorization is flawless, would this make the process acceptable?

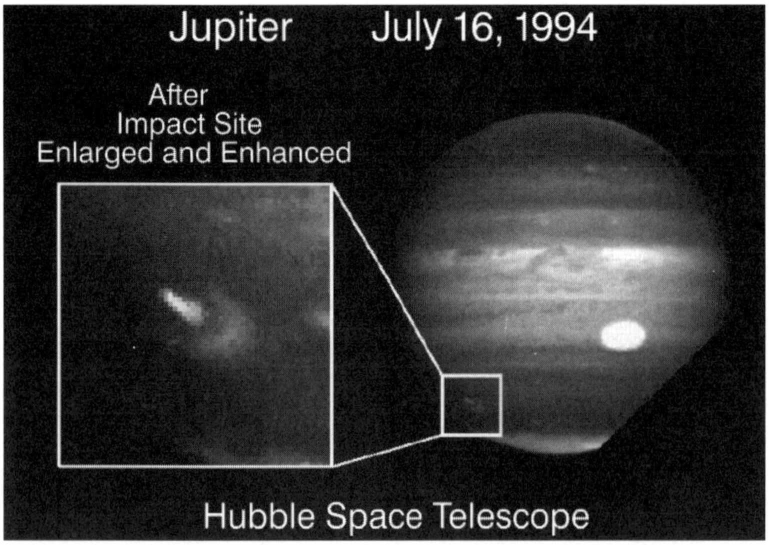

Figure 16.1

The Shoemaker-Levy/Jupiter collision. Computers play a central and critical role in producing images of celestial events that we can subsequently view and study. (Courtesy of NSSDC/HST Science Team.)

IMAGE MANIPULATION

Computers have also been used to manipulate advertising and news images. The problem with the second scenario is the public's perception. People generally believe a news photograph portrays reality, an event as it occurred. By altering the image, the public's trust could be broken.

But image editing and manipulations are not new phenomena. The photographic process itself, in which you select a specific lens and compose a shot, is a form of editing. Image manipulations are also established darkroom fare.

Nevertheless, the new electronic systems raise the latter process by a notch. It is now easier to manipulate an image, and the prerequisite tools, such as a PC and graphics packages, are readily available.

In one example, you can take a picture of a site and electronically add a proposed building. In another example, you can take a picture of a swamp and electronically alter it to make it look like prime real estate. Although each manipulation produces a new image, there are differences in the context of our discussion. If the first picture is created as part of an environmental impact study, and is clearly labeled as a simulation, not many people would have a problem with it. In fact, it can actually provide a service. But if you do not label the swamp for what it is, that's when problems start cropping up.

Newspapers and magazines with ready access to even more sophisticated systems have been guilty, at times, of essentially turning swamps into valuable land. In two examples, *National Geographic* and the *St. Louis Post-Dispatch*, respectively, moved a pyramid in one scene and eliminated and replaced a can of soda in another.[6] On face value, a can of soda may not be that important. Yet in the context of this photograph, it had a specific meaning with regard to the individual pictured in the photograph.

It also raised other questions. What else can you remove, add, or replace in other images? Will a government manipulate a picture to portray an event not as it occurred, but as it wants it to appear? What about abuses with legal evidence and the potential manipulation of video imagery?

This situation is further exacerbated by the emergence of electronic still photography (silverless photography). Instead of using film, in which the original image is preserved as a negative or a slide, an image is electronically captured. A picture can subsequently

Jupiter   July 16, 1994

After
Impact Site
Enlarged and Enhanced

Hubble Space Telescope

be reproduced as a hardcopy with a printer and/or can be relayed over a telephone line to another location. It can also be fed to a computer for desktop publishing or video applications. This same capability, though, may make it easier to manipulate an image. It is already in a malleable form, and you do not have to bother with a print or negative.[7]

How do you prevent these abuses? One possible and partial solution may be the use of a labeling system to indicate a photograph has been altered. Another answer may lie in the field's ethical underpinnings. Because a digital manipulation could be very hard to detect, if at all, an individual's ethical standards may be one of the only safeguards you have and can trust.[8]

Finally, note that professional and consumer electronic cameras are available. The capabilities vary, as do the complementary applications. As indicated, they range from desktop publishing to more traditional photographic tasks.

Image quality also greatly improved in the late 1990s as prices dropped. In one example, a Sony camera, which recorded images on a floppy instead of the more typical (and expensive) memory card, cost less than $500. Equipped with a zoom lens and electronic viewfinder, you could generate pictures that could be input to your computer and subsequently saved and printed. For considerably more money, you could purchase a professional camera that could rival traditional film systems.

ELECTRONIC MUSIC

The proliferation of electronic musical instruments has led to its own controversy. Musicians, for instance, used a *New York Times's* ad to protest a budgetary and/or profit-related move in which eight string players were replaced by a synthesizer. The musicians had played for a Broadway production, and the ad read, in part:

If the producers . . . really believed that a synthesizer sounds as good as real string instruments,

they could have used the synthesizer to play the string parts from the beginning! But they knew that real strings sound better, so they hired eight top-notch string players to enhance the production—that's what they wanted theater critics to hear![9]

This incident is not isolated, and other professionals could be similarly affected.

The key points for our discussion are the sound's quality and the human presence. In the former, the musicians argued that real strings sound better than a synthesizer. A similar argument has been waged between CD (digital) and high-end LP (analog) proponents.

In the latter, we all have expectations for different events. When you see a play where an orchestra is central to the production, the expectation almost calls for a conventional orchestra. It is part of the ambience, at least, as we currently perceive it.

Yet this perception could change. Although we may object to electronic instruments in certain circumstances, this situation may be the norm for the next generation.

Until that time, technological changes will continue to conflict with traditional standards. But a balance may eventually be reached. In the music world, both electronically and traditionally generated music have value. They can coexist, when used appropriately.

COPYRIGHT

The music industry has also been affected by sampling. It is possible to electronically copy and use part of an artist's work without permission. The same scenario applies to images. You can use a scanner and image editing software to copy and alter a graphic.

In both cases, the artist's rights are violated. But the PC's ubiquitous nature has made it almost impossible to protect the artist. This problem is only magnified in light of contemporary storage and distribution systems. For example, it is possible to gain access to an image library through a single CD-ROM.

Figure 16.2

Computer manipulation, in this instance, is performing a valuable service: revealing the potential impact of a proposed physical plant alteration. Note the top photo of the existing conditions and the bottom photo that depicts the proposed changes. (Courtesy of David C. Young/Young Associates-Landscape Architecture.)

Photo of Existing Conditions

Simulation of Action - Proposed Stack and Plume

New York State Electric & Gas - Clean Coal Technology Project at Milliken Station
Prepared by: Young Associates - Landscape Architecture

Related issues are as follows:

1.   Once information is committed to an electronic form, it becomes widely susceptible to illegal copying.

2.   Multimedia and desktop video productions are built on digital pictures as well as audio and video clips. There's a great demand for this raw material.

3.   Simply keeping track of information is problematic. CD-ROMs and other optical media can accommodate megabytes of information. If you are creating a compilation of pictures for distribution, you must be sure the individual or organization providing a picture has the "full, clear, and unquestioned legal right" to the image.[10] If not, you could commit a copyright violation.

4.   Enforcement raises serious questions. You can legislate against illegal copying, but how do you enforce the law? Are you going to differentiate between a hobbyist who unwittingly uses a copyrighted image in a newsletter and a multimedia producer? Or are they going to be treated equally? If an exception is made for the hobbyist, does not the copyright owner still pay the price, since he or she is not compensated for the work?

5.   As new applications unfold, new copyright challenges emerge. For example, in the late 1990s, MP3, a compression system "that allows people to pass near-CD quality music files easily" over the Internet, became popular.[11]

Some music companies and organizations were alarmed that the Internet's ubiquitous presence would make music piracy even more rampant. Others took a different approach and believed the exposure would actually boost their CD and cassette sales. This was particularly true for newer and alternative bands.

This situation is not unique, and different protection schemes have and will continue to be introduced to protect copyrighted information. For the Internet, this may include adopting standards, such as the Secure Digital Music Initiative, that would allow music distribution but protect copyright owners.[12]

6.   The copyright issue must be examined from a system's perspective since it cuts across different fields, including, as indicated, the Internet. Organizations and individuals are establishing their own web sites on the Internet, personalized information pools with links to other sites. But like the optical disk and computer software fields, how do you protect intellectual property, especially when new sites are sprouting up every day? How do you monitor legal and illegal usage?

This environment also raises another question. As outlined by Lawrence Magid in an *Information Week* article:

It's possible for me, or anyone who runs a web site, to point to an object [e.g., a photograph] at another site so that the object is displayed automatically. The object would appear to reside at my site, even though it remained at its original site. I would not be copying the image—that is clearly illegal—but I would be representing it as mine.[13]

Consequently, what are the implications of this capability? While Magid declined to adopt this approach to his work, will other people follow suit? Even if this type of operation is legal, is it ethical, without some type of explanatory notation?

There are, however, legal ways to use preexisting information, whether it is a sound, image, or digital clip. You could obtain the proper clearances from the copyright owner. Or you could buy an image and/or sound collection.

7.   Legislative and judicial action have also been taken to help protect materials that are available via the Internet. These include the No Electronic Theft Act and the Digital Millennium Copyright Act.

The No Electronic Theft Act, signed into law by President Clinton in December 1997, closed a loophole in the illegal distribution of software. Criminal infringement would occur when a party infringed on a copyright in a willful manner for commercial advantage or private financial gain.[14] But this was not a prerequisite for infringement. Infringement would also occur if, during any 180-day period, one or more copies of a copyrighted work having a retail value of more than $1000 were illegally reproduced or distributed via electronic means, regardless of whether the party committing the action gained from the infringement.[15]

The Digital Millennium Copyright Act was created in the late 1990s. Two important sections for our discussion are the

- World Intellectual Property Organization (WIPO) Copyright Treaties Implementation Act
- Online Copyright Infringement Liability Limitation Act.[16]

The WIPO section of the Digital Millennium Copyright Act guarantees adequate legal protection and legal penalties against circumvention of effective technological measures employed by authors to control the use of their work. Manufacturing and trafficking in such technology is prohibited as well.[17]

The WIPO can trace its roots to international treaties ratified in 1998 by the United States and 120 other United Nations countries. The treaties were written to protect copyright works fixed in traditional and nontraditional mediums. The goal? To further develop rights previously established by the terms of the Berne Convention and a World Trade Organization agreement.[18] The new act would more securely protect a copyright owner's right to reproduce, communicate to the public, and adapt their work. Copyright owners were also granted the right to prohibit the commercial rental of their computer programs and musical recordings and to make their work available online, among other stipulations.[19]

The Online Copyright Infringement Liability Limitation section of the act was designed to encourage service providers and copyright owners to work together to discover and deal with online copyright infringement. The act also gives service providers a clearer picture of their legal responsibilities.

Service providers are also protected from copyright infringement liability if they simply transmit information over the Internet. However, they are required to remove material from a web site if copyright infringement is apparent.[20]

This section of the act also contains strong language to protect copyright owners against potential abuses that could arise from the latest technological advancements. For example, it criminalizes any interference with antipiracy measures built into commercial software.[21]

8.   Case law has also been decided based on copyright infringement issues. In *Playboy v. Frena* and *Playboy v. Hardenburgh*, for instance, bulletin board services (BBSs) illegally posted copyrighted photographs from *Playboy* magazine. A BBS is an electronic communications system that generally predates web sites. They are used for distributing information and can serve as communications conduits for people who dial-in to their sites via modems.

In each instance, the service provider had not obtained copyright clearance to post the photos. *Playboy* sued and the courts ruled that *Playboy* had the exclusive rights to display the works. Therefore, copyright infringement had occurred.[22]

In another case, *Sega v. MAPHIA*, a BBS owner knowingly and intentionally offered subscribers access to Sega video games and the hardware to copy the games. Based on these facts, the court found in favor of Sega, holding that copyright infringement had occurred.[23]

In a different vein, a student who operated his own BBS made software, valued in the millions of dollars, available to his subscribers. He was sued in federal court by the government. But the court ruled he could not be held legally liable for copyright infringement since he had not charged his subscribers for the software. The aforementioned No Electronic Theft Act, which was passed at a later date, has since eliminated this loophole.[24]

Finally, in *Tasini et al. v. New York Times et al.,* the court examined the posting of copyrighted articles of freelance writers on electronic databases and/or CD-ROMs without an author's permission and/or additional payments. The case was heard at the U.S. District Court and the Court of Appeals level with two different outcomes.

The U.S. District Court ruled in favor of the *New York Times* and the other publishers. The court indicated the additional publication of the works did not constitute copyright infringement. However, on appeal, the circuit court overturned this decision holding that posting the works via the Internet and/or on CD-ROMs was an additional use. Without the author's consent, it constituted a copyright infringement.[25]

To sum up, the No Electronic Theft Act and the Digital Millennium Copyright were written, in part, in response to the new technologies and their applications. It is also an evolving field, as electronic services, including those on the Internet as described in a later chapter, continue to evolve.

THE DEMOCRATIZATION AND FREE FLOW OF INFORMATION

The production technologies have contributed to the democratization of information. With the correct outfit and skills, you can publish a newspaper, magazine, video, or multimedia project. This capability implies that information cannot be controlled by only a select group of people or by the government.

Established media organizations could be censored or even shut down. But it may be impossible to completely shut off the flow of information. There may be too many desktop publishing and video systems, copy machines, VCRs, and other production and distribution mechanisms.

This concept has been put to the test on different occasions, including the aborted August 1991 coup in the Soviet Union when Mikhail Gorbachev was deposed. Boris Yeltsin and his support group used desktop publishing and other communications tools to keep the Russian citizens and the world community apprised of events. The free flow of information, during this time, contributed to the coup's eventual collapse.[26]

The democratization of information also has a personal connotation. Using sophisticated communications tools, you could launch your own company for, say, desktop publishing or serving Internet customers.

You could also present your personal ideas in the open marketplace via a letter, political broadside, video production, or multimedia presentation. You may only attract a small audience. No one may even agree with you. But in the context of our discussion, is this important?

CONCLUSION

To sum up, all five topics have common threads. They are the offspring of production technologies, they have influenced society, and they have raised important questions. The most pressing issues are related to news image manipulation and the democratization of information.

For the democratization of information, the communication revolution has presented us with powerful tools. We have become information producers and are now more ensured of the free flow of information. You can censor a newspaper, but can you shut down thousands of DTP/video systems?

There is also the old joke about the censorship police knocking on your door. But with the new technologies, there may now be too many doors.

For the news media, can people still trust your reporting? How do you know what is real and what is fabricated? Once you lose the public's trust, can it ever be fully recovered?

This has also raised an incident with another application. As outlined in the following editorial from *Broadcasting & Cable*, it has serious implications. "Box of Virtual Chocolates," by John Eggerton, deputy editor, and Harry Jessell, editor, originally appeared in the January 7, 2000, issue, p. 150, of *Broadcasting & Cable*.[27] Reprinted with permission by *Broadcasting & Cable*.

On the subject of believing. It sure isn't seeing anymore.

When *Forrest Gump* first came out, some of us didn't know actor Gary Sinise. The filmmakers didn't show him from the waist down until after his character's legs had been amputated, so we though the actor was likely an amputee. He isn't, of course. It was computers that removed his legs, just as they had put Tom Hanks into George Wallace's schoolhouse door speech and Kennedy's White House. In an age when computers can put anything into a picture or take it out, the distinction between reality and special effects is as blurry as the virtual images are sharp. That is why we reacted so viscerally to the quote from the director of the *CBS Evening News* to the effect that virtual insertion technology has "applications that I think are very valid and lend themselves perfectly to news, such as *obscuring things you don't want in the frame* [my emphasis]."

Now CBS News executives are saying that they are going to be careful in how they use the video insertion technology and so far they have been. They have used it only to post their logo in strategic places. . . . They haven't distorted events in any substantive way, although they have the distinction of being the only news organization to show Times Square as it wasn't at the dawning of the millennium.

Nonetheless, we believe that the best policy regarding video insertion is not to use it during newscasts. Viewers must be able to trust that video, especially live video, is the real thing. . . .

No less a CBS executive than former president Frank Stanton had this to say on the subject, on the occasion of presenting a First Amendment award to Walter Cronkite in 1995:

"Digital technology opens a Pandora's box. The options are startling and tempting. No longer will it necessarily be that what you see is what you get. Ultimately it will be easy and inexpensive to fake the picture. And virtually undetectable. Consider the temptations and the burden these developments will put on the producer, the television reporter and his organization. The Forrest Gumps of the evening news could have a field day. And the public could be the loser. The audience will not know who or what to believe."

We join Mr. Stanton in urging news departments everywhere to consider the temptations and their consequences. It may be impossible to put the digital genie back in the bottle, but when the news business becomes about obscuring things, we're all in trouble.

REFERENCES/NOTES

1. The shades look too similar.
2. The processed image is a false color image.
3. An analogy can be drawn with the Sistine Chapel restoration. Some critics claimed this process altered Michelangelo's original work; others stated it was a cleaning and restoration process.
4. Steve Ciarcia, "Using the ImageWise Video Digitizer: Part 2: Colorization." *Byte* 12 (August 1987): 117. *Note:* As part of the process, a new print has been made, which may actually help preserve a copy of the film.
5. You may notice a label that appears on a colorized product: "This is a colorized version of a film. . . . It has been altered without the participation of the principal director, screenwriters, and other creators of the original film." From *Village of the Damned,* aired on TBS, 1995.
6. Jane Hundertmark, "When Enhancement Is Deception," *Publish* 6 (October 1991): 51.
7. Ibid.
8. Don Sutherland, "Journalism's Image Manipulation Debate: Whose Ethics Will Matter?" *Advanced Imaging* 6 (November 1991): 59.
9. Advertisement, "Grand Hotel: The Rip-Off," *New York Times* (November 3, 1991): Section 4, 7.

10. Henry W. Jones, III, "Copyrights and CD-ROM," *CD-ROM Review* (November/December 1987): 64.

11. Christopher Jones, "Digital Media Is Spinning Toward a Distribution Revolution," *NewMedia* 9 (June 6, 1999): 28.

12. Ibid.

13. Lawrence Magid, "Defining Ethics On the Net," *Information Week* 547 (October 2, 1995): 126.

14. No Electronic Theft Act. Pub. Law 105-147. 111 Stat. 2678. 1997, downloaded from www.GSE.UCLA.Edu.

15. Ibid.

16. H.R. 2281. Bill Summary and Status from the 105th Congress. October 28, 1998.

17. Digital Millennium Copyright Act. Pub. Law 105-304. October 1998, downloaded from thomas.loc.gov.

18. Kent Middleton et al., *The Law of Public Communication,* 5th ed. (Reading, Mass.: Addison-Wesley-Longman Publishing, 2000), 224.

19. Ibid, 225.

20. Digital Millennium Copyright Act. UCLA, Online Institute for Cyberspace Law and Policy, downloaded from www.gse.UCLA.edu.

21. Digital Millennium Copyright Act.

22. *Playboy Magazine v. George Frena.* 839 F. Supp. 1552.1993. *Playboy Magazine v. Russ Hardenburgh.* 982 F. Supp. 503. 1997.

23. *Sega v. MAPHIA.* 857 F. Supp. 679. 1994.

24. *United States v. David LaMacchia.* 871 F. Supp. 535. 1994.

25. *Jonathan Tasini et al. v. New York Times et al.* 972 F. Supp. 804. 1997. *Jonathan Tasini et al. v. New York Times et al.* 192 F. 3d 356. 1999.

26. See Richard Raucci, "Overthrowing the Russians and Orwell," *Publish* 6 (November 1991): 21, and Howard Rheingold, "The Death of Disinfotainment," *Publish* 6 (December 1991): 40–42, for additional information about this event and similar incidents, such as the 1989 Tiananmen Square uprising. Rheingold's article also touches on other implications.

27. John Eggerton and Harry Jessell, "Box of Virtual Chocolates," *Broadcasting & Cable* (January 7, 2000): 150. Reprinted with permission.

SUGGESTED READINGS

Colorization/Image Manipulation

Baxes, Gregory A. *Digital Image Processing.* New York: John Wiley & Sons, 1994. An excellent look at image processing.

Berry, Richard. "Image Processing in Astronomy." *Sky and Telescope* (April 1994): 30–36; Scher, Chris and John O'Farrell. "Improve Your Astrophotos by Combining Images." *Sky and Telescope* (October 1999): 135–139. Overviews of image processing, PC-based astronomical applications, and relevant software.

Chouthier, Ron. "Glossary of Image Processing Terminology." *Lasers and Optronics* 6 (August 1987): 60–61. A glossary of image processing terms.

Hardin, R. Winn. "Biometric Recognition: Photonics Ushers in a New Age of Security." *Photonics Spectra* 31 (November 1997): 88–100. Interesting look at identification systems that tap imaging techniques.

Higgins, Thomas V. "The Technology of Image Capture." *Laser Focus World* 30 (December 1994): 53–60; Klare, Matthew. "Still Life in Pixels." *Interactivity* 4 (September 1998): 11–21; Wheeler, Michael D. "Information Processing: Law Enforcement Uses Digital Imaging and Storage to Track the Criminal." *Photonics Spectra* 32 (November 1998): 107–111. Electronic camera overview and applications.

Larish, John L. *Electronic Photography.* Blue Ridge Summit, Penn.: Tab Professional and Reference Books, 1990. Electronic photography, including still camera systems and image manipulation.

Wels, Susan. *Titanic.* New York: Time Life Books, 1997. Examines the history/fate of the *Titanic;* includes information about imaging techniques used to capture pictures of the ship.

Copyright

Brown, Peter. "Is Hollywood Holding Its Breath on Copyright Protection?" *Digital Television* 2 (May 1999): 44; Haar, Steven. "Online Copyright Battle Shapes Up." *Inter@active Week* 3 (January 29, 1996): 16; Meeks, Brock N. "Clinton Seeks Copyright Law Tweak." *Inter@ctive Week* 2 (September 11, 1995): 13. Copyright issues and new technologies.

Jones, Christopher. "Digital Media Is Spinning Toward a Distribution Revolution," *NewMedia* 9 (June 6, 1999): 26–34. A comprehensive look at copyright issues and the Internet.

Labriola, Don. "Getting Through the Media Maze." *Presentations* 8 (September 1995): 23–29; Weiss, Jiri. "Digital Copyright; Who Owns What." *NewMedia* 5 (September 1995): 38–43. Copyright hints (for example, how to obtain images).

Roberts, Jon L. "A Poor Man's Guide to Copyrights, Patents and Trade Secrets." *Advanced Imaging* 9 (September 1994): 13–14. Copyright/patent/trademark overview.

GLOSSARY

Colorization: The process by which color or "colorized" versions of black-and-white movies are produced.

Democratization of Information: New equipment (for example, PCs and desktop publishing/video systems) contribute to the free flow of information in society. The implications? More people can now be information producers, and it may be harder to institute censorship on a broad scale.

Image Processing: The field and technique in which an image (for example, created by a video camera) is digitized and manipulated. Typical operations include image enhancement and correction.

MP3: A compression tool primarily used to distribute music over the Internet. It has copyright/legal implications.

Photograph and Image Manipulation: The manipulation of photographs for, in one example, electronic retouching (used in advertising). A potential problem is the alteration of news images.

17 Information Services: Interactive Systems

The next two chapters cover information services. Chapter 17 focuses on America Online (AOL) and other interactive, computer-based information systems. Chapter 18 is devoted to the Internet.

INTERACTIVE SYSTEMS

Interactive information systems can meet your individual needs—you can request and receive specific information. This is in contrast to another type of information service, a *teletext* magazine.

In brief, a teletext magazine is an electronic publication delivered to a television set via a television signal.[1] It is composed of pages of text and graphics, and like a print publication, can feature news, sports, and other stories. Subscribers generally receive the same information, and the only option is the page you may want to view at any given time.

Teletext services only achieved limited success in the United States. Competing technical standards, cost factors, and other issues contributed to a flat market.[2] They were, however, more successful in England and other countries.[3]

In contrast with a teletext service's *one-way information stream*, an interactive system performs as the name implies. You can interact with the system to request and receive specific information and services.

This family can also be divided into two general classifications: *videotext* and *PC-based* operations. For our discussion, PC-based systems describe both *dedicated* and more *consumer-oriented*, AOL-type operations. The former also targets narrow subscriber groups with specialized information pools.

Videotext Operations

A videotext service is a graphics-oriented, interactive operation. You could tap a database of thousands of frames or pages of information through its two-way capabilities and a telephone line, the prevalent communications channel for two-way systems.[4]

Videotext service has been distinguished by an easy-to-use interface and hookup. Consumers have also been a primary target, and different terminals were developed for this connection. A typical configuration consisted of a keypad, decoder, and television set.[5]

Yet like the teletext industry, the U.S. videotext industry failed to mature. The systems found wider acceptance in other countries, however, and they did help pioneer a simplified user interface and other features adopted by their more successful PC-based counterparts.

Prestel, Telidon, and Viewtron. The world's first public videotext system, Prestel,

Figure 17.1
An index page of a teletext magazine. (Reprinted by permission of ELECTRA, a part of Great American Broadcasting; Electra.)

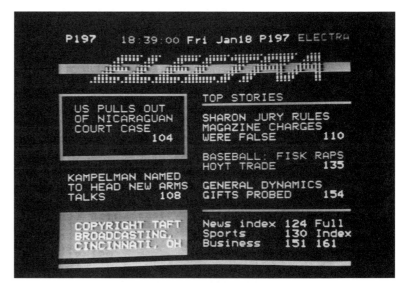

was launched in England in 1979. Unlike free teletext information, a subscriber was assessed a fee. This included potential information provider (IP) charges, a general videotext practice.

An IP is an organization that contributes information to a database. An airline IP, for instance, could provide flight, seating, and pricing information. IPs have also supplied newspaper stories as well as financial and other assorted data.

Prestel incorporated a series of indexes and menus to facilitate an information search. They guided you from broad to narrow subject areas. The service also featured a gateway option to gain access to computers outside of the host system. It is, to all intents, a gate, a link, to these other operations. In one example, travel agents used this feature to interact with airline computers.[6]

Telidon, for its part, can be viewed as a videotext system and standard. The Telidon project was the brainchild of the Canadian Department of Communications. One designer goal was to create a videotext system free from hardware constraints. Older terminals would be capable of accessing "future command formats," while newer terminals would be compatible with old or preexisting data.[7]

Telidon supported an alphageometric coding scheme and alphageometric graphics. Picture description instructions (PDIs) essentially defined geometric shapes as they would appear on the screen—a line command specified two endpoints subsequently connected by the terminal to form the line. This approach enhanced the graphics creation and information relay processes.[8]

Telidon also contributed to the development of the North American Presentation Level Protocol Syntax (NAPLPS) standard. Based on AT&T and Telidon concepts, it was selected to drive various videotext services.[9]

In use, the Telidon standard served as host for individual videotext systems throughout Canada. These ranged from Grassroots, a Manitoba project that provided farmers with commodity price and local information, to Teleguide, a tourist resource system accessed through public terminals.[10]

Viewtron, launched in Florida in 1983, serves as a representative example of a U.S. videotext system. It offered subscribers weather, marine forecasts, and games. Viewtron also carried the *New York Times*, other IPs, and a gateway option for electronic banking and shopping.

A local business hooked into the system could also send a subscriber an electronic receipt for an order.[11] This capability highlighted one of Viewtron's strengths. Besides supporting electronic transactions, it could be integrated in a community.[12] Despite its features, consumers did not support Viewtron. The initial high cost for its Sceptre terminal and user fees were two contributing factors. An attempt was subsequently made to expand its customer base to PC owners across the country—to make it more of a national system. These actions failed, however, and the service was terminated in early 1986 after a financial loss.[13] Another consumer-oriented videotext project, Times Mirror's Gateway, similarly folded.

The closing of both systems highlighted a trend. It appeared that U.S. videotext services were not viewed as priority items. Although they did not target all consumers, they nonetheless failed to attract enough individuals who could afford or wanted to become subscribers.

In contrast, their contemporary PC-based counterparts were more successful. A partial explanation may lie in the targeted user groups. For example, earlier PC-based systems were geared toward more focused audiences, people who were willing to pay for a service for specific reasons. You did not just get movie reviews and weather information. This principle was particularly true for dedicated systems that were designed for lawyers and other select groups with desirable information pools.

Teletel. One of the most successful interactive information systems to date has been France's Teletel. To promote this national service and to lower printing costs, the telephone directory was installed on the network. The government subsequently distributed free

Minitel terminals to French citizens. These factors contributed to its impressive growth.

The system has subsequently supported numerous information services. It has also compiled an impressive array of user statistics. Highlights, based on 1994 data, include the following:

1.   There were more than six million Minitel terminals. PC owners could also tap into the system with Minitel emulation software.

2.   As a reflection of its wide acceptance, thousands of supporting services sprung up. You could shop at home and even locate consultants in different fields.

3.   In 1994, 1913 million calls were made.[14]

4.   International network access was provided by the Internet.

In addition, MinitelNet traffic dramatically rose over seven years. As represented in hours:[15]

1988	1,000
1989	3,000
1990	152,390
1991	312,245
1992	506,241
1993	931,323
1994	1,344,461

In sum, the French network emerged as one of the world's premiere information and communications systems. Through the initial distribution of terminals, and other factors, it became an integrated member of France's overall communications infrastructure. It is also flexible, and in one example, has ties to the Internet.

PC-Based Operations

When first launched, PC-based companies attracted computer-literate users. You had to set up a modem and communications software, and the on-screen information was text based.

But as the systems evolved, the connection process was simplified and graphical on-screen representations became popular. In one example, taking a page from a videotext environment, you were and are greeted by a graphically rich display and interface. Spearheaded by Prodigy and later by AOL, subscribers can point and click their way to different databases and electronic services. Consequently, operations are now easier to use, especially for newer PC owners.

An AOL-type system also seemingly cuts across the traditional U.S. videotext and the PC-based markets. General consumer and PC user needs are addressed.

Finally, during the mid- to late 1990s, various companies were slated to move to the Internet as a distribution vehicle.[16] Some

Figure 17.2
The typical connection between a home user and an information company is the telephone line.

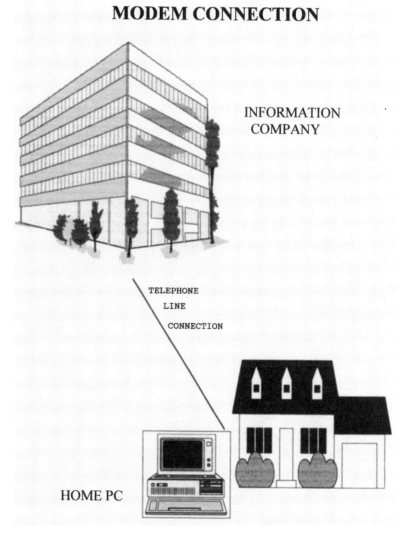

MODEM CONNECTION

INFORMATION COMPANY

TELEPHONE LINE CONNECTION

HOME PC

planned, however, to maintain their traditional services as long as there was sufficient demand. In other cases, a company may have retained its proprietary system but now offered an Internet link to its subscribers.

Consumer-Oriented Systems: AOL. AOL serves as an illustration for a contemporary consumer-oriented operation, in contrast with the dedicated systems described in the next section. The typical connection is by a PC, modem, and telephone line.

America Online has dominated the consumer-oriented end of the industry. Like Prodigy, it also helped pioneer a user-friendly interface. It is graphical and AOL's sign-on procedure and other processes are somewhat automated.[17]

By using AOL, you can speed up certain tasks. In one application, you may want to buy stock. But depending on where you live, it may take several or more days to establish an account. The solution? Connect with an AOL-type system and sign on with an electronic broker. You could potentially set up the account and buy the stock in the same day.

Industry has also been supported by AOL and other systems. Business-related databases and services have been featured and there has been strong interest in electronic and financial transactions.[18]

AOL also features an array of online databases. Online implies, in this situation, information you receive while you are linked to and are interacting with the company's computer. Sample databases include national and international news, travel guides, and software libraries you can tap to download a variety of programs.

One of AOL's most popular options has been its chat rooms, essentially electronic meeting places. Think of a topic, and a chat room has probably covered it. You could join an existing room or start your own. Part of a chat room's popularity, or variations found on other systems, has been its anonymity. You could meet new people and choose to respond or not.

Some individuals believe this communication form is dehumanizing and could lead to a further erosion of the socialization process. Others view it as simply another communications tool. This subject resurfaces in Chapter 19.

In sum, AOL-type systems have responded to specific consumer and business needs. The rich assortment of services and information, in addition to the PC's proliferation at work and home, have contributed to their overall growth. They are also cost effective and typically provide an Internet hook.

Dedicated Information Retrieval Operations. Dedicated information retrieval systems are generally geared for smaller and more specialized subscriber groups. Such systems have offered legal, medical, and other unique information pools. They are also differentiated from consumer-oriented services, for our purpose, by the targeted user groups and databases. Three members of this classification have been Lexis, Nexis, and NewsNet.

Lexis, a legal database, has supported legal citations, printouts of court cases, and other functions. It has also covered elements of other countries' legal systems, an important consideration in light of the international community's impact.

Nexis has been adopted by the media and other users as a news retrieval service. It has operated as a broad-based information resource composed of more than 100 newspapers, magazines, and specialized publications and transcripts.

NewsNet, for its part, has been particularly salient for the communications industry. Topics ranged from satellites to the FCC to optical disks. It also supported, as have some other services, an electronic clipping function. Rather than tediously searching through a database to spot specific information, the system could be set up for this purpose. A subject area is defined. Next, relevant news, which is constantly updated, is collected and stored in an electronic mailbox. It is subsequently retrieved by the subscriber.

Dedicated systems have also adopted the Internet for distribution. In some cases, it has helped simplify the user interface since you can now use a Web browser, which is described in the next chapter, to search and gain access to information.

Summary.   PC-based systems offer their subscribers a broad range of information and electronic services. They also continue to flourish despite the Internet's explosive growth.

On one hand, they support proprietary information and services that may not have been available on the Internet. This is particularly true of the dedicated information companies. They are also well established and even feature an Internet connection. Thus, PC-based systems and the Internet are somewhat complementary.

In this light, the previously described move to the Internet could be viewed as a change of venue. As a dynamic and popular environment, the Internet is *the* place to be. On the Internet, a company can still sport proprietary information pools. But now it can increase its audience, open new revenue channels and options, and take advantage of the Internet's other features.

There is a challenge, though, for consumer-oriented services. Since the Internet offers so much free information and services, a company may have to develop specialized and attractive information pools and services to retain its customer base. Value-added services are also becoming more important. For example, you may spend hours searching the Internet for specific medical information. If a company provides this service, and points the way to these and other sites, would you remain a customer?

The dedicated systems are somewhat more protected from this scenario. Their highly focused information pools and/or services may not be duplicated on the Internet.

APPLICATIONS

When viewed in its entirety, the computer-based communications universe is an important resource. You can send e-mail to a friend, search for a book, and communicate with other people in real time.[19] You can also join special interest groups, conduct research, buy and sell stock, retrieve electronic journals, and tap into the Internet.

Interactive systems have also contributed to the birth of an electronic democracy. They can promote the free flow of information and open a dialogue between a government and its citizens, as stated in a 1990 report, *Critical Connections: Communication for the Future*. In fact, during a 1992 presidential primary, former governor Jerry Brown used two-way services to communicate with voters during his campaign bid.[20]

But this tool's emergence in the political arena has implications. Will network access be limited only to those who can afford it? Or could the system be used to promote a specific political agenda rather than the free exchange of ideas?

In another twist, interactive systems have supported virtual communities. Groups of people who may live in different geographical regions, but share a common interest, can congregate electronically.[21]

Education has also been served. In one case, numerous colleges are now offering courses and degrees online, typically through an Internet link. A key factor is geographical freedom: Unlike traditional schools, you are not bound to a specific locale to take a course.

One of Thomas Jefferson's sentiments was that "no republic could maintain itself in strength without the broad education of its people."[22] Interactive systems may help sustain Jefferson's ideas through its diverse data pools, virtual communities, and distance education schemes.

IMPLICATIONS

Paperless Society

The systems that make up the interactive universe, including the Internet, have contributed to the growth of a paperless society. In such a society, information is increasingly created, exchanged, and stored in an electronic form. This subject is covered in Chapter 22.

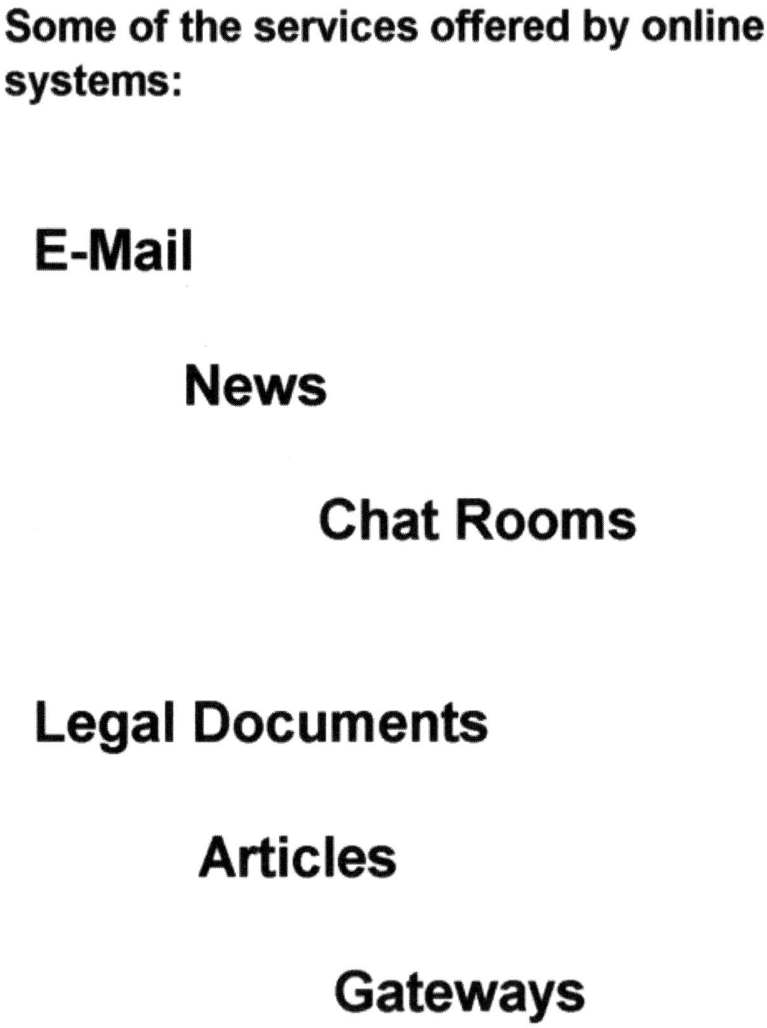

Some of the services offered by online systems:

E-Mail

News

Chat Rooms

Legal Documents

Articles

Gateways

Figure 17.3
Sample of the services offered by computer-based information systems.

Privacy and the First Amendment

Privacy, especially with regard to e-mail, is another pressing issue. Electronic messages can be relayed on a local area network within an organization and to outside locations through different conduits. But there has been a problem with the perception of this class of communication. As stated in a *Byte* editorial:

Thinking of electronic information differently from, say, a personal letter at the post office is natural. The intangible character of electronic information makes the attributes of privacy and ownership somehow seem less real. So we often take for granted how, for example, we might react to someone reading private e-mail. Yet

doing so is no different from opening a sealed envelope that's been stolen from the post office.[23]

The last line is particularly salient. While reading private e-mail is conceptually no different from opening a traditional letter, people may not view it in this light. More important, what you think is private mail may not be legally defined as such.

Another concern is First Amendment rights. Without supporting legal mandates, the electronic information and communications systems would be open to censorship and a potential chilling effect. The latter could lead to *de facto* censorship. This topic and that of e-mail are explored in greater depth in later chapters.

Legal Issues

Computer Security. Individuals have also been embroiled over legal issues. In one example, Craig Neidorf, a publisher of the electronic magazine *Phrack*, was arrested and his operation was raided by the government for publishing a telephone document that purportedly contained sensitive material. The information was actually obtained from a telephone company's computer by another individual, and Neidorf was "indicted on felony charges of wire fraud and interstate transportation of stolen property."[24] But when it was revealed that the same information was publicly available from the telephone company for a small fee, the prosecution dropped the case.

Neidorf's and related cases, though, brought the opinions of two opposing camps into focus. On one side, law enforcement agencies have upheld their duty to prosecute anyone who illegally gained entry into a computer system, which in Neidorf's case, ultimately resulted in the electronic publication of a stolen document. On the flip side, defenders of both Neidorf and individuals involved in similar cases, indicated that publishers and their electronic publications should be afforded the same protection as the traditional print media. Furthermore, while illegal computer entry should be prosecuted, the punishment should fit the crime and the individual's intent. It should not be disproportionate to the true extent of the damage an illegal intrusion may have caused.[25]

Wiretapping and Encryption—An Introduction.   Two other technological/legal issues are wiretapping and encryption. In the past, the structure of the telephone system made it easier to tap someone's line. But the introduction of digital communication, fiber-optic systems, and encryption techniques, where information is rendered unintelligible without the proper decoding mechanisms, have made electronic eavesdropping more complicated. For most people, the added security is welcome news. But for law enforcement agencies, their job has become more complicated. Consequently, different legislation had been floated to ensure an agency's continued access to this information, potentially through an electronic back door, which would allow an agency to gain access to the targeted information.[26]

Elements of the business community and other organizations have been opposed to this type of legislation on privacy and economic grounds. Although a court order must be obtained to initiate wiretapping, as has been the practice, legislation may make the communications network more susceptible to illegal tapping. The system would also be more open to potential government and nongovernment abuses. Ultimately, instead of using technology to build a more secure communications system, which could carry voice as well as computer transactions, it could potentially be compromised.

A network could also be more expensive to build, and a manufacturer would have to contend with more red tape. Other concerns center on the stifling of technological developments and international implications.[27] As other nations developed more secure networks, U.S. systems could be burdened by this organizational hierarchy and a communications structure that could be vulnerable to government and potentially illegal eavesdropping.

The government, for its part, has defended its stance in the name of law enforcement and national security. The prosecution and

conviction of a criminal/terrorist could hinge on, for instance, recorded telephone calls. The inability to tap this and other forms of information would hinder such an operation.

Finally, friction over encryption entered the international market. Tight export standards made it difficult for U.S. companies to compete in this arena. Legislative initiatives, including those by Senator Leahy and his colleagues, sought a balance between individual/business rights and law enforcement/national security concerns.[28]

Wiretapping and Encryption—Legislation and Cases.   Two examples of legislative action are the Communications Assistance for Law Enforcement Act (CALEA), signed into law in 1994, and different versions of the Security and Freedom Through Encryption Act (SAFE) from the 106th Congress. The CALEA was partly born from a government concern—as stated, sophisticated communications technologies would significantly hinder the government's ability to conduct legal surveillance activities. But the American Civil Liberties Union, the Electronic Privacy Information Center, and other public interest groups, questioned the CALEA's potential Fourth Amendment privacy implications.[29]

In brief, telecommunications carriers must guarantee they can isolate and intercept all wire and electronic communication, once proper legal authority has been obtained.[30] The CALEA also requires carriers to deliver intercepted communications and call identifying information to the government. The FCC also has the authority to create rules establishing technical standards in accordance with the act's assistance capabilities requirements.[31] Note also that the Department of Justice and FBI have petitioned the FCC to expand their capabilities to gather surveillance information. This request has been strongly opposed by public interest groups. As of this writing, the FCC is considering the legal arguments.[32]

The 106th Congress, for its part, drafted different encryption bills. HR 850 is representative of the group.

Sponsored by Representative Bob Goodlatte (R-Va.) and Zoe Lofgren (D-Calif.), HR

850 legalizes the use of any type of encryption within the United States, permits the sale of any type of encryption product in interstate commerce, and prohibits the government from requiring any kind of key escrow (an "unlocking" mechanism the government would possess).[33]

Internationally, HR 850 amends the Export Administration Act of 1979. The new legislation would grant the Secretary of Commerce jurisdiction over all encryption products not designed or modified for military purposes. Furthermore, following a one-time 15-day technical review by the Secretary of Commerce, an export license would not be required for the following types of information:

- Generally available encryption software and hardware products.
- Generally available products containing encryption or with encryption capabilities.
- Technical assistance and data used to install or maintain such products.
- Products not used for confidential purposes.[34]

HR 850 also permits the export of custom-designed encryption products and those with encryption capabilities, if they are used by banks or if comparable products can be commercially obtained in other countries.[35] The act also mandates that an encryption product that had not required an export license in the past would not need one in the future. Finally, the president would retain the right to prohibit the export of encryption products to terrorist nations, among other groups.[36]

Besides legislative initiatives, cases concerning the constitutionality of U.S. encryption policy have been heard, including the following:

1.   *Karn v. Department of State.* The District Court examined if the State Department had jurisdiction over the export of a diskette containing cryptographic software. The court held that under the terms of the Arms Export Control Act, Karn was not entitled to a judicial review of the legality of State Department

jurisdiction over such software. The court further held that even if the cryptographic software was considered "speech," controlling its export would not violate the First Amendment because the regulation was content neutral, narrowly tailored, within the government's legitimate interest to control the export of defense articles, and was based on a rational premise.[37]

2.   *Bernstein v. United States.* The case was heard on appeal from the U.S. District Court for the Northern District of California to the Ninth Circuit. The case examined the constitutionality of a Department of Commerce policy that made it illegal to export encryption programs of a certain sophistication without first securing a license from the department. The policy also prohibits the posting of such programs on the Internet because foreign nationals could subsequently download them.

The Ninth Circuit, in a 2–1 decision, upheld the lower court decision. Math professor Bernstein could legally post his encryption program's source code, which comprised the program, on the Internet without obtaining permission. Senior Ninth Circuit Judge Betty Fletcher concluded by ruling that code is free speech and should be afforded the same full First Amendment protections guaranteed for traditional language.[38] However, in October 1999, it was determined this case be heard *en banc* by all of the judges in the Ninth Circuit.[39]

In sum, while legislative initiatives continue to establish a legal framework for surveillance and encryption law, case law will be heard in an effort to interpret the intricate details of such legislation in new and legally unique circumstances. As in other situations where technological advances come into contact with the law, numerous legal questions remain.[40]

CONCLUSION

The systems explored in this chapter have contributed to the growth of our information and communications infrastructure. They have launched new applications and have brought data pools and electronic transactions to our fingertips.

But as discussed, important issues remain to be resolved that cut across the legal, political, and social arenas. Other such topics are discussed at different points in the next chapter.

REFERENCES/NOTES

1. They have been distributed as part of a television signal's Vertical Blanking Interval (VBI). *Note:* During the television scanning cycle, the electron beam initially scans the odd-numbered lines and when it returns to scan the even-numbered lines, a period of time elapses in which the beam is blanked or turned off. This is known as the VBI. All the lines in a television frame do not contain picture information. Teletext magazines are inserted in a portion of the approximately 21 "empty" lines, the VBI, of a television signal.

2. Joseph Roizen, "Teletext in the USA," *SMPTE Journal* 90 (July 1981): 603. U.S. standards were World System Teletext (WST), based on British operations, and the North American Broadcast Teletext Standard (NABTS) developed through the combined efforts of CBS, the Canadians, and the French. Domestic supporters included Electra and Keyfax (WST) and EXTRAVISION (NABTS). The British services were Ceefax and Oracle.

3. Magazines have also been used for education. According to Dr. Goldman, one such project director, a teletext magazine may serve as an active teaching tool. See Ronald J. Goldman, "KCET and Broadcast Teletext: Why We Came and What We've Done," *Communication Options* (April/May 1983): III, C1.

4. As of this writing, the telephone system also serves as the primary consumer conduit for Internet access.

5. Stand-alone terminals were also developed.

6. Prestel, press release, August 29, 1983.

7. C. D. O'Brien, H. G. Brown, J. C. Smirle, et al., "Telidon Videotext Presentation Level Protocol: Augmented Picture Description Instructions," CRC Technical Note 709-E (Ottawa, Canada: Department of Communications, 1982), 1.

8. Andrej Tenne-Sens, "Telidon Graphics and Library Applications," *Information Technology and Libraries* 1 (June 1982): 101. *Note:* It also helped ensure future compatibility.

9. NAPLPS was the joint creation of the Canadian Standards Association and the American National Standards Institute.

10. Keith Y. Chang, "Videotext: A Pillar of the Information Society," Department of Communications, Canada.

11. "The Electronic Mall Arrives; All Under One Roof (Yours)," *Viewtron Magazine and Guide* 1 (November 1983): 6.

12. Public institutions (for example, a library) could also join the network.

13. "Knight-Ridder Pulls Plug on Viewtron," *Broadcasting* (March 24, 1986): 45.

14. French Telecom Intelmatique, "Overview of Minitel in France," downloaded from www.minitel.fr, May 1996.

15. French Telecom Intelmatique, "France French Telecom Intelmatique," downloaded from www.minitel.fr, May 1996.

16. "Prodigy Acquired," *Internet World* (August 1996): 20.

17. But you paid for this enhanced interface with a slower system. You had to wait until graphics were drawn and/or downloaded for subsequent recall.

18. These include buying and selling stock; Dow Jones News/Retrieval specialized in these types of operations.

19. Real time implies, in this context, that the conversation is instantaneous. It is the written equivalent of the telephone.

20. Steve Higgins, "'Electronic Democracy' Wins Votes," *PC Week* 9 (May 18, 1992): 19. The use of new technologies in politics is not a new idea. For more details about such applications, see Robert G. Meadow, ed., *New Communication Technologies in Politics* (Washington, D.C.: The Washington Program of the Annenberg School of Communications, 1985).

21. U.S. Congress, Office of Technology Assessment, *Critical Connections: Communication for the Future* (Washington, D.C.: U.S. Government Printing Office, 1990), 190.

22. Noble E. Cunningham, Jr., *In Pursuit of Reason: The Life of Thomas Jefferson* (Baton Rouge, La.: Louisiana State University Press, 1987), 337.

23. Dennis Allen, "Editorial: Ethics of Electronic Information," *Byte* 17 (August 1992): 10.

24. Howard Rheingold, "The Thought Police on Patrol," *Publish* 6 (July 1991): 46. *Note:* The document was concerned with the 911 emergency system.

25. Mitchell Kapor, "Why Defend Hackers?" *Effector* 1 (March 1991): 1.

26. ABC News, "FBI Pushing for Enhanced Wiretap Powers," transcript from Nightline, show #2870 (May 22, 1992), 3. *Note:* The collected information, such as a telephone conversation, could subsequently be used as evidence in court.

27. Sam Whitmore, "Drop a Dime and Stop Some Spooky Legislation," *PC Week* 9 (May 18, 1992): 100.

28. U.S. Senator Patrick Leahy, "Leahy Introduces Encryption Communications Privacy Act," March 5, 1996. p. 1.

29. Surreply Comments of the ACLU, The EPIC, The EFF, and Computer Professionals for Social Responsibility, Before the Federal Communications Commission, CC Docket No. 97-213. downloaded from www.aclu/Congress.

30. CALEA. 47 USC 1001-1010. *Tech Law Journal,* downloaded from www.techlawjournal.com, February 2000.

31. FCC. FCC Proposes Rules to Meet Technical Requirements of CALEA, CC Docket No. 97-213, downloaded from www.fcc.gov, October 1998.

32. Surreply Comments of ACLU, EPIC, EFF, and Computer Professionals.

33. "Summary of Encryption Bills in the 106th Congress," *Tech Law Journal,* downloaded from www.techlawjournal.com, November 1999.

34. Ibid.

35. Ibid.

36. Ibid.

37. *Karn v. the Department of State.* 925 F. Supp. 1. 1996.

38. *Bernstein v. United States.* 97-16686. CV-97-00582. 1999.

39. Brenda Sandburg, "9th Circuit Set to Review Encryption Case En Banc," *The Recorder/Cal Law,* downloaded from www.lawnewsnetwork.com, October 1, 1999.

40. In another case, the court examined the constitutionality of State Department regulation controlling the use and transfer of sophisticated encryption software. The court ruled that according to the language of the Export Administration Regulations, the State Department's policy did not abridge the First Amendment since it only controlled the distribution of the encryption software itself and did not impact on ideas concerning encryption. See *Junger v. Christopher/Junger v. Daley.* 8 F. Supp. 2d 708. 1998.

SUGGESTED READINGS

Armon, Carl, Dan Glisson, and Larry Goldbeg. "How Closed Captioning in the U.S. Today Can Become the Advanced Television System of Tomorrow." *SMPTE Journal* 101 (July 1992): 495–498. Overview and standards for closed captioning.

Banisar, Dave. "EPIC Analysis of New Justice Department Guidelines on Searching and Seizing Computers." Downloaded from www.epic.org, December 1999. Legal implications surrounding computer seizures.

Connelly, Terry. "Teletext Enhances WKRC's Local News Image." *Television/Broadcast Communications* (October 1983): 52–58. An overview of the Electra teletext magazine.

France Telecom Intelmatique. Teletel traffic data. Downloaded from France Telecom Intelmatique, 1996.

Harrison, David. "Lexis-Nexis Moves to the World Wide Web." *Inter@ctive Week* 4 (October 13, 1997): 42–43; Jones, Kevin. "The French Evolution Comes to the Web—Slowly." *Inter@ctive Week* 5 (November 30, 1998): 25. Two traditional services and Internet implications.

Kening, Dan. "A Connected Electorate." *CompuServe Magazine* 11 (September 1992): 35. An earlier use of an interactive system (CompuServe) as an information clearinghouse for the 1992 presidential candidates.

Martin, James. *Viewdata and the Information Society.* New York: Prentice-Hall, 1982. An excellent source about early interactive/videotext systems.

McCabe, Kathryn. "Step Up to the Modem: Baseball Goes Online." *Online Access* 7 (Spring 1992): 14–15, 17. A baseball lover's dream come true: using an online service to manage a team you assemble while using statistics from real players.

Morgenstern, Barbara, and Michael Mirabito. "Educational Applications of the Keyfax Teletext Service." *Educational Technology* 8 (August

1984): 46–47. Examines some of the educational applications of a teletext magazine.

Pavlik, John V. *New Media Technology.* Needham Heights, Mass.: Allyn & Bacon, 1996, 268–275. Excellent overview of the encryption/clipper chip issue.

Rayers, D. J. "The UK Teletext Standard for Telesoftware Transmissions." In International Conference on Telesoftware. London: The Institution of Electronic and Radio Engineers, 1984. Explores telesoftware technical issues—using a teletext magazine to distribute computer software.

U.S. Congress, Office of Technology Assessment. *Critical Connections: Communication for the Future.* Washington, D.C.: U.S. Government Printing Office, 1990. Chapter 6 of the report explores communication and the democratic process. The topics range from the use of remote sensing satellites to computerized information systems.

GLOSSARY

Computer Forum: An electronic meeting place. A forum can support a special interest group.

Consumer-Oriented PC System: A two-way interactive system designed for broader consumer needs.

Decoder: The device that strips the teletext information from the VBI and displays the information.

Dedicated Information Retrieval Operations: A two-way interactive system that supports specialized information pools and narrow subscriber groups.

Gateway: An electronic "gate" used to gain access to external computer systems.

Information Provider (IP): An organization that provides information carried by an interactive service.

Interactive System: An information system you can interact with to request and subsequently retrieve specific information.

Teletext Magazine: An electronic magazine that appears on a television screen. It is a spectrum-efficient communications service.

Videotext System: A videotext system is a dedicated and interactive information service. Generally geared toward consumers; business options could be supported.

18 Information Services: The Internet and the World Wide Web

Jules Verne's *Around the World in Eighty Days* describes Phineas Fogg's momentous journey. Today, we can complete the same trip in seconds via the Internet. In brief, the Internet can be described as a global data highway. You can travel on this electronic road to exchange information with sites scattered across the globe.

The Internet can trace its roots to the late 1960s. It started as a U.S. government project, the Advanced Research Projects Agency Network (ARPANET). Designed, in part, to experiment with and to demonstrate decentralized computer networking, ARPANET eventually evolved into the Internet structure.[1]

The Internet also remains a decentralized entity. It can be viewed as a collection of independent computer systems that no one individual or organization owns. It is almost like the Wild West; an information frontier without boundaries that is primarily governed by technical standards.[2]

The Internet is also an evolving network. New information pools become available as additional computers are linked to it. Internet contributors include government agencies, profit and nonprofit organizations, educational institutions, and individuals. You can read government documents, send and receive e-mail, browse through the Library of Congress, join discussion groups, and visit personalized information sites.

Internet-based operations have also virtually eliminated geographical and time-based constraints. You can retrieve information from around the world, 24 hours a day. This capability was vividly demonstrated in the summer of 1994 when the comet Shoemaker-Levy 9 collided with Jupiter. This once in a lifetime event created a scientific and public sensation that was particularly felt on the Internet.

As comet fragments slammed into the planet's atmosphere, images were made available and were retrieved from Internet sites. The demand was so great that primary and even mirror sites, which stored duplicate image files, were overwhelmed by requests. One estimate placed the number of downloaded images at the two million mark in a relatively short time.[3]

Prior to the Internet, this level of accessibility was not possible. But now you can assemble your own image library and participate in the event in almost a real-time sequencing, that is, as it unfolds.

The collision also demonstrated how the Internet complemented more traditional communications venues. The PBS television station WHYY ran a live program that tied an Internet link with "satellite dishes and a videophone to bring together multiple images and experts for an interactive program shared with other PBS stations around the country."[4] The different media provided viewers with a broad overview of this event in addition to expert commentary.

The Internet also complements AOL-type services. For example, AOL supports proprietary information pools. But information may be available via the Internet that is not accessible through AOL. Consequently, you may use both to meet your information needs.

The Internet's popularity also made it the center of a fierce competition. In the mid-1990s, AOL and other companies rushed to offer their subscribers Internet access, especially access to the World Wide Web (WWW), or web, for short.

An electronic cottage industry was born in the same general time frame. Entrepreneurs recognized that some people wanted Internet access, but not the other offerings provided by AOL-type organizations. Thus, they launched companies that primarily served as communications conduits: If you owned a PC and modem, they provided the hookup and necessary software.

These companies also opened up cyberspace to more people. *Cyberspace*, a term popularized by William Gibson's science fiction work *Neuromancer*, can be described as a computer-generated environment or world where you can interact with other people and work and play.

The creation of cyberspace has spawned, in turn, a collection of other words and complementary activities. A March 1994 *CompuServe Magazine* article presents such a list, most of which are self-explanatory. A sample includes cyberart, cybergames, cybersex, and cyberschool.[5]

When you enter cyberspace, either through the Internet, an AOL-type organization, or through another means, you have a universe of information and activities at your fingertips. Some have positive connotations, while others are more questionable. An example of the latter is the potential illegal distribution of someone else's work.

WORLD WIDE WEB

The web is a product of the Swiss-based CERN research center. It was pioneered by Tim Berners-Lee to help facilitate the exchange of information, and by the mid-1990s, its popularity had soared. As outlined by Eric Richard, the web could be described as "a collection of protocols and standards used to access the information available on the Internet . . . [which is] the physical medium used to transport the data." The web has been primarily defined by three standards:

URLs (Uniform Resource Locators), HTTP (HyperText Transfer Protocol), and HTML (HyperText Markup Language). These standards are used by WWW servers and clients to provide a simple mechanism for locating, accessing, and displaying information available through other common network protocols. . . . However, HTTP serves as the primary protocol used to retrieve information via the web.[6]

In essence, the web can be viewed as an overlying net that, as described by Richard, is used to mine or gain access to the Internet's information pools.[7] You actually travel across the web via a *browser*, software used to visit different sites and to conduct other activities. These include engaging in business, searching online catalogs, conducting financial transactions, retrieving scientific data, and reading movie reviews.

Prior to the web, you typed a series of commands to gain access to information and to conduct other Internet-based operations. In the case of the File Transfer Protocol (FTP), you initiated a sessions by typing *ftp*, followed by an electronic address (e.g., ac.anyschool .edu). The last three letters specify, in this case, an educational (edu) rather than a commercial (com) or other type of organization/institution. Once connected, you could use keywords to retrieve files, such as the aforementioned comet images. Additional tools include telnet and gopher.[8]

For some, these tools made the Internet a readily traversable environment. For others, the Internet was still a hostile territory that you could only cross with a handful of instruction books, a series of arcane commands, and a prayer.

But the second perception greatly improved with the web's growth. Unlike most traditional Internet tools, you could explore the information universe with software that sported an intuitive interface. As a typical user, you could throw the instruction books away.

HTML and URL

The HTML specification is used for document formatting and design. It is an ASCII-based system that employs tags to format text and to delineate a document's appearance. In two examples, the <P> and <H> tags, respectively, mark a new paragraph and a text heading level (on-screen text size).

HTML is also used to embed links to other documents and Internet services.[9] These links, and the web's underlying structure, are based on a concept you are already familiar with from an earlier chapter, hypertext. You navigate across the web, retrieving information and visiting new sites, via this tool.

There are also basic aesthetic guidelines for document design. For example, if you use the same heading level on every line of text, what should a reader focus on? Is everything equally important? Are the textual and graphical elements too tight—is there enough blank space for eye relief and design purposes?

The URL, for its part, is an addressing system. It can identify a document's location. A typical URL is http://www.ac.anyschool.edu/newcomm.html. When you activate its link, you will retrieve the newcomm.html document at the ac.anyschool.edu site.[10] As you create a web page, you can forge links to other information by using URLs embedded in your document.

Browsers

As indicated, you explore the web with a browser, which for most users is a graphical rather than a text-based tool. With a text-based system, the on-screen information is limited to text.[11]

Graphical browsers are analogous to GUIs. You use a mouse to point to and select different functions. But with the web, you can retrieve information or travel around the world. It is a visual interface that also supports graphics and other visual information.

Different graphical browsers have been released. Mosaic, a product of the National Center for Supercomputing Applications (NCSA), helped define this software field. It also helped make the web the Internet's hot spot.

Netscape, another product, subsequently captured some 70 percent of the market by 1995. When its stock went public, its initial $28 per share price quickly rose to a $75 per share.[12]

Regardless of the system you use, they generally share some common elements:

- Drop-down menus with commands.

- Hypertext link ID system. Hotspots or links are color coded for easy identification.
- A point-and-shoot interface like GUIs. To complete an operation, move the cursor and click a button.
- Navigational buttons you can activate to go, for instance, back to the previous link (for example, a different document and/or site).
- An option to view a page's underlying HTML code. This is a good way to learn how to create your own documents.
- Extensibility. As the web evolves and new media types are introduced (e.g., audio format), your browser should be able to accommodate them through "plug-ins"—software that extends your browser's capabilities.

As you publish on the web, you should also be aware of potential browser conventions. At one point, Netscape used formatting extensions that other browsers may not have supported. Consequently, a document may not have appeared as originally designed if viewed with a different browser.[13]

You must also pay attention to other technical issues, such as the size of files tied to a document. You may design your home page so a visitor is greeted by an aesthetically

Figure 18.1

Microsoft's Internet Explorer (browser). Note the navigational buttons located at the top of the window. (Courtesy of Microsoft Corporation; Internet Explorer.)

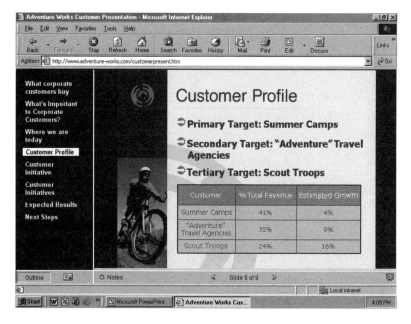

pleasing graphic. But if the file is too large, it may take too long to fully appear on the screen. This is a reflection of the typical telephone line/modem connection. The design constraint should become less of a problem as faster lines are adopted.[14]

Information Management: Search Mechanisms and Agents

You can electronically wander around the web for days looking for specific information. You can also drown in the mountains of e-mail you may or may not want to receive. Two tools may provide some help and relief.

The first, a web search mechanism, is used to identify and locate web-based resources. You can find government documents or locate a site that covers Phil Ochs or another folk singer.

One of the most popular search mechanisms has been Yahoo. A comprehensive service, Yahoo features a keyword search option as well as broad subject categories that can facilitate this operation.

The second tool, an agent, is able to sort through mountains of e-mail and sift the junk mail from important messages. By observing your work patterns, the software will also "learn" how to work for you.[15] This capability is critical. The web's once small trick-

le of information has become a raging torrent. If you work in the computer industry, how do you keep track of key developments? How do you separate important from trivial information?

Agents may play a role in finding solutions to these and other questions. They may help make the torrent more controllable and information management more transparent for us.

But agents raise their own questions. Agents may find themselves engaging in the electronic version of "doing lunch" with each other as they roam through cyberspace. Basically, one agent could query other agents for information. If some of the data are personal, there may be privacy implications. Other information, which could detail your preferences for news or electronic shopping, would be a boon to manufacturers and service providers.[16] Consequently, software safeguards must be written to prevent unauthorized access to private and proprietary information.

INTERNET AND WEB GROWTH

Some Factors

Several reasons for the Internet's growth are listed in the following subsections. They also provide a framework and an historical perspective for tracking this infrastructure's technological maturity.

Interface

The web has provided users with a familiar and intuitive interface. Many, if not most, PC owners are already familiar with GUIs and, as indicated, graphical browsers extend this metaphor to the Internet.

Browsers also help make Internet access transparent. Like a telephone, you do not have to think about or fully comprehend the web's underlying technologies or structure to use it.

Real-Time Interaction

The web's real-time interaction can save time. Prior to the web, you generally relied on file descriptions to identify data. The problem? If this information was inaccurate

Figure 18.2

An example of a web page design. Its underlying HTML code is shown in Figure 18.3. Programs such as PageMill have extended this programming and design capability to a broad user base. Much like a DTP program, you design the page(s) and the program generates the code. (Software courtesy of Adobe Systems, Inc.; PageMill.)

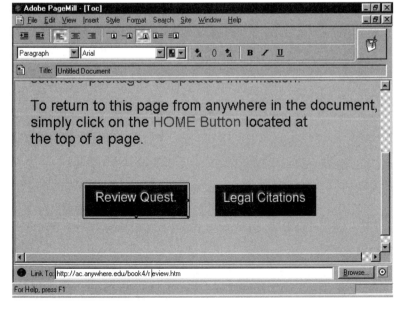

or imprecise, you may have spent valuable time retrieving a file that was actually not suited to your purposes.

With the web, you may be able to listen to and view specific data types, such as a graphic, before downloading. Most sites use small, low-resolution *thumbnails*, pictures that are linked to larger, higher resolution images, as visual guides. You look at a thumbnail, and if it suits your needs, you retrieve its counterpart with the click of a mouse button.

Similarly, audio and video retrieval became a real-time event. Real time, for this discussion, signifies you do not have to download an entire file before you can hear or see it (e.g., video).[17]

Various products support this "streaming" of video and audio information over the Internet. Xing Technology, for one, offered this capability as well as an enhanced version for the broadcast industry. RealNetworks also became a dominant player in this application.[18]

Regardless of the system, the capability to deliver real-time audio and video is an important achievement. Numerous applications can be accommodated, including the launching of Internet-based radio stations.

Commercial radio stations, for instance, can simultaneously distribute their programming via the web. Individuals can also open their own stations for a minimal investment and without an FCC license or spectrum allocation.

Although the audience is limited to online users, you can reach the international community with this mechanism. Web surfers, for their part, can pick and choose the materials they want to hear and view.[19]

The technology could also support real-time audio and video depositories. Information could be retrieved and viewed on demand. Other applications include advertising, running movie and television promotions, online demonstrations, training, delivering electronic news, and entertainment.

The teleconferencing field could also receive a boost. As described in a later chapter, teleconferencing makes it possible for two or more individuals, located in different sites, to hold an electronic meeting. These real-time

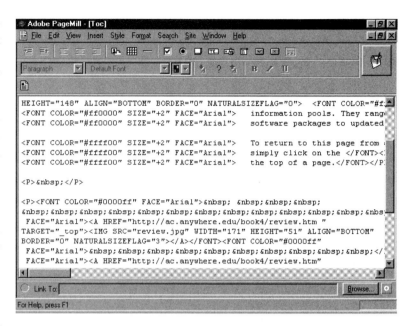

Figure 18.3

The underlying HTML code for the web page shown in Figure 18.2. (Software courtesy of Adobe Systems, Inc.; PageMill.)

audio and video tools could help transform the Internet into the world's largest teleconferencing network.

In a related application, the Internet was tapped for "telephone" conversations. You could talk with a friend with your PC, microphone, and special software, all without long-distance telephone charges. Although this capability was a boon for many users, telephone companies, as you might guess, were not as enthusiastic about the application.

Note also the parallel between audio and video distribution via the Internet and early PC-based digital movies. Some video producers may have initially wondered what the fuss was all about with QuickTime and Video for Windows. The on-screen size was small and the picture quality was poor.

However, as the technology base matured, they became important video and multimedia production tools. The same scenario may play out for "real-time" web services. Even though the video quality is limited, as of this writing, when a typical modem/telephone line connection is used, the potential remains. High-speed communications channels will accelerate this trend.

Finally, as briefly covered in a later section, an intellectual property question remains with regard to using audio and video clips (for example, Internet radio and television stations). How do you monitor the legal use of this information?

HTML Authoring and Java

Most people are able to create HTML-based documents. If you're comfortable with macros for spreadsheets and have access to a word processing program with an ASCII option, you can create web publications. All you need is the prerequisite Internet link.[20]

New tools have also simplified web page construction. Dedicated authoring programs have been introduced and modules have been released for desktop publishing and presentation programs. Essentially, with the right software, your project can be published and distributed across the print and electronic domains.

Additional Internet enhancements, including those spearheaded by Sun Microsystems, Apple, Microsoft, Netscape Communications, and others, have also made the web a more flexible environment. One development has been Sun Microsystem's Java programming language.

Using Java, you can write applets, programs that "can be included in an HTML page, much like an image can be included. When you use a Java compatible browser to view a page that contains a Java applet, the applet's code is transferred to your system and executed by the browser."[21] Applications range from animations to interactive ads.[22]

Although Java may have required programming skills when introduced, newer products have simplified the development process. Java and related products also signaled an important stage in the Internet's maturation. Starting as a text-based information and communications channel, it gradually emerged as a robust and flexible system. Much like the first PCs, which evolved into today's sophisticated machines, the Internet has evolved into a sophisticated multimedia engine.

Virtual Reality Modeling Language

The Virtual Reality Modeling Language (VRML) introduced virtual reality applications to the web. As will be discussed, virtual reality systems enable you to enter into a computer-generated environment. You can explore computer-generated cities, another planet, the human body, or a museum's collection.

Geared for limited-capacity communications channels, VRML extended this capability to the Internet and the web. While there were limitations versus a more conventional setup, it provided the web with a third dimension; you can now literally travel through cyberspace.

This capability can accommodate different applications. Real estate agents can take prospective clients on a tour of a building. You can also set up a museum with links. As you pass through a door to another room, you could be hyperlinked or connected to a new document, and for this type of application, a new virtual environment.[23]

Exhibits could also be rotated to highlight different pieces, and new "rooms" could be added. Unlike a conventional museum, which has limited exhibition space, a virtual museum could readily extend its display capacity.

A VRML-type interface, when combined with the web's ubiquitousness, could also support other operations. In one potential application, photographs from a future space probe could be used to produce a flight over Mars or other planetary body. Using a VRML-type system, these data could be quickly released to thousands of interested armchair astronomers and explorers via the web. VRML and similar systems can also help make the Internet a richer multimedia environment.

Cottage Industry, Advertising, and the Democratization of Information

The web has promoted the growth of an electronic cottage industry. Participants include service providers who may offer web access, web page designers, and advertisers.

As a visually driven environment, the web can be an ideal advertising platform. Small ads may greet you when you navigate to a site, and an ad could tap the web's interactive capabilities. In the latter, a computer company's home page may serve as a launching point to retrieve additional product data. You could also tap Java, VRML, and other capabilities to produce exciting ads—you are not restricted to a static, print layout.

Other benefits include support for a deeper information pool, potentially reaching a larger audience, and setting up feedback mechanisms through forms. So, if you are the Yankees, besides listing home games, you can present team information and audio-video sequences.

Yet despite these benefits, the web does pose a unique challenge. Unless an ad is incorporated on a regularly viewed page, like an ad placed on a search mechanism's home page, people may not get to see it. When you read a magazine, you may scan the ads as you flip through the pages. The same scenario generally does not play out on the web. You typically have to seek a site. Thus, one suggestion for advertisers has been to make their sites interesting stopping points and to offer interactive games and other incentives.[24]

If you are running a for-profit web site, another potential problem is generating enough advertising revenue to cover your operating costs, which include telecommunications, hardware/software, and personnel fees.[25]

For newspaper and magazine publishers, lower cost web publishing is offset by lower advertising incomes. While "countless magazines leaped onto the web, their advertisers haven't always leaped with them."[26] Basically, much like earlier videotext operations, you may lose money. The solution? You may have to run the web operation, at a loss, until your Internet-based advertising becomes more established. Other options include charging a subscription fee, using an information meter described in the next section, or providing advertisers with value-added incentives. The latter may include bundling traditional print and Internet ads in a package.[27] Or the web site could prove valuable as a public relations tool and a resource to attract new and to retain current print subscribers.

On a more positive note, the web does enable large and small companies to engage in this form of advertising. This can help level the playing field for reaching potential clients. But the playing field can also shift with the ebb and flow of technology.

Larger companies also have a funding advantage. They can load their web sites with the latest technological advancements. More in line with traditional advertising, they can also place ads on other, strategic web pages and hire the best graphic artists.

But a key fact remains unchanged: The Internet provides individuals with a platform to launch and advertise their own companies. Like other fields, technological developments can also trickle down to make them more accessible to the average user.

This same capability also extends the concept of the democratization of information. For a modest fee, or even for free, as has been the case with various "hosting" companies, you can establish a personal web site. Whether or not someone actually visits is another matter. It depends on your site's contents and perceived value. Even if it only has a few "hits" or visitors, you can still tap this communications and information tool to present your ideas.

OTHER CONSIDERATIONS

The Internet and web have raised additional considerations. These range from heavy traffic concerns to security to censorship.

Channel Capacity and Implications

The availability of high-speed communications lines is a critical issue for Internet users. They are generally tied to a standard modem/telephone line connection. This has implications.

Relay times are increased. While surfing the web, you may find a QuickTime file you want to retrieve. You click a mouse to start the procedure, and with a typical modem, you can probably finish lunch by the time the file is downloaded. A faster modem might help, but depending on the file's size, you may still have time for coffee.

There's also an impact on the type of data you can use. As outlined, the web supports

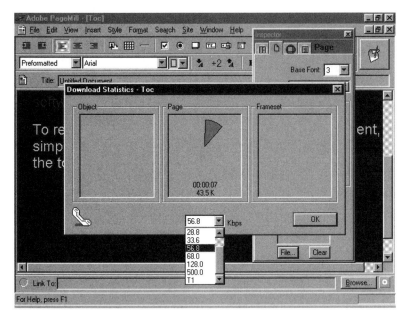

Figure 18.4

HTML design programs may also support other features—in this case, the estimated time to download your project based on the user's set-up. (Software courtesy of Adobe Systems, Inc.; PageMill.)

real-time audio and video. But the communications channel still imposes restrictions, including a reduced frame rate and a lower quality image. You could alleviate these problems with a DSL or other, high-speed line.

A second option is a satellite relay. A modem and telephone line are generally used for the information requests. The data are then relayed via a satellite's wide communications channel.[28]

Another possibility is using a cable system for the Internet connection, via a cable modem. One test site was established in Elmira, New York, in the mid-1990s. The system delivered a variety of information and provided Internet access, the latter at a rate that far exceeded a conventional modem connection.[29] As of this writing, cable modems have also entered the more mainstream market and are offered by a number of companies.

But despite this advantage, questions were initially raised about cable-based ventures. First, had web participation reached a critical mass—were there enough interested users? Second, could a cable system, especially an older system, accommodate a high data rate? Third, when first launched some individuals expressed concern that cable modem systems were ". . . primarily one-way. Whereas the Internet was intended to be a collaborative

communication medium, cable modem technology would continue the Net's 'decline' into a broadcast metaphor."[30] In essence, if a telephone line was used for the return link from a subscriber to the company, it would be limited in its capabilities. Much like a conventional modem/telephone line connection, a subscriber's information options would be curtailed.

But the third scenario's impact, which could also affect a satellite operation, could be handled through a higher speed return or upstream link. Individuals could interact with the system and relay their own information at a higher rate (for example, engage in a teleconference). For cable companies, this could also be an option. If so equipped, a cable company could offer its subscribers two options: using the telephone line for a slower upstream link, or for a higher fee, tap the cable system for a higher speed return.[31] This same scenario is playing out in the satellite field where two-way satellite links are established.

The typical user's needs also merited consideration. At this time, a high-speed return link may not have been as critical as having access to a fast Internet connection.

By the mid-to-late 1990s, you may also have experienced a general Internet slowdown. The cause? Overburdened equipment and communications lines, more frequent graphical and audio-based relays, and an expanding user base.

As outlined in a 1995 report, it was estimated that Internet traffic exceeded 30 terabytes of data a month, equal to "30 million 700 page novels."[32] The end result was longer retrieval times and possibly not being able to gain access to specific sites. Various solutions were offered. These ranged from the construction of new communications channels to the creation of a two-tiered Internet structure, the focus of this discussion.

A first tier member would pay a fee that guaranteed a certain level of Internet access. Historically, you generally roamed the Internet for "free," other than telecommunications and service provider costs. An access fee would reverse this trend.[33]

Free implies you do not pay for services that may be borne by an institution. So even though you may not incur any direct costs, the institution does (for example, network charges and equipment).

The second tier would be comprised of users who would not pay a fee and would operate under current or future, potentially restricted conditions. This may include access to only specific sites.

The development of a pay-for-use scheme and, in fact, more expensive Internet access, raised two questions. Much like any other commodity, should we be willing to pay for Internet service? Or in a modified structure as just presented, should we be willing to pay for a certain level of performance? It is a complicated issue that became more complicated with the 1996 Telecom Act's passage. Various factors play a role in this scenario.

First, if Internet use becomes too expensive, what are the economic, social, and political (for example, uneducated electorate) implications? According to some critics, a pay-for-use option would actually create a third tier—people who might be willing to pay, but could not afford to do so.

Second, the 1996 Telecom Act mandated that advanced information and communications services should be nationally accessible. As stated in the act,

ACCESS TO ADVANCED SERVICES—Access to advanced telecommunications and information services should be provided in all regions of the Nation.

ACCESS IN RURAL AND HIGH COST AREAS— Consumers in all regions of the Nation, including low-income consumers and those in rural . . . and high cost areas, should have access to telecommunications and information services . . . that are reasonably comparable to those services provided in urban areas . . . at rates that are reasonably comparable to rates charged for similar services in urban areas.[34]

The definition of what constitutes advanced services would be flexible and could change with new technological developments. The act also supported schools and select agencies through lower rates and other, special considerations.

A goal was to extend the Communications Act of 1934. At that time, a universal service concept was devised to help ensure telephone access "at reasonable rates" to all Americans.[35]

Yet while universal service, and for the purpose of our discussion, Internet access may be important, some questions were raised about the act's provisions. In one example, you may have access to a service. But if you cannot afford to buy a computer or other device, who picks up the bill?[36]

Even though inexpensive systems geared for Internet access may become widely available, they will still be too expensive for many users and poorer school districts. "Reasonable" service rates may also be beyond someone's financial reach. Will subsidies be provided for both?[37] Some individuals are opposed to subsidies for economic and philosophical reasons.[38] Others support them to extend Internet access.

An unknown factor, particularly for individual use, is advertising and advertising revenue. Like commercial broadcasting, could the Internet evolve into a commercially sponsored communications system? Ads are common, but could they generate enough revenue to keep the Internet free?[39]

In fact, free Internet service became a reality by the late 1990s. You signed on with a company, and when connected, would be greeted by a banner ad(s)—typically a rectangularly shaped ad situated on the bottom of your screen. You might also be required to visit an advertiser's site during your hook-up.

Consequently, you could tap into the Internet for what, many considered to be, a minor inconvenience. But potential restrictions remained: Would there be adequate technical support? Would you be able to tap the Internet's latest technological features? Would you be greeted by frequent busy signals when you tried to connect with the service?

Another element may be the widespread adoption of an information meter. A small fee would be charged for accessing specific information and other functions. By charging only a nominal amount, it could help keep Internet costs down and within the reach of most Internet users.[40] We may also see a combination of solutions, including an information meter/advertiser supported operation.

An information meter may also promote copyright protection since you could receive a small fee for the use your work. Analogous structures already exist in the music/broadcast industries.

But there may be problems. How do you contend with sites run by thousands of different people? If personal radio and television stations become an Internet fixture, how do you monitor "broadcasts" for copyright violations? How do you ensure that an information meter is used?

Security, Sales, and E-Commerce/E-Business

Security is another important Internet issue. This has encompassed providing security for companies to prevent break-ins and for financial transactions.

In one example, firewalls and other techniques can help safeguard your data. A firewall is "a barrier placed between your network and the outside world to prevent unwanted and potentially damaging intrusion of your network."[41] Firewalls can block out unwanted visitors and help maintain a system's integrity. But, like any other protection scheme, it should be viewed as only a component of a broader security system. You may also determine who gets access privileges to specific data pools.[42]

Firewalls also protect a company from the potential intrusion of its own employees. BellSouth, for one, set up an intranet environment, an internal Internet-like system that could be used for data sharing. To help protect this data, firewalls have separated its different operating units.[43] Note also that intranets became a growing market in their own right.

Besides firewalls, organizations face another security breach. While not new, the Internet community received a warning shot across it electronic bows in early 2000. Some of its "hottest" sites were temporarily disabled.

Individuals using special software essentially overloaded different computer-based systems with information, thus causing the downtime. Legitimate users could not enter the sites, causing revenue losses, including advertising revenue. This is analogous to losses experienced by a radio or television station if it goes off the air. In response, companies tightened their security and the FBI was called in to help.[44]

This type of security breach had and has enormous repercussions—by the late 1990s, the Internet was a commercial boom town. It was a time when electronic business ventures exploded. What would happen if they, as well as communication exchanges, were paralyzed? What would happen if sensitive, computer-based information was stolen?

Commercial Transactions

For our discussion, the Internet serves as an infrastructure that can support transactions ranging from stock trading to sales. It is also the world's largest shopping mall and information/communications network.

Large and small companies offer Internet users a vast array of products and services. The Internet's accessibility, ubiquitous nature, cost effectiveness, and freedom from geographical and time constraints play key roles in this development.

Banks also became players in this field. A traditional bank could reduce its overhead and reap substantial profits by creating a virtual bank, in essence, an electronic bank or a bank without walls. But individuals who conduct business over the Internet have to be ensured that their credit information will remain secure and private.[45] Other issues include digital cash and signature verification schemes.

Internet-based auction companies, such as eBay, also flourished. As a subscriber, you could electronically bid on products ranging from cars to ancient coins to toys. You could also sell your own products for a fee. The cottage industries described in the chapter are also part of this emerging world.

Two terms are associated with the use of the Internet for business, *e-commerce* and *e-business.* E-commerce "generally refers to buying and selling over the Internet. . . . E-business has a wider meaning, encompassing e-commerce but generally using the Internet and the technology behind it to connect business processes over the Internet."[46] Some relevant activities include

- Providing service
- Promoting brand awareness
- Extending market reach
- Distributing information
- Delivering distance learning
- Managing business partners
- Launching a web-based business[47]

Besides selling services/products to individuals, which is the way we normally think about Internet business (for example, ordering a camera from an online catalog), operations include company-to-company services—companies selling their services/products to other companies.

The overall growth of the Internet-based business market was partly reflected by a popular culture phenomenon, the 2000 Super Bowl. As you watched commercials between the plays, you probably noticed that many spots were paid for by Internet-based companies. It is a rapidly growing sector of the international financial market, and one in which revenues are expected to be measured in the trillions of dollars during the early years of the twenty-first century.[48]

But as indicated, maintaining a secure system is crucial for this success. In fact, one tool may be *biometrics*, the "automatic identification of an individual based on [his/her] physiological . . . traits."[49] Example systems range from fingerprint ID to voice recognition to matching your eye's pupil. It would be analogous to computer vision systems where information is compared to a stored image or template. In one example, a special keyboard may scan your fingerprint that is subsequently compared against the data stored by, for example, a bank's computer.[50]

The thrust behind the technology is primarily twofold—to preserve privacy and for security. Biometric applications are typically more secure than PIN or password-based systems. So when you complete an online transaction, the other party can be pretty confident *you* are completing the transaction, not another individual. Different companies have and are adopting such applications.[51]

One question remains, however. By adopting this more secure system, could we actually be further limiting our privacy? Besides traditional information, companies could have other, more personal information stored about each of us.

Monty Python, Censorship, and Privacy

Another Internet item may be familiar to Monty Python fans: spam, or more precisely, spamming. One of the online universe's greatest attractions has been Usenet newsgroups, basically special interest or discussion groups. By 1995, there were approximately 14,000 such groups, and the Internet has provided a door to these discussions.[52]

In April 1994, a law firm posted an ad for its services to thousands of newsgroups, the practice of spamming. According to many users, this action violated Internet etiquette or "Netiquette." This triggered an avalanche of flames or hostile electronic responses.

Spamming had and still has a number of implications. First, the messages can contribute to the ever-growing Internet traffic jam. Second, it has led to the development of cancelbots, programs that automatically erase messages that are posted to multiple newsgroups. Cancelbots, in turn, have raised First Amendment questions. Does anyone have the right to curtail spamming, or as some individuals have claimed, to hamper free speech?[53]

Third, the 1994 incident highlighted other Internet-related issues. These have included the use of mail bombs, large messages designed to clog-up your electronic mailbox, and other censorship topics.

Censorship became particularly important with the passage of the 1996 Telecom Act. Opponents of one of the act's provisions indicated that First Amendment rights would be violated. To protest the law's passage, different

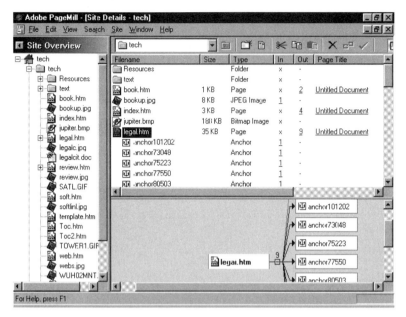

Figure 18.5

PageMill supports a web site management option. As illustrated in this shot, you can visually scan the documents and other information that comprise your site. Links are also graphically displayed (bottom). (Software courtesy of Adobe Systems, Inc.; PageMill.)

web sites colored their pages black. This included Senator Leahy's (D-Vt.) official site.[54]

Privacy issues also became a primary concern. How do you ensure that other people are not monitoring the sites you visit? One solution has been to subscribe to a company that provides an Internet link but protects your identity. Legislative initiatives have also been proposed. Various privacy and censorship issues are covered in more depth in later chapters.

The Internet remains an evolving system. It is also a challenging topic to cover since it literally changes every day. Taking these factors into consideration, several generalizations can still be drawn.

CONCLUSION

The Internet will continue to grow in importance as a premiere information and communications system. It has provided us with a tool to conduct research, to launch new business ventures, to exchange information, and to participate in newsgroups.

Its continued growth should also be fueled by faster communications channels. This enhanced capability will enable us to communicate faster with a wider variety of data.

It is also important to view the Internet as a focal point for the convergence of different communications companies/channels and technologies. The former include the telephone, satellite, and cable industries. The latter range from streaming audio and video data to VRML to applications that have not yet left the drawing board. It is this last feature that makes the Internet so exciting. As its driving technologies are refined, it becomes possible to support new applications.

It may also be possible to receive and view your television programming via the same communications channel and device that you use to surf the Internet. As an element of this diverse information mix, the Internet may contribute to the creation of an integrated information and entertainment system.

In one example, the Internet may become the world's largest distribution center for music and video. As described, MP3 and other compression schemes have made it possible to relay music over the Internet. As these technologies are refined, we may be ordering our music via the web in addition to going to music stores. The same scenario may play out for movies—they may be retrieved over the web and/or other distribution tools. But on the flip side, this has led, to an extent, to illegal pirating.

Another factor, covered in Chapter 21, is the Internet and censorship. While the Internet is the world's largest information conduit, censorship in the United States and abroad could limit the access to certain types of information. In the United States, for example, provisions in the 1996 Telecom Act would have had an impact on the information we could receive. Cuba, China, and other countries, for their part, have tight Internet controls in place.[55]

REFERENCES/NOTES

1. See Howard Rheingold, *The Virtual Community* (New York: Addison-Wesley Publishing Co., 1993), for an in-depth look. *Note:* ARPANET was also a reflection of the Cold War. The United States was concerned that a nuclear conflict would sever traditional

communications/information ties, thus the impetus for an ARPANET-type environment.

2. It could almost be categorized as an organism or artificial life. It is growing and is host to other organisms (for example, the cancelbots described in a later section).

3. J. Kelly Beatty and Stuart J. Goldman, "The Great Crash of 1994: A First Report," *Sky and Telescope* 88 (October 1994): 21.

4. Adam Gaffin, "Comet's Impact Felt on Internet," *Network World* 11 (July 25, 1994): 65.

5. Cathryn Conroy et al., "All Things Cyber," *CompuServe Magazine* (March 1994): 10–21. *Note:* Cybersex encompasses a range of activities, including interactive conversations with other people.

6. Eric Richard, "Anatomy of the World-Wide Web," *Internet World* (April 1995): 28.

7. Peter Kent, "Browser Shootout," *Internet World* (April 1995): 46.

8. The telnet tool enables you to log on to a remote computer. You're not limited to file transfers and can fully interact with the system as though you're on site. Gopher is a menu-driven system that combines, in part, FTP and telnet capabilities. Many people still use FTP and telnet as stand-alone applications (outside of the web).

9. Robert Jon Mudry, *Serving the Web* (Scottsdale, Ariz.: The Coriolis Group, 1995), 109. *Note:* These other services have included FTP, gopher, and telnet. The protocols that drive the web also have applications on other types of networks.

10. The http designation indicates the document is on a web server. Other addresses might be, for example, for FTP sites.

11. Educational institutions, which may have computer labs equipped with "dumb" terminals, could fall under this category. Lynx has been a popular program in this environment.

12. "Bull Rush Greets Netscape," *PC Catalog* (September 22, 1995): 12.

13. There's a wealth of software you can use for web page design and site maintenance.

14. Other connections are discussed in a later section.

15. Kurt Indermaur, "Baby Steps," *Byte* 20 (March 1995): 99. *Note:* This topic was first introduced in the multimedia chapter.

16. Evan I. Schwartz, "Software Valets that Will Do Your Bidding in Cyberspace," *New York Times* (January 9, 1994): Section F, 11.

17. Real time also implies that a radio station's programming, for example, is simultaneously distributed and can be listened to over the air and on the web. The term *live* may also be associated with the function.

18. Xing Technology Corp., press release dated October 17, 1995, downloaded from www.xing.com, November 14, 1995. *Note:* This technology could also be used on other types of networks.

19. A web surfer is analogous to a cable television surfer.

20. For more sophisticated operations, though, you may need programming expertise. For example, you can create a document with a form, but you may have to write a cgi script to actually tap this information. Or you may be able to use a premade script that could be downloaded from the Internet.

21. 1995 Sun Microsystems web site, downloaded from http://java.sun.com, November 1, 1995.

22. See Dan Amdur, "The Scene is Set for Multimedia on the Web," *NewMedia* 5 (November 1995): 44, for such an example (fantasy baseball).

23. A paper by David Raggett, Hewlett-Packard Laboratories, "Extending WWW to Support Platform Independent Virtual Reality," downloaded November 4, 1995.

24. Ron Maclean, "New Game Rules," *American Advertising* 2 (Spring 1995): 9.

25. These include web page designers.

26. Eric Brown and Dan Amdur, "The NewMedia 500," *NewMedia* 6 (August 12, 1996): 38.

27. Ibid.

28. Frank Beacham, "DirecTV Offers On-Ramp to the Net," *TV Technology* 13 (September 1995): 1.

29. Interview with Hal Rubin, Trigger, Inc., May 30, 1996.

30. Eric Brown, "Cable Modem Access Moves Up to 30Mbps," *NewMedia Today* (October 24, 1995), downloaded on November 4, 1995.

31. Adelphia Cable offered this type of option as of this writing (www.adelphia.net).

32. Mark John, "Cybergridlock Feared," *Communications Industries Report* 12 (August 1995): 9.

33. Ibid.

34. Telecommunications Act of 1996, 104th Congress, 2nd session, January 3, 1996.

35. Dow, Kohnes & Albertson, Memorandum, "Implications of Telecommunications Act of

1996 for Colleges and Universities," February 13, 1996, 1.

36. Charles S. Clark, "Regulating the Internet," *CQ Researcher* 5 (June 30, 1995): 573.

37. For a closer look at this topic prior to the act's passage, see Thomas J. DeLoughry, "Colleges and Telephone Companies Battle Over Future of Internet," *Chronicle of Higher Education* (May 19, 1993): 25.

38. The philosophical issue focuses on essentially using your money (for example, higher telephone bills to help pay for someone to use, in this case, advanced telecommunications services).

39. Mark Berniker, "Internet Begins to Cut into TV Viewing," *Broadcasting & Cable* (November 6, 1995): 113.

40. Jeff Ubois, "Augering the Matrix," *Internet World* (November 1995): 69.

41. John Bryan, "Build a Firewall," *Byte* 20 (April 1995): 91.

42. Peter Kolsti, "Firewalls Keep Users from Kicking Your Apps," *Network World* 12 (November 13, 1995): 61.

43. Steven Vonder Haar, "Firewalls Address the Threat Within," *Inter@ctive Week* 3 (January 29, 1996): 23.

44. Rutrell Yasin, "Cyberterrorists Crash Web Party," *InternetWeek* (February 14, 2000): 66.

45. Peter Noglows, "Virtual Tellers: Internet Bank to Open," *Inter@active Week* 2 (October 9, 1995): 53.

46. Scott Bury, "The e-Business Explosion," *Electronic Publishing* 24 (February 2000):24.

47. Martin C. Clague, "Understanding e-Business," Appears in *Renewing Administration* (Boston, Mass: Anker Publishing Co., Inc., 1999), 47.

48. Ibid, 59.

49. Anil K. Jain et al, "Biometrics: Promising Frontiers for Emerging Identification Market," *Computer* 33 (February 2000), online download.

50. Jesse Berst, "Web Sites Want Your Fingerprints (How It Could Save You Money)," downloaded from ZDNET, March 6, 2000.

51. Ibid.

52. Joel Furr, "The Ups and Downs of Usenet," *Internet World* (November 1995): 58.

53. Laurie Flynn, "Spamming on the Internet," *New York Times* (October 16, 1994): F9.

54. Reuters New Media, "Parts of Internet Go Black in Protest," downloaded February 14, 1996.

55. Michael Hatcher, Jay McDannell, and Stacy Ostfeld, "Computer Crimes," *American Criminal Law Review* (Summer 1999) (36 Am. Crim. L. Rev. 397), downloaded from Lexis, May 2000.

SUGGESTED READINGS

American Advertising. Vol. 2 (Spring 1995). This issue covers advertising and the Internet. Includes Maclean, Ron. "New Games New Rules," 9–12.

Baldasserini, Mario. "Playing with Fire." *Computer Shopper* (June 1995): 606–608; Bryan, John. "Build a Firewall." *Byte* 20 (April 1995): 91–96; Vonder Haar, Steven. "Firewalls Address the Threat Within." *Inter@ctive Week* 3 (January 29, 1996): 23. Coverage of firewalls. The last article examines firewalls within an organization.

Coco, Donna. "A Web Developer's Dream." *Computer Graphics World* (August 1998): 54–58. Presenting *A Midsummer Night's Dream* over the Internet via VRML.

Cosper, Amy C. "One-Way Data Traffic over DBS Satellites." *Satellite Communications* 21 (May 1997): 32–34. Satellite data relays (including the Internet).

Coursey, David. "So What's an Intranet." *Inter@ctive Week* 3 (January 15, 1996): 28; Schwartz, Jeffrey and Richard Karpinski. "Intranets Grow Up." *InternetWeek* (December 21, 1998): 17. Intranets.

Cox, John. "It May Be Immature, But Fans Claim Java's No Jive." *Network World* 13 (March 18, 1996): 1; 14–15. A look at Java and an early evaluation.

Dobson, Rick. "Dynamic HTML Explained, Part 1." *Byte* 22 (November 1997): 53–54. HTML issues. There's also a wealth of information available on the web.

Dodge, John. "Can the Internet Let Freedom Ring in China?" *PC Week* 16 (June 14, 1999): 3; Pappas, Leslie. "The Long March to Cyberspace." *The Industry Standard* 3 (February 21, 2000): 134–139. Internet issues and China. *Note:* Censorship questions, including the Internet and the United States, are covered in Chapter 21.

Flynn, Laurie. "Spamming on the Internet." *New York Times* (October 16, 1994): F9. Spamming and a Q/A session with Laurence Canter and Martha Siegel; Gaffin, Adam, and Ellen Messmer. "Censors Hit Cyber-Speech." *Network World* 11 (October 31, 1994): 1; 79; Marcus,

Joshua A. "Commercial Speech on the Internet: Spam and the First Amendment." *Cardozo Arts and Entertainment Law Journal* 1998. (16 Cardozo Arts & Ent Lj 245). Downloaded from Lexis. Censorship and the Internet. Includes a look at spamming, mail bombs, and cancelbots.

GartnerGroup Insight, 1 (October 1999). The GartnerGroup issues numerous informational pieces focusing on e-commerce/e-business.

The Industry Standard covered the "attacks" on Internet sites. Two articles from the February 21, 2000, issue are Abreu, Elinor. "The Hack Attack," 66–67 and Sprenger, Polly. "Go Hack Yourself," 68–69.

John, Mark. "Cybergridlock Feared." *Communications Industries Report* 12 (August 1995): 9; Meeks, Brock. "Commerce Dept. Study Highlights Unequal Access." *Inter@active Week* 2 (July 31, 1995): 12. The Internet and too much traffic, possible two-tier system, and access issues.

Krivda, Cheryl. "Data-Mining Dynamite." *Byte* 20 (October 1995): 97–103; Indermaur, Kurt. "Baby Steps." *Byte* 20 (March 1995): 97–104; Watterson, Karen. "A Data Miner's Tools." *Byte* 20 (October 1995): 91–96. Intelligent agents and various data mining techniques.

Lewis, Peter. "The World Wide Web: Open for Business." *Computer Shopper* (February 1995): 600–601; Messmer, Ellen. "Firms Face Business Pitfalls on the 'Net." *Network World* 12 (October 16, 1995): 35; Singleton, Andrew. "Cash on the Wirehead." *Byte* 20 (June 1995): 71–78; Singleton, Andrew. "The Virtual Storefront." *Byte* 20 (January 1995): 125–132; Streeter, April. "Easing into E-Commerce." *EmediaWeekly* (October 5, 1998): 18–21. The Internet and commerce issues. The first and last articles also cover setting up your own business in this environment.

McKeon, John. "Examining the Net's Effect on the '96 Elections." *Communications Industries Reports* 12 (August 1995): 1, 6. The Internet and political applications/implications.

Schwartau, Winn. "Striking Back." *Network World* 16 (January 11, 1999): 1, 33–35. Breaking into/disrupting online operations and security reactions/precautions.

Venditto, Gus. "Instant Video." *Internet World* (November 1996): 85–101; Waggoner, Ben. "Streaming Media." *Interactivity* 4 (September 1998): 23–29. Media streaming and earlier systems.

GLOSSARY

Browser: Software used to visit different web sites.

Cyberspace: A term popularized by William Gibson's science fiction work *Neuromancer,* cyberspace can be described as a computer-generated environment (or world) where you can interact with other people and work and play.

File Transfer Protocol (FTP): The process that allows a user to transfer files to and from remote sites.

Firewall: A barrier typically placed between your network and the outside world to prevent unauthorized entry.

HyperText Markup Language (HTML): The basic language for creating web documents. If you are comfortable with macros for spreadsheets and have access to a word processing program with an ASCII option, you can create web publications.

Java: A language that extends the web's capabilities.

Real-Time Audio and Video Retrieval: The ability to hear/see a file without having to completely download it.

Spamming: If you are a Monty Python fan, a funny skit (spam, spam, spam . . .). If not, the act of posting an ad, for example, to thousands of newsgroups. This can trigger an avalanche of flames—hostile electronic responses.

Telnet: Telnet enables you to log onto a remote computer.

Virtual Reality Modeling Language (VRML): Introduced virtual reality applications to the web.

Web Search Mechanisms: Software tools used to identify and locate web-based resources (for example, keyword search).

World Wide Web (WWW): A product of the Swiss-based CERN research center, it was pioneered by Tim Berners-Lee to help facilitate the exchange of information. You use the web to travel across the Internet and to retrieve and exchange information ranging from text to digital video.

19 Teleconferencing and Computer Conferencing

TELECONFERENCING

In Chapter 1 you wrote a note to a friend about the New York Yankees. Now, you want to celebrate a World Series win. Do you write another note, use the telephone, or hold a personal videoconference so you can also see each other?

The latter falls under the umbrella term *teleconference*. A teleconference is an electronic meeting between two or more sites. It can range from an audioconference, where you hold an interactive conversation, to a videoconference, where video information can also be exchanged.

Another type of meeting, a computer conference, is an extension of the systems presented in Chapters 17 and 18. A computer conference can range from an exchange of pictures between two people to dedicated networks. It is placed in this chapter for organizational purposes, and for our discussion, the focus is on its electronic meeting characteristics.

The Codec

The codec (coder/decoder) helped teleconferencing's growth. In a videoconferencing environment, codecs "digitize the analog television signal and compress it so that it can be transmitted at a variety of data rates. . . ."[1] This information can then be placed on lower capacity and less expensive communications channels. A codec at the end of a relay reverses this process.[2] A range of codecs are available. They have supported a 1.544 megabit/second rate as well as higher and lower rates.

The compression, though, may affect a relay's perceived quality, especially at the lower transmission rates. In one example, small picture changes or motions could be accommodated, but rapidly gesturing hands and similar movements may not be smoothly reproduced. Thus, there's a trade-off between reduced transmission speeds and costs versus the picture quality.

Codecs are also used in multimedia and web-based production, among other areas. When you create a QuickTime movie, for instance, you select a codec that is the most suitable for the task at hand.

VIDEOCONFERENCING

There are different types of videoconferencing. In a two-way videoconference, you can hear and see each other through cameras, monitors, microphones, and speakers. In a one-way videoconference, the information exchange is a one-way audio and video delivery system. But there's an option for interaction through a telephone or a fax connection. In a typical meeting, a party receiving the information could ask questions by telephone.[3]

Videoconferences have also been either motion or nonmotion. Motion implies movement, but the capabilities vary. You can appear on a television screen in a lifelike manner. Or your motions may be jerky, and the picture quality may be poor.

In a nonmotion videoconference, a series of still images is relayed. Even though the visual element is not lifelike, an audio hookup could support a conversation.

Regardless of its form, a meeting can be held over satellites, telephone lines, and other communications channels. A company

Figure 19.1
*A teleconferencing setup.
Note participants in this
room (foreground) and
the shot of a conferee at
another location
(background screen).
(Courtesy of US Sprint.)*

may also own a private teleconferencing network, or a public room could be rented for the occasional meeting.

Two-Way Videoconferencing
In a two-way videoconference, the various parties can see and hear each other, and its primary advantage lies in the replication of a face-to-face meeting. The participants or conferees can react to each other's body language, valuable visual clues in interpersonal transactions.

A room can also be designed with this goal in mind. For example, the table where the conferees sit can be shaped to maximize their view of the room's monitors. The object is to promote good eye contact between the different parties.[4] Strategically placed cameras can produce various shots of the conferees, and an additional camera(s) can shoot graphics and other supporting materials. Proper lighting and acoustics are also important.

Before the development of dedicated facilities, a meeting may have been hampered by inadequate equipment and design considerations. A large office may have been converted into a videoconferencing room by simply adding the necessary communications equipment. Thus, lighting and other production considerations, the human element that would have made people more comfortable, may have been overlooked.

Another past and current problem has to do with the notion of appearing on camera. Some people get nervous, and yet others think a videoconference is an unnatural way to communicate.

Finally, two-way videoconferences had generally been used for point-to-point meetings, uniting two sites. But advances have made it possible to take advantage of a multisite capability. This has also become a popular option.

One-Way Videoconferencing
The one-way videoconference, which has also been called business television, is typically relayed via satellite from one location to multiple sites (point-to-multipoint). A one-way videoconference has generally been analog, but newer digital compression techniques, when combined with satellite delivery, can make for an effective relay. In this situation, though, the picture quality may be more of a concern than with a two-way setup, where a lower quality image, but a higher compression ratio may be more acceptable.[5]

In a typical application, the audio and video information can be a one-way stream from a corporate headquarters to its branch offices. These sites may communicate back with a telephone or other hookup for question-and-answer periods and discussions.

In light of its point-to-multipoint capabilities, the one-way videoconference has prompted the development of dedicated videoconferencing networks.[6] It has also been supported by organizations that lease satellite time and the prerequisite facilities. One-time or occasional videoconferences, also called special events or ad hoc videoconferences, have also been held.

This electronic meeting's power has been vividly demonstrated in medical videoconferences. Through this real-time process, doctors can view a new surgical technique performed at another site.

Nonmotion Videoconferencing
In a nonmotion videoconference, still images are delivered over voice-grade or faster

channels. The relay time can vary, depending on the communications line and the image's resolution, and either dedicated or PC-based configurations can be used.

A nonmotion videoconference has some advantages. A system can be relatively inexpensive, and the transmission cost can be reduced to that of a telephone call. It is also suitable for situations where visual information must be exchanged, but not necessarily moving pictures. These have included X-rays and illustrations for the medical and educational fields.

AUDIOCONFERENCING

An audioconference can be thought of as an extended telephone conversation. But instead of talking with only one person, you may be talking with several or more people. In this operation, multiple sites can be connected through a teleconferencing bridge.

An audioconference is a satisfactory communications tool in many situations. It is relatively inexpensive and is supported by its own array of equipment, including a microphone that may serve several people through a 360-degree pickup pattern.

Audioconferencing can also handle numerous applications. It can create an inexpensive communications link between a journalism class and a sports reporter, who may live and work 200 miles away. It can also support business meetings and national political campaigns.

The primary criterion for adopting audioconferencing, or one of the visual systems, is an organization's needs. In many cases an audioconferencing link may suffice.

OTHER CONSIDERATIONS

The teleconferencing field has grown over the years. In fact, it surged in the early 1990s when product and service sales for the videoconferencing sector alone rose from $127 million to $510 million in 1987 and 1991, respectively.[7] An accelerated growth pattern has been projected through the twenty-first century. Some of the reasons for this growth are discussed in the following subsections.

Standards
A standard, H.261, also known as Px64, has resulted in a level of compatibility between manufacturers to help facilitate the flow of information.[8] Other standards promoted videoconferencing's growth over communications channels ranging from the ISDN to the "plain old telephone system" or POTS.[9] Even though we're hurtling toward a digital world, the analog telephone infrastructure is still important. By videoconferencing over standard telephone lines, you could tap the large, existing network.[10]

Transmissions
The industry has benefited from lower transmission and equipment costs. It is less expensive to hold a videoconference, and the quality has improved. Basically, newer codecs can support an enhanced relay at lower transmission speeds.[11]

It is also easier to hold a meeting. New equipment, standards, and an improved communications infrastructure have made this application more user accessible and transparent.

Flexibility
Teleconferences can handle more types of information. In one case, computer and conventional videoconference capabilities have been married. In a multimedia videoconference, graphics and computer data can be exchanged, documents can be electronically annotated, hardcopies can be printed, and simplified interfaces can be set up.[12] The goal is to provide users with the same tools they might use in a face-to-face meeting.

Teleconferencing is also becoming more flexible as the industry continues to mature. Organizations can rent public rooms to experiment with videoconferencing or for occasional use. Satellite and fiber-optic relays are now widely available and networks accommodate domestic and international relays.[13]

A company can also use a portable and less expensive roll-about unit to hold a videoconference. This unit can be moved from room to room and eliminates the need for a dedicated facility. However, its aesthetic elements could be enhanced if the unit is used in a board or conference room with the proper lighting and acoustical treatment.

The Internet

As covered in Chapter 18, the Internet is expanding at a dizzying pace. This environment has also supported teleconferencing, including meetings held across the Internet as well as those within an organization through an intranet.

One pioneering product, CU-SeeMe, offered users a low-cost videoconferencing option via a PC.[14] The quality was poor, compared to television standards, but it did provide the Internet community with an inexpensive teleconferencing tool.

More important, this type of product pointed the way to more sophisticated applications. In short, the Internet could emerge as the world's largest teleconferencing environment. This is particularly true as new techniques that support real-time audio-video relays are developed and widely adopted.

Note also that teleconferencing could be accommodated on a local area network. Even though the process may not be as simple as plugging cameras in PCs, it may serve as an existing teleconferencing foundation.[15]

PERSONAL VIDEOCONFERENCING

Videoconferencing has also come to the desktop. In the early 1990s, a number of desktop configurations were introduced. These devices can trace their origins to AT&T, which experimented with the Picturephone transmission standard and the Picturephone, its hardware component.

The Picturephone was a compact desktop unit that incorporated a small television monitor, camera, and audio components. Even though it was a pioneering service, it did not

catch on in either the business or private sectors. The transmission and hardware costs, a Picturephone user's possible reluctance to appear on camera, and the inadequate reproduction of printed documents hampered sales.

Since 1964, when AT&T publicly demonstrated a video telephone at the World's Fair, newer products have been introduced.[16] These include consumer-oriented nonmotion units, which relayed still black-and-white images over telephone lines, and newer motion systems.

The latter have ranged from PC-based to stand-alone desktop models. Analog and digital transmission schemes have been supported, and consumers and businesses have been targeted.

While personal videoconferencing may be inevitable, the time frame for its broad acceptance is unknown. The quality, as of this writing, may still be unacceptable to some users, and the human factor must be considered.[17]

New transmission lines, as outlined in previous chapters, may support faster relays, and equipment costs may continue to fall. But systems geared for the consumer may run into the same problem faced by the original Picturephone: Many people may not want to appear on camera. Think of the way you use the telephone. When talking, are you always attentive? Or are you also working on a computer, eating, watching TV, or as indicated in research that evaluated the Picturephone, reading, doodling, or even yawning?[18] While you could use a system in a voice-only mode, would the person you're talking to be insulted?

This type of system calls for a different mind-set. It may only be fully accepted when a generation of children actually grows up with the technology, and it is a part of their everyday lives. At that time, appearing on camera will just be old hat.

ADVANTAGES OF TELECONFERENCING

As a communications tool, teleconferencing has advantages, regardless of its form:

1. It can promote productivity. By linking an organization's offices, meetings can be held either as needed or on a regular basis. With a private setup, a company could also meet on short notice to respond to crisis situations.[19]

2. Employee morale can be boosted. Management can use a teleconference to keep employees informed about relevant events.

3. Guest lecturers have electronically met with classes while businesses have engaged consultants, especially those who may be unable to travel to a given site.

4. Time and money can be saved by reducing travel.

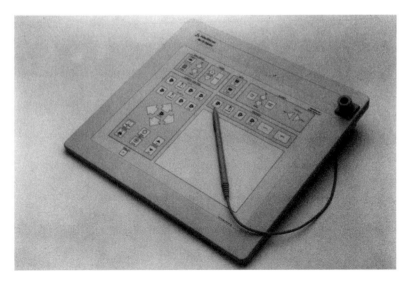

Figure 19.2
More transparent and easier-to-use control interfaces, such as the one depicted, have simplified teleconferencing operations. (Courtesy of VideoTelecom Corp.)

Your job may call for extensive traveling, which normally entails a drive to the airport, waiting and flight time, another drive to the meeting point, and the return flight. This itinerary may then be repeated. In today's world, this wasted time and energy are two resources that can never be recouped. The financial cost can also be prohibitive.

Teleconferences can help alleviate these problems. If an electronic meeting could be arranged between East and West Coast offices, the same information could potentially be exchanged through a teleconference, rather than a series of trips.

But teleconferencing will not eliminate all travel. Face-to-face communication is still important. For example, if you are learning how to use a complicated piece of equipment, an on-site seminar may be appropriate. Some people also prefer face-to-face meetings. Or at the very least, the first meeting is conducted in person, which could be followed up by teleconferences.

Another advantage is that, as with business travel, you could save time and money by conducting a job interview via a teleconference. You may also be able to interview more candidates.[20]

Another possible advantage is that an organization could rent its facilities, during downtimes, to generate additional revenues.

In sum, the teleconference is emerging as another tool that organizations and individuals can tap to improve their communication capabilities. These advantages also extend to the international market. Because many companies are now global in scope, teleconferencing can help unite these geographically distant sites. This capability cannot be overemphasized because international considerations continue to play an integral role in domestic operations.

The development of PC-based and inexpensive teleconferencing outfits, which may become standard office and home fixtures, may also change this meeting from a special to an everyday event. Wide and cost-effective communications channels, enhanced display systems, and more powerful processors may accelerate this trend.

COMPUTER CONFERENCING

The term *computer conferencing* can describe a computer-based meeting. For our discussion, a conference can range from an exchange of pictures to dedicated networks that may link multiple domestic and international users (for example, the Internet).

In one application, a PC-based configuration could unite national real estate offices. An office in Denver could send pictures of houses to the New York branch for review by a client. A telephone conversation could supplement this exchange.

Some of the computer-based systems described in Chapter 17 can also support conferencing. An advantage of this operation is its availability. Scattered groups of people can gain access to the system, and the company provides the prerequisite organizational and communications infrastructures.

In a different setting, an organization may opt to create its own internal network. Meetings can be rapidly convened and a variety of information can be exchanged.

In one growing application area, individuals can collaborate on a document through data conferencing. People share the same file for editing. When working on complex projects, this capability could speed up and help support the writing process.[21] In this light, it could be viewed as an interactive extension of e-mail.

Hardware and software have also supported computer conferencing in real- and non-real-time situations. Real time, in this context, implies that messages can be sent and received as you interact with the system and other people. The non-real-time elements may encompass a series of longer messages, a central database of information, and a record of comments all the conferees can see.

Telecommuting

Computer conferencing can also be held between an individual and an organization, as is the case with telecommuting. As defined by Osman Eldib and Daniel Minoli, telecommuting is "the ability of workers to either work out of their homes or to only drive a few minutes and reach a complex in their immediate neighborhood where, through advanced communication and computing support . . . they can access their corporate computing resources and undertake work."[22] As further described by Eldib and Minoli, telecommuting can also be associated with the virtual office, covered in Chapter 9.

For our present discussion, if you are a telecommuter, you can work at home and maintain contact with the office by computer. You can be an employee, a consultant, or a chief executive officer.

Telecommuting is becoming a popular work option. It can, for example, be cost and time efficient. Think about your own commute, especially if you have to travel on an overextended highway or mass transit system.[23] Telecommuting would reduce the frequency of, or possibly eliminate, this daily subway, bus, or car trip.

Telecommuting also gives you more options. You could choose to live further away from work or take a geographically inaccessible job.

Telecommuters may also be more productive and experience a higher level of job satisfaction. There are also corresponding social benefits, such as fuel savings, air-pollution reduction, and child care improvements.[24]

Several factors have combined to make telecommuting a viable alternative. These include faster communications lines, the proliferation of PCs, and the information age. A job in the information sector could potentially be completed at home.

Yet everyone is not happy with this development.

You may be uncomfortable with this work environment, and there is a potential for abuse—you could be treated as a contract worker instead of a salaried employee with its attendant benefits. Lower wages may also be earned. You could be cut off from the "informal communication channels" and be "less well integrated into an organization's structure and culture."[25]

Another issue has flared up concerning work-at-home safety conditions. The U.S. Labor Secretary wrote a letter to an employer stating that "companies have the same responsibilities for employees working at home as in the office."[26] The letter was withdrawn after criticism from the business community. But it did draw into focus two contrary viewpoints:

1. An employer should be liable for safe, working conditions in an employee's home. The goal is to essentially provide these employees with the same rights as those who work in a traditional setting.

2. Because an employer does not have direct control over an employee's home,

such a regulation would be unfair and impossible to enforce. It could also curtail the growth of telecommuting with its attendant benefits.[27]

As of this writing, the issue is still under debate. What might actually develop is a blending of both viewpoints. Employees would be afforded certain protections while employers would be free from the more intrusive regulations.

A final question focuses on another issue—does telecommuting help reduce human communication to a machine-dominated format? This could have an adverse effect on the socialization process since the interpersonal relationships we develop at work may play an important role in our lives.

Isaac Asimov, the late science fiction writer and science authority, paints an interesting portrait of such a society. Communications technologies have contributed to the birth of a rigid social structure where personal human contact is avoided. In Asimov's world, even casual meetings are conducted through lifelike 3-D visual relays.[28]

Could our own society be similarly affected? What if telecommuting is weighed with other technologies that enable us to shop at home, to create sophisticated home entertainment centers, and to view a museum's collection via the web?

Education and Medicine

The concept of telecommuting has also been extended to education and medicine. For education, you can register for computer-based and -delivered courses as a form of distance education. Like "real" school, you are assigned a teacher and textbooks. But instead of taking notes in class, lectures can be downloaded, and your teacher may be post electronic office hours.

Students can also expand their course options because they are not limited by geography and time. If you do not live near a school, and work during the day, it may still be possible to pursue a degree. Teachers, for their part, can reach a broader student body and can tap the Internet and other electronic resources when designing a course.

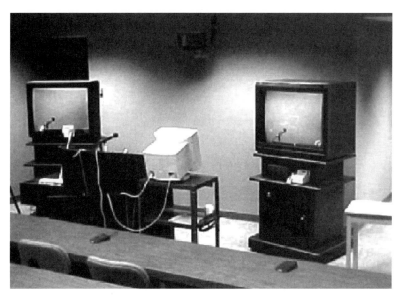

Figure 19.3
An example of a video-conferencing room used for applications ranging from holding courses to interviews to meetings. The monitors are used to display the participants at other sites; microphones are placed on the tables for each participant. (Courtesy of Marywood University.)

Similarly, the Internet has emerged as an important conduit for this application. Various tools, including the WebCT software package, support this application.[29]

Note also that courses can be delivered by compressed video. This setup more accurately duplicates a classroom situation since you can see and interact with an instructor. But it does carry a price. Much like a videoconferencing room, you have to set up a suitable environment with monitors, microphones for student responses, and possibly, a document camera and computer. An instructor may also have to learn how to teach to a camera/monitor in addition to other prep work.

For medicine, it is possible to use telecommunications tools for distant diagnostic purposes, among other applications. In one case, electronic meetings can be held between medical personnel at sea and on land. Consequently, even if you are situated at a remote site, you could tap someone's medical expertise via teleconferencing tools. The military is also experimenting with systems, originally pioneered by NASA, to remotely monitor the medical status of its personnel.[30]

CONCLUSION

Teleconferences and computer conferences have emerged as potent communications tools. They serve organizations, and through personal videoconferencing, may have an impact on our daily lives.

We may also have direct contact with such systems through telecommuting and by taking electronic courses. Although both applications are not without their faults, they may enable us to work at a geographically removed site and to attend a class that may, just a few years ago, have been impossible to take.

REFERENCES/NOTES

1. "Understand the Options," *Networking Management—Videoconferencing and Teleconferencing Directory,* Mid-November 1991, 43.

2. Codecs are also used in other fields: in telephone operations and software codecs in PC-based video/multimedia.

3. Bill Dunne, "Screening the Confusion Out of the Different Types of Videoconferencing," *Communications News* (November 1985): 34.

4. David B. Mensit and Bernard A. Wright, "Picturephone Meeting Service: The System," in *Teleconferencing and Electronic Communications: Applications, Technologies, and Human Factors* (Madison, Wis.: Regents of the University of Wisconsin, 1982), 174.

5. Clarke Bishop, Ken Leddick, and Jim Black, "Compressed Digital Video for Business TV Applications," *Communications News* (December 1991): 40.

6. A network can support educational and business applications, among other things. See the International Teleconferencing Association, "ITCA Teleconferencing Definitions," Guide to Membership Services brochure.

7. The International Teleconferencing Association, "Business Television Private Network Market Reaches $606 Million," press release, February 1992, from a survey by TRI and IFC Resources, Ltd., of 100 companies responding to the impact of videoconferencing on their operations.

8. "Video Communications Comes of Age with H.261 Standard," *Communications News* (February 1991): 10.

9. Andrew W. Davis, "Videoconferencing via POTS Now: Proprietary Codecs and Emerging Standards," *Advanced Imaging* 10 (June 1995): 32.

10. The quality may be lower.

11. Robert Lindstrom, "Shrinking of the Globe," *Presentation Products* 6 (June 1992): 22.

12. Conversation with VideoTelecom Corporation, Summer 1992, developer of such a teleconferencing system.

13. Sprint Communications, "Sprint Meeting Channel-Rates and Services," brochure.

14. Davis, "Videoconferencing Via POTS Now," 36.

15. This capability also expands videoconferencing's potential user base.

16. Patrick Portway, "The Promise of Video Telephony Made in 1964 Finally Is Being Fulfilled at Reasonable Prices," *Communications News* (February 1988): 35.

17. Elliot M. Gold, "Trends in Desktop Video and Videophones," *Networking Management* (May 1992): 46.

18. Howard Falk, "Picturephone and Beyond," *IEEE Spectrum* (November 1973): 48.

19. The International Teleconferencing Association, "Business Television Private Network Market," brochure.

20. Since you saved time and money.

21. Herman Mehling, "The Ties that Unbind," *Information Week* (November 6, 1995): 58.

22. Osman Eldib and Daniel Minoli, *Telecommuting* (Norwood, Mass.: Artech House, 1995), 1.

23. Ibid, 4.

24. Since a parent works at home.

25. Robert E. Kraut, "Telecommuting: The Trade-Offs of Home Work," *Journal of Communication* (Summer 1989): 43.

26. Fran Calpin, "Work at Home Injuries," *The Sunday Times,* (February 20, 2000): C1.

27. Ibid.

28. Isaac Asimov, *The Naked Sun* (New York: Ballantine Books, 1957), 46.

29. WebCT is used to create interactive, online courses.

30. John Rhea, "Telemedicine Ties into Tactical Communications Infrastructure," *Military & Aerospace Electronics* 8 (April 1997): 11–13.

SUGGESTED READINGS

Concelmo, Peter A. "Business Television Blossoms with Distance Learning." *Satellite Communications* 21 (October 1997): 20–30. Satellite-delivered business television.

Doyle, Bob. "How Codecs Work." *NewMedia* 4 (March 1994): 52–55. Technical description of codecs/compression.

Gold, Elliot M. "PC Graphics Add a New Dimension." *Networking Management* (April 1992): 38–41; Halhed, Basil R., and Lynn D. Scott. "Making Multimedia Conferencing Work for You." *Video Systems* 17 (December 1991): supplement, 2–4. An examination of the use of graphics in videoconferencing and multimedia conferencing.

Halhed, Basil R., and Lynn D. Scott. "Videoconferencing Terminology." *Video Systems* 18 (May 1992): supplement, 8–10. A list and accompanying definitions of the more important terms in the videoconferencing field.

Jerram, Peter. "Videoconferencing Gets in Sync." *NewMedia* 5 (July 1995): 48–55; Sherman, Lee. "New Systems Provide Virtual Meetings at Your Desktop." *NewMedia* 4 (November 1994): 49–54; Yager, Tom. "Better than Being There." *Byte* 18 (March 1993): 129–134. Desktop videoconferencing and data collaboration.

Kames, A. J., and W. G. Heffron. "Human Factors Design of Picturephone Meeting Service." A paper presented at Globecom '82, sponsored by the IEEE, in Miami, Florida, November 29 to December 2, 1982. A review of an AT&T videoconference service. The paper also discusses the tests that were conducted to judge the system's effectiveness as a communications tool.

Mehling, Herman. "The Ties that Unbind." *Information Week* 552 (November 6, 1995): 56–62. Dataconferencing: applications, advantages, and drawbacks.

Nisenson, Kyle. "Tune in to IP Videoconferencing." *Network World* 15 (September 21, 1998): 43–45; Olgren, Christine. "Trends in the Use of Teleconferencing Illustrate the Wide Variety of Applications That It Offers." *Communication News* (February 1988): 22–26. Two articles about teleconferencing; the second article presents an interesting view of the state of the industry in the late 1980s.

Schooley, Ann K. "Allowing FDA Regulation of Communications Software Used in Telemedicine: A Potentially Fatal Misdiagnosis?" *Federal Communications Law Journal* (May 1998). (50 Fed. Com. L.J. 731). Downloaded from Lexis; Smith, Delbert D. "Distant Doctors." *Satellite Communications* 22 (May 1998): 32–40. Telemedicine and legal implications.

GLOSSARY

Audioconference: A form of a teleconference. In an audioconference, individuals at two or more sites can speak to and hear each other.

Codec: A combination of the words coder/decoder. A codec converts an analog signal into a digital format. In a teleconferencing environment, it also compresses the signal so the information can be relayed on lower capacity and less expensive communications channels. A codec at the end of the relay converts the signal back into an analog form.

Computer Conference: A meeting conducted through computers that can support the exchange of information ranging from text to graphics.

Electronic University: A school that offers courses via computer.

Motion Videoconference: A motion videoconference can duplicate a face-to-face meeting. In a two-way, motion configuration, for example, the participants can see and hear each other.

Nonmotion Videoconference: A nonmotion videoconference supports the relay and exchange of still pictures.

Personal Videoconferencing: Personal videoconferencing is a general term that describes desktop videoconferencing systems.

Telecommuter: An individual who works at home and maintains contact with the office by computer.

Teleconference: An umbrella term for the various categories of electronic meetings and events, ranging from audioconferences to videoconferences.

Videoconference: A teleconference wherein, as implied by the name, video or visual information is exchanged.

20

E-Mail and Privacy

The United States has a rich tradition of protecting an individual's privacy rights. This protection can be traced to the Fourth Amendment, which guarantees

the right of the people to be secure in their persons, houses, papers, and effects, against unreasonable searches and seizures, shall not be violated, and no warrants issued, but upon probable cause, supported by Oath or affirmation, and particularly describing the places to be searched, and the persons or things to be seized.[1]

California, Washington, and some other states also provide privacy guarantees within their constitutions. In California, for example, "All people by nature are free and independent and have inalienable rights. Among these are enjoying and defending life and liberty, acquiring and possessing, and protecting property, and purchasing and obtaining safety, happiness, and privacy."[2]

Based on the intent found within such documents, legislation has been enacted to protect our privacy rights. But during the past several years, new communications technologies have led to a potential deterioration of these privileges. In response, federal and state courts have been asked to examine and interpret relevant law. Of particular concern is the area of privacy and human-to-computer and computer-to-computer communication, which includes e-mail privacy.

FEDERAL CASE LAW

The Fourth Amendment provides protection to individuals in narrowly defined situations, but it does not guarantee employee privacy in the workplace or in their e-mail. Fourth Amendment protection is only afforded if the employee has a reasonable expectation of privacy. This expectation must be balanced against the employer's right to control, supervise, and maintain workplace efficiency.

According to the reasonableness standard, the search and seizure of an employee's private workplace property must also be reasonable at its inception and in scope. An e-mail search is justified at inception if there is a reasonable belief the search will provide evidence of employee workplace misconduct or if the search is required to retrieve noninvestigatory work-related material—for example, your supervisor accesses some routine work reports from your e-mail on a day you are sick.[3] An e-mail search is reasonable in scope if "the measures taken are reasonably related to the objectives of the search."[4]

Two-Prong Test

A two-prong test must also be conducted. First, a subjective expectation of privacy must exist. This is your belief that you are entitled to a reasonable expectation of privacy with regard to your briefcase, personal mail, and other workplace personal items, including e-mail. Second, there must be an objective expectation of privacy. Society must agree that your perception is reasonable.[5]

The reasonableness of a search must also be judged by the search's intrusiveness. In one example, the federal courts have ruled that no reasonable expectation of privacy exists in workplace situations where there are extensive security precautions (for example, at a military site).[6] In contrast, a warrantless body cavity search for suspected workplace misconduct is not reasonable because it is very intrusive.[7] Therefore, a search of employee items, including their e-mail, is a complex issue.

Precedent and *Stare Decisis*

The paucity of federal e-mail cases has led privacy advocates to cite other, relevant Fourth Amendment cases.[8] The cases discuss whether employees have a reasonable expectation of privacy in their desks, file cabinets, wastebaskets, and similar workplace items. If they are afforded a reasonable degree of privacy protection, this could establish precedent for e-mail.

Precedent is defined as an established rule of law set by previous courts, deciding a case involving similar facts and legal issues.[9] Another legal concept that comes into play is *stare decisis*. *Stare decisis* ensures that similar cases will be decided in a similar way.[10] So precedent sets the standard—your private wastebasket is protected by privacy law. *Stare decisis* then guarantees that similar wastebasket privacy cases will be decided in the same way.

Based on these two legal concepts, e-mail privacy advocates claim the following:

- A line of Fourth Amendment cases serve as precedent for extending workplace privacy protection to an employee's e-mail.
- Workplace e-mail should be afforded the same privacy protections other workplace items currently enjoy.

Cases

O'Connor v. Ortega **(480 U.S. 707.1987).** One such case, which could serve as a precedent, was *O'Connor v. Ortega.* The U.S. Supreme Court ruled that Dr. Ortega, a physician who was a state hospital employee, had a reasonable expectation of privacy in his desk, file cabinets, and office. They were not shared with other employees and only contained personal items.

The Court decided this case on Fourth Amendment grounds. It ruled that state hospital administrators had violated Ortega's privacy rights. They failed to meet the requirements of the reasonableness standard while investigating possible work-related misconduct.[11]

In brief, hospital administrators were concerned about Ortega's management of the hospital's psychiatric program. They subsequently placed him on leave and launched an investigation. At this time, the administrators searched Ortega's office, removed some items, and subsequently used them in an administrative hearing that resulted in Ortega's firing. He responded with a lawsuit against the hospital.

In deciding the case, the Court recognized the need to fairly balance the employee's legitimate right to privacy against the employer's need to supervise, control, and maintain the workplace's efficient operation. However, the court also ruled that a search must be reasonable at its inception and in scope. In the *Ortega* case, the Court ultimately ruled that Ortega had a reasonable expectation of privacy in his workplace items.[12]

Following this logic, an employer would violate an employee's rights if private, workplace e-mail was accessed without any prior suspicion of work-related misconduct or if the search was more intrusive than necessary. Any evidence of work-related misconduct/criminal behavior, which had been obtained in this manner, would be inadmissible in an administrative or judicial proceeding (for example, to fire the employee).

Richard Schowengerdt v. General Dynamics et al. **(823 F. 2d. 1328. 944 F. 2d. 483.1987).** Richard Schowengerdt was a civil engineer employed by the U.S. Navy to work on high-security weapons and projects. In light of their sensitive nature, random workplace searches and other extensive security precautions were taken. Nevertheless, following a routine search of his office and desk, Schowengerdt filed suit claiming he had an objective reasonable expectation of privacy in his workplace.

The court disagreed stating that employees knew about the "tight security procedures and the ongoing surveillance."[13] Therefore, the measures taken to guarantee the facility's security would negate any objective expectation of privacy.

Based on this case, it appears that if the employer has a policy informing employees that their e-mail may be read, accessing it would only be legal for monitoring purposes. It would be illegal in narrow circumstances, such as access by an unauthorized person.

Michael A. Smyth v. The Pillsbury Company (914 F. Supp 97. 1996).

In 1996, the federal courts finally heard cases specifically involving workplace e-mail privacy. In *Smyth v. Pillsbury,* the U.S. District Court for the Eastern District of Pennsylvania ruled that an employee who used the company's private, internal e-mail system to communicate "inappropriate and unprofessional comments" to his supervisor, did not have a reasonable expectation of privacy in his e-mail messages. He had been terminated for this action.[14]

The court also ruled that the company's right to prevent inappropriate, unprofessional, or illegal comments and activity, outweighed the employee's workplace e-mail privacy interest.[15] This decision serves as a particularly strong endorsement of employer rights—Pillsbury's company policy states that employee e-mail communication is confidential and privileged. Nevertheless, the court still held that Pillsbury had the right to monitor internal employee e-mail communication.[16]

John Bohach et al. v. The City of Reno (932 F. Supp. 1232. 1996).

Another federal case specifically involving workplace e-mail privacy is *Bohach v. The City of Reno.* Unlike *Smyth v. Pillsbury,* the Bohach case involved a public employer, the Reno (Nevada) Police Department.

Two police officers used the department's Alphapage message system to exchange information that was later used in an internal investigation against the officers.[17] They filed suit against the police department in federal court on the grounds that the storage, retrieval, and disclosure of the messages violated their right to workplace e-mail privacy.[18]

The court ruled that while the two officers may have believed their messages were private, no reasonable expectation of privacy in their workplace e-mail messages existed.[19] This decision was based on the following:

1. The system automatically recorded and stored all messages.
2. Officers were informed that all messages would be logged and that some types of messages were prohibited on the system.
3. Message recording was considered a part of the "ordinary course of business" for a police department.
4. As the service provider, the police department was free to access its system messages and use them as they saw fit.[20]

Thus, the court held the officers did not have a reasonable expectation of privacy in their public, workplace e-mail messages.

Case Law Summary as It Relates to E-Mail

The courts provide the employer with certain advantages in the employer–employee relationship. If an employer can demonstrate that the place of business, including office files and computer systems, is easily accessible to coworkers, customers, and others, the employee does not enjoy a reasonable expectation of workplace privacy. This includes e-mail.

However, if an employer accesses an employee's e-mail, this action would constitute a search and, thus, the employer must meet the reasonableness standard. Because an employer has the right to supervise a business and see that it is efficient, a search warrant does not have to be obtained to gain access to routine, noninvestigatory business information.

In essence, although the Fourth Amendment provides protection in certain narrowly defined situations, it does not guarantee your privacy in the workplace, or as of this writing, your e-mail.

FEDERAL LEGISLATION

Beyond federal case law, privacy advocates have been concerned about legislative initiatives. One problem has been the legal system's inability to keep pace with technological developments.

For example, Congress enacted Title III of the Omnibus Crime Control and Safe Streets Act in 1968 to protect certain privacy rights. But it became apparent, especially in the 1980s, that the act was not comprehensive enough. It only provided protection for wire and oral communication and the aural interception of voice communication. Human-to-computer communication and computer-to-computer communication were not protected.[21]

Recognizing the gap, Congress passed the Electronic Communications Privacy Act of 1986 (ECPA) to eliminate some of the loopholes.[22] But while the ECPA does apply to public e-mail, private e-mail is not protected.

A public e-mail system functions much like a public telephone company and enables you to send and receive messages.[23] If such a company illegally discloses a message's content to unauthorized individuals, it can be sued. In contrast, a private, internal e-mail system, such as one set up by a company, only has to guarantee that "unauthorized parties cannot gain access to your e-mail."[24]

STATE LEGISLATION AND CASE LAW

A body of state legislation and case law exists that specifically addresses e-mail workplace privacy.[25] In California, for instance, section 630 of the California Penal Code states the following:

The Legislature hereby declares that advances . . . have led to . . . new devices and techniques for the purpose of eavesdropping on private communications and that the invasion of privacy resulting from the continual and increasing use of such devices . . . has created a serious threat to the free exercise of personal liberties and cannot be tolerated in a free and civilized society.[26]

E-mail that was illegally accessed could fall into this category. Code violations would render any illegally obtained information inadmissible. However, the California code supports the legitimate use of the latest technologies, which could include accessing an employee's e-mail, if a legal search warrant had first been obtained.[27]

With regard to case law, one relevant example is *Alana Shoars v. Epson America* (No. 073234. Super. Ct. No. SWC 112749.1994) A court examined state law and the state constitution to determine if an employer could legally access, print out, and read an employee's workplace e-mail.[28]

Alana Shoars was hired by Epson to teach employees how to use the company's e-mail system to communicate with external parties via an MCI setup. Shoars issued personal passwords to each employee and told them their e-mail would be considered private communication. But Epson management did not share Shoar's viewpoint about confidentiality. Employee e-mail was frequently accessed, printed out, and read.

When discovered, Shoars confronted her supervisor and demanded the practice be stopped. He threatened to have her fired if she continued to pursue the matter. Shoars refused and was fired. She then filed suit claiming that Epson's actions violated the California Penal Code and the California State Constitution.

Epson management claimed they did not violate the law. Employees were provided with computer tools to tie their internal e-mail system with MCI's operation. This meant that messages were automatically logged and downloaded to Epson's computers (for example, for troubleshooting).

More pointedly, Epson claimed its supervisors had to scan and read through this information as a part of the support process. Thus, Epson argued the law had not been violated.

After hearing the arguments, the court ruled in Epson's favor. The company had not engaged in illegal tapping or unauthorized connections with a telephone line. The court further held the act of downloading messages, as practiced by the company, did not constitute a violation. It similarly ruled in favor of Epson on other grounds.

A more recent state workplace e-mail privacy case, *Restuccia v. Burk Technology, Inc.* (5 Mass. L. Rptr. 712. 1996), was heard by the Massachusetts State Court in 1996.[29] This case involved two employees who were fired by a private employer, Burk Technology. The company contended the firings were based on the content of their internal workplace e-mail messages and for the excessive use of the company's e-mail system.[30] The employees subsequently filed suit in state court on the grounds that accessing and reading their workplace e-mail messages violated the Massachusetts wiretapping statute.[31]

The Burk Technology e-mail system had been created for internal communication between employees and required the use of a password. Employees had not been told their messages were automatically saved or that e-mail messages could be accessed by company personnel. Furthermore, Burk Technology did not have a company policy prohibiting communication of nonbusiness messages via the e-mail system. But it did have a written policy prohibiting excessive system use.[32]

The Massachusetts state court eventually ruled that an employer's accessing and reading of e-mail messages, which had automatically been saved, did not violate the state wiretapping statute. The court further held this practice fell under the "ordinary course of business" exemption clause and as such, constituted protected company action.[33]

CONCLUSION

Online privacy is afforded protection under various federal and state constitutions and legislation, but it is not fully extended to the workplace. Once you become an employee, much of this protection is lost or at least significantly diminished.

For example, most employer privacy guidelines prohibit "unauthorized parties" from accessing and disclosing employee e-mail and other personal workplace effects. The employee is also protected from an un-reasonable and warrantless search. An employer, however, has sweeping powers to monitor and disclose an employee's e-mail without prior knowledge or consent.

While an employee's expectation of e-mail privacy is shaky at its best, two avenues of protection may be passwords and addresses. In the former, employees are assigned private passwords, thus the expectation of privacy exists. In the latter, if e-mail is addressed to a specific individual, the sender believes the message will be sent to that person alone.

But employers have disputed these contentions. The hardware and software required to use e-mail are company owned—these tools are subject to company review. An employer also has a legitimate right to supervise employee work, including e-mail, for quality control.

Federal and State Response

E-mail privacy supporters have filed suit in federal and state courts on the grounds that employer monitoring of e-mail violates the law. Specifically, federal courts have been asked to decide if this practice violates the Fourth Amendment.

To date, the federal courts have declined to extend such protection to workplace e-mail. Most constitutional scholars predict that future Fourth Amendment protection would only be extended to workplace e-mail in cases where obvious abuses exist.

State courts have generally followed suit. Like federal court, most scholars maintain that privacy protection would only be afforded in extreme circumstances. Workplace privacy advocates also claim that federal and state legislation should offer some protection. However, as of this writing, they have provided limited protection at best.

Federal and state law have also made a distinction between a public employer (for example, the government) and a private employer. Generally speaking, if you work for a public employer, you have slightly more privacy protection. A private employer has more leeway and can provide significantly less protection.

Employer Code

If an employer has expectations for employee use of company property, including e-mail, formal guidelines should be drafted and distributed (for example, each party's legal/ethical responsibilities). When designing such policy, the employer should also define the workplace privacy that will be afforded to the employee, regardless of the communication form.

For example, does existing company policy provide privacy protection to an employee's paper files, desk drawers, and voice mail? If so, how does it compare with the privacy afforded to e-mail? If e-mail has been singled out for special treatment, the employee should be informed.[34]

Workplace privacy should also meet the needs of both the employer and the employee. But it can be a very fine line. An employer has a right to monitor a business. But the employee has a right to expect a positive and secure workplace. If, for instance, e-mail is going to be monitored, the employee should be warned. The least intrusive method of obtaining the desired information should also be used. This access should be limited in scope, and the employer should preserve as much of the employee's privacy as possible.

In conclusion, individuals have privacy protection on the federal and state levels. But in reality, much of the protection is illusionary once you become an employee. Until a body of law is created that is more favorable to employee privacy rights, your workplace e-mail will be subject to employer monitoring.

REFERENCES/NOTES

1. Constitution, amend. IV.
2. California Constitution, Article I, Section I.
3. *New Jersey v. TLO.* 469 U.S. 325. 1985.
4. *United States v. Charles Katz.* 389 U.S. 347. 1967.
5. Ibid, 359.
6. *Richard Schowengerdt v. General Dynamics et al.* 823 F. 2d. 1328. 944 F. 2d. 483. 1987.
7. *Gerald W. Shields v. David Burge et al.* 874 F. 2d. 1201. 1989.
8. Cases have, however, been heard (which was not the case at the time of the book's 3rd edition).
9. *Black's Law Dictionary* (St. Paul, Minn.: West Law Publishing, 1979), 1059.
10. Ibid, 1261.
11. *O'Connor v. Ortega.* 480 U.S. 717.
12. Ibid, 717.
13. *Richard Schowengerdt v. General Dynamics et al.* 823 F. 2d. 1328. 944 F. 2d. 483. 1987.
14. *Michael A. Smyth v. The Pillsbury Company.* 914 F. Supp. 97. 1996, downloaded from www.loundy.com.
15. Ibid.
16. Ibid.
17. *John Bohach et al. v. The City of Reno.* 932 F. Supp. 1232. 1996, downloaded from Lexis.
18. Ibid.
19. Ibid.
20. Ibid.
21. Steven Winters. "Do Not Fold, Spindle or Mutilate: An Examination of Workplace Privacy in Electronic Mail," *Southern California Interdisciplinary Law Journal* 85 (Spring 1992): 45–46.
22. Electronic Communications Privacy Act of 1986. 100 Stat 1848. Section 2510, 1.
23. Electronic Communications Privacy Act. viii.
24. Ibid.
25. California Constitution, Article I, Section I.
26. California Penal Code, Section 630.
27. California Penal Code, Section 631.
28. *Alana Shoars v. Epson America, Inc.* No B 073234, Calf. Super. Crt. No. SWC112749.
29. Jenna Wischmeyer. "E-Mail and the Workplace," 1998, downloaded from raven.cc.ukans.edu, December 1999.
30. Ibid.
31. Ibid.
32. Ibid.
33. Ibid.
34. Electronic Mail Association. *Access to and Use and Disclosure of Electronic Mail on Company Computer Systems: A Tool Kit for Formulating Your Company's Policy* (September 1991): 1.

SUGGESTED READINGS

Case Law

Bonita Bourke v. Nissan Motor Corporation in U.S.A. (B 068705. Super. C. 1993).

Bonita Bourke v. Nissan Motor Corporation in U.S.A. (No. YC003979).

Cameron v. Mentor Graphics (No. 716361. Cal. Spr. Ct.).

Francis Gillard v. Schmidt (570 F. 2d. 825. 1978).

Lacrone v. Ohio Bell Co. (182 N.E. 2d. 15.1961).

National Treasury Employee Union v. William Von Raab (489 U.S. 656. 1989).

O'Connor v. Ortega (480 U.S. 709. 1987).

Ray Oliver v. Richard Thornton (466 U.S. 170. 1984).

Rhodes v. Graham (37 S.W. 2d. 46. 1931).

Roach v. Harper (105 S.E. 2d. 564. 1958).

Alana Shoars v. Epson America, Inc. (No. 073234. Super. Ct. No. SWC 112749. 1994).

Samuel Skinner v. Railway Labor Executives Association (489 U.S. 602. 1989).

Watkins v. Berry (704 F. 2d. 577. 1983).

Federal and State Legislation

Computer Security Act of 1987. 101 STAT. 1724. P.L. 100-235. 1988.

Electronic Communications Privacy Act of 1986. 100 STAT. 1848. P.L. 99-508. 1986.

Privacy for Consumers and Workers Act. Hearing. S. 984. 1993.

Title III Wiretapping and Electronic Surveillance. 82 STAT. 211. P.L. 90-351. 1968.

Journals, Magazines, and Online Documents

American Management Association International. "Electronic Monitoring and Surveillance: 1997 AMA Survey." 1997. Downloaded from www.cybersquirrel.com, November 1999.

Barlow, John Perry. "Electronic Frontier: Bill O' Rights. *Communications of the ACM* (March 1993): 21–23.

Baumhart, Julia. "The Employer's Right to Read Employee E-Mail: Protecting Property or Personal Prying," *LABLAW* 8 (Fall 1992): 1–65.

Boehmer, Robert. "Artificial Monitoring and Surveillance of Employees: The Fine Line Dividing the Prudently Managed Enterprise from the Modern Sweatshop." *Dickinson Law Review* (Spring 1992): 593–666.

"Current Legal Standards for Access to Papers, Records, and Communications." 1999. Downloaded from www.cdt.org, February 2000.

CyberSpace Law Center. "Workplace Privacy Bibliography." 1998. Downloaded from www.cybersquirrel.com, November 1999.

Electronic Mail Association. *Access to and Use and Disclosure of Electronic Mail on Company Computer Systems: A Tool Kit for Formulating Your Company's Policy.* September 1991.

Electronic Mail Association. *Protecting Electronic Messaging: A Guide to the Electronic Communications Privacy Act of 1986.*

Tuerkheimer, Frank. "The Underpinnings of Privacy Protection." *Communications of the ACM* (August 1993): 69–73.

Witte, Lois. "Terminally Nosy: Are Employers Free to Access Our Electronic Mail?" *Dickinson Law Review* (1993): 545–573.

GLOSSARY

Electronic Communications: Any transfer of signs, signals, writings, images, sounds, or data transmitted via wire, radio, electromagnetic, photoelectronic, or photo-optical systems.

Electronic Communications Privacy Act (ECPA): This act was passed in 1986 to close some regulatory loopholes that existed in the Omnibus Crime Control and Safe Streets Act. This protects the acquisition of the contents of any wire, electronic, or oral communication via an electronic or mechanical device.

Objective Expectation of Privacy: Society agrees with an employee that he/she has a reasonable expectation of privacy in workplace items.

Precedent: An established rule of law set by a previous court deciding a case involving similar facts and legal issues.

Stare Decisis: Similar cases should be decided in a similar way.

Subjective Expectation of Privacy: The right to expect that personal workplace items such as a briefcase, personal mail, desk, files, and wastebaskets will not be subject to an unreasonable employer search.

21 The New Technologies and the First Amendment

The new communication technologies and their applications have raised constitutional questions, including the First Amendment implications of regulating online obscenity and libel. Relevant legislation and case law concerning these issues are discussed in this chapter.

For online obscenity, the main focus is the Communications Decency Act and the Child Online Protection Act. Background information is presented first, followed by an examination of relevant case law.

THE COMMUNICATIONS DECENCY ACT

In early 1995, U.S. Senators James Exon (D-Nebr.) and Slade Gorton (R-Wash.) sponsored the Communications Decency Act (CDA) to regulate online sexual content. According to Senator Exon, this legislation, which was later incorporated in the 1996 Telecom Act, was necessary. Some individuals were using online services to communicate with children about inappropriate sexually oriented topics. Some adults were also offended by the sexually explicit material that was available over the Internet.[1]

In the past, sexually explicit speech was regulated on the basis of consent. For example, the 1934 Communications Act only regulated nonconsensual sexual speech that was unwanted by one of the parties.[2] But the CDA made no distinction between consensual and nonconsensual speech. It criminalized all obscene, lewd, lascivious, filthy, and indecent communication distributed via a telecommunications device. This law also placed the regulatory burden on providers and services.[3]

The CDA's Constitutionality

The CDA was challenged in court on constitutional grounds. In 1996, two lower courts held portions of the CDA unconstitutional. CDA opponents indicate the act would have a *chilling effect* on computer communication. Chilling implies that a threat of regulation may make you think twice about providing what may be inappropriate information. In this case, you may not post some questionable material online because you thought it might be illegal—communication is stopped before it can actually take place. In this light, it could be viewed as a form of censorship.

Critics also contend the CDA would create a new "measuring stick" to determine if a message was obscene. Under the CDA, only material judged appropriate for children would be permitted online.[4]

Another concern was the targeting of obscene as well as indecent speech. Indecent speech is defined as nonobscene language that is sexually explicit or includes profanity.[5] Although most First Amendment scholars

Figure 21.1
The ability to convert one media type to another has enhanced the communication process. But our new capabilities have also led, at times, to legal situations (as discussed in this chapter). In this shot, PageMaker, a desktop publishing program, can export a file to an HTML format. (Software courtesy of Adobe Systems, Inc.; PageMaker.)

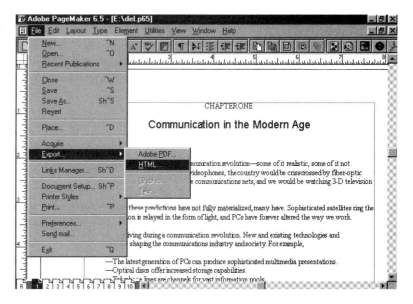

agree that obscene speech is not protected, there is no precedent for denying constitutional protection for indecent speech. In fact, the Supreme Court has ruled that indecent speech is protected.

In the precedent-setting case, *Miller v. California*, nonobscene speech was ruled constitutional. Only language that appeals to prurient interests, is patently offensive, and lacks serious literary, artistic, political, and scientific value is prohibited.[6] Indecent speech does not fall within this definition and has been protected by the First Amendment.

Furthermore, in *Sable Communications of California Inc. v. Federal Communications Commission*, the court ruled that sexual expression, which is indecent but not obscene, was protected. For example, the ability of a child to potentially retrieve a sexually explicit telephone message does not render such speech unconstitutional.[7]

Internet Regulation and the Broadcast Model

When a new communication form is born, Congress and other bodies must determine if and how it will be regulated. In this process, established communication systems may be examined as regulatory models.

For the Internet, the broadcast industry served as one example. Broadcast regulation has been based, in part, on the spectrum scarcity theory—since the spectrum is finite, only a limited number of individuals could receive a license. Thus, FCC involvement in this arena could, depending on your viewpoint, be considered a legitimate use of power.

But when applied to computer communication, critics contend the scarcity argument is not valid. As described in previous chapters, the Internet and online services constitute a vast and rapidly growing infrastructure. Essentially, "whenever you add a computer to the Internet, you increase the Internet size [sic] and capabilities."[8]

A similar principle applies to the pervasive medium theory, which argues that the broadcast industry could be regulated since messages are "thrust" on a sometimes unsuspecting audience. But like spectrum scarcity, this regulatory stance may not be valid in computer-based systems.

Computer communication on the Internet, for instance, requires you to take a number of affirmative steps to retrieve information. "Exposure to the content is primarily driven by choice with little risk of unwitting exposure to offensive or indecent material."[9] Consequently, the broadcast model may not be a valid Internet regulatory model even though it was used to explore such options.

Internet Concerns

One key factor, which provided impetus for Internet regulation, was a concern that some Internet material may be inappropriate for children. But CDA critics indicated the online community was developing blocking and screening software and options. Parents, teachers, and other responsible adults would be able to control the material children could retrieve.[10]

In one case, Cyber Patrol functioned by producing a "CyberNOT" list of objectionable sites, and it organized questionable content into categories. In the violence and profanity entry, obscene words and communication that encouraged extreme cruelty against any animal or person were included. Sites with these characteristics would be placed on the list and minors would be blocked from entering.[11]

The Case: *American Civil Liberties Union v. Attorney General Janet Reno*

The CDA gave the U.S. attorney general the authority to investigate and prosecute any party in violation of the act's indecency and patently offensive provisions. Specifically, the indecency provision (Section 223) states:

. . . any person in interstate or foreign communication who, 'by means of a telecommunications device,' 'knowingly . . . makes, creates, or solicits' and 'initiates the transmission' of 'any comment, request, suggestion, proposal, image or other communication which is obscene or indecent, knowing that the recipient of the communication is under 18 years of age,' shall be criminally fined or imprisoned.[12]

The patently offensive provision (Section 223)

. . . makes it a crime to use an 'interactive computer service' to 'send' or 'display in a manner available' to a person under age 18, 'any comment, request, suggestion, proposal, image, or other communication that, in context, depicts or describes, in terms patently offensive as measured by contemporary community standards, sexual or excretory activities or organs, regardless of whether the user of such service placed the call or initiated the communication.'[13]

In response, the American Civil Liberties Union (ACLU) and other plaintiffs filed suit in federal court to halt the provisions' enforcement. They requested a temporary restraining order (TRO), and Judge Ronald Buckwalter of the U.S. District Court for the Eastern District of Pennsylvania granted a limited TRO.

The grounds? The terms "indecent" and "patently offensive" were unconstitutionally vague and required interpretation. Following this action, a panel of three judges from the Third Circuit Court of Appeals was convened to examine this case. Attorney General Janet Reno agreed not to investigate or prosecute violators until the judges rendered their decision. The court said Reno could file charges for violations in the future, if the provisions were found to be constitutional.[14]

Some case highlights are as follows:

1.   The government said the ACLU was overreacting to the CDA's language since speech would only be restricted in narrow circumstances. The government cited supporting factors. For example, a telecommunications facility/service would not be in violation of the CDA if it acted in good faith and took "reasonable, effective, and appropriate actions to restrict or prevent access" to inappropriate materials by a minor.[15]

2.   The ACLU did not dispute that obscene, pornographic, and harassing material was inappropriate on the Internet. It argued the CDA reached too far and violated the First Amendment since it "effectively bans a substantial category of protected speech from most parts of the Internet."[16]

3.   The government indicated that "shielding minors from access to indecent materials is the compelling interest supporting the CDA."[17] According to the court, "whatever the strength of the interest the government has demonstrated in preventing minors from accessing 'indecent' and 'patently offensive' material online, if the means it has chosen sweeps more broadly than necessary . . . , it has overstepped onto rights protected by the First Amendment."[18] The court then had to answer the question: Did the government overstep it bounds?

4.   What may be objectionable to one person in one city may not be objectionable to another person in another city. Angels in America, for instance, a play about sexuality and AIDS, may be unacceptable in certain communities, but the play won two Tonys and a Pulitzer Prize, and some adults could view it as appropriate material for older minors.[19]

If introduced online, however, this same material would fall under the CDA's purview. Other nonobscene material, ranging from certain National Geographic photographs to sculptures to textual descriptions, could follow suit and would be subject to the "CDA's criminal provisions."[20]

5.   The ACLU had to demonstrate it was likely to prevail on the case's merits. If the case went to court, the ACLU would most likely win.

6.   The ACLU had to convince the court that if the TRO was not granted, irreparable harm would occur. Citing past case law, the ACLU claimed halting protected speech, even for a brief time, constituted irreparable harm.

7.   Past courts have ruled that criminalizing protected speech has a chilling effect on the speaker's freedom of expression, also resulting in irreparable harm.

8.   The ACLU had to establish that a TRO was in the public's best interest. Citing past case law, it was alleged that an open marketplace of ideas and the preservation of protected speech is always in the public's interest.

The Ruling

The court eventually ruled the CDA was overly intrusive, and the government's claim of a compelling interest in restricting these materials from children was outweighed by the right to free speech.[21]

Criminal prosecution resulting from a violation could also subject the content provider to public damnation and would create a financial hardship. In fact, content providers could back away from questionable material out of fear of prosecution if they make a mistake as to what constitutes indecent and/or patently offensive material.[22]

The terms *indecency* and *patently offensive* are also inherently vague. The government could not identify the specific community standards that would be used to judge questionable online material.[23] The court further noted that Congress could have supported the development of blocking software, thus placing the decision-making process in the hands of parents and educators.

Congress could have also followed a well-traveled path by looking toward the print media for guidance. Nonobscene, but indecent and/or patently offensive books and magazines, are available everywhere.[24] The government, however, is not primarily responsible for controlling a minor's exposure to these materials. The decision making rests squarely on the shoulders of adults.

The court also indicated that the Internet "may fairly be regarded as a never-ending worldwide conversation," and the government cannot use the CDA to "interrupt that conversation. As the most participatory form of mass speech yet developed," the Internet merits the highest level of protection.[25]

Furthermore, while the lack of government regulation "has unquestionably produced a kind of chaos, . . . The strength of the Internet is that chaos. Just as the strength of the Internet is chaos, so the strength of our liberty depends on the chaos and . . . the unfettered speech the First Amendment protects."[26] Thus, the court held the indecency and patently offensive provisions of the CDA unconstitutional.[27]

The court granted the TRO and instructed Attorney General Janet Reno, and other parties, that they were "enjoined from enforcing, prosecuting, investigating, or reviewing any matter" concerning these matters.[28]

The attorney general subsequently appealed the case and in June 1997, the U.S. Supreme Court heard *Reno et al. v. American Civil Liberties Union et al.* (117 S.Ct. 2329. 1997). In a 7–2 decision with Justice John Paul Stevens writing for the majority, the Supreme Court upheld the lower court, finding the indecency and patently offensive portions of the CDA violated free speech protected by the First Amendment.[29]

Justice Stevens wrote, "We are persuaded that the CDA lacks the precision that the First Amendment requires when a statute regulates the content of speech. In order to deny minors access to potentially harmful speech, the CDA effectively suppresses a large amount of speech that adults have a constitutional right to receive and address to one another."[30]

The Court found those portions of the CDA unconstitutionally vague, overbroad, and not sufficiently narrowly tailored based on the fact that, ". . . the scope of the CDA is not limited to commercial speech or commercial entities. Its open-ended prohibitions embrace all nonprofit entities and individuals posting indecent messages or displaying them on their own computers in the presence of minors. The general, undefined terms 'indecent' and 'patently offensive' cover large amounts of nonpornographic material with serious, educational or other value."[31]

CHILD ONLINE PROTECTION ACT

Despite this strong endorsement of free speech, Congress passed the Child Online Protection Act (COPA), also known as "CDA-II," in October 1998. COPA was more narrowly drawn than the CDA. Nevertheless, free speech advocates still opposed it on First and Fifth Amendment grounds.

COPA made it a crime to permit anyone under the age of 17 to access materials that were construed as harmful to minors from the web. Any communication, image, recording, or information that was obscene, or met other criteria, would fall under this aegis. The latter included materials that

- "the average person, applying contemporary community standards, would find pandering to . . . prurient interests";
- depict ". . . in a manner patently offensive with respect to minors, an actual or simulated sexual act . . . or a lewd exhibition of the genitals . . .";
- "taken as a whole, lacks serious literary, artistic, political, or scientific value for minors."[32]

Additionally, an individual who "devotes time, attention, or labor to such activities [for example, making such materials available to minors], as a regular course of . . . trade or business, with the objective of earning a profit . . ." would be liable.[33]

An individual, however, who acted in good faith to restrict minors from accessing proscribed online material would be exempt from prosecution. This would require the use of a credit card, adult access code, or other reasonable identification scheme.[34] In sum, COPA would apply in these situations:

1. When material harmful to minors is knowingly communicated.
2. When the communication is made for commercial purposes.
3. When the material is communicated via the web.
4. When you are the person actually posting the illegal material.

Note also that COPA would still permit adults to receive "illegal" materials. This would be possible through the use of one of the aforementioned methods of identification.[35]

Objection

Despite COPA's built-in exemptions, the ACLU, the Electronic Frontier Foundation, and other parties filed suit before the U.S. District Court for the Eastern District of Pennsylvania in November 1998. They requested a TRO to halt enforcement of COPA. Much like the CDA, U.S. District Court Judge Lowell A. Reed, Jr., had to answer pertinent questions, including the following:

- Based on the presented information, was it likely that the ACLU and the other suing parties would win if the case went to trial?
- Would less harm result by granting the TRO or by denying it?[36]

After deliberating, Judge Reed granted a preliminary TRO against government enforcement of the COPA. The TRO was initially granted for 14 days and then extended for an additional 2 months until relevant testimony could be heard.[37]

The Case

The ACLU argued COPA was unconstitutionally vague under the First and Fifth Amendments. It was also overbroad, created an economic and technological burden for online speakers, would chill speech, and violated speech protected for minors. Attorney General Reno and the Justice Department countered and argued that COPA was constitutional. It was narrowly tailored, only targeted materials harmful to minors, and would not prevent adults from gaining access to such materials through a technologically and economically feasible identification system.[38]

Eventually, Judge Reed ruled the government had failed to show that COPA required the least restrictive means available to prevent access to materials that were harmful to minors. Furthermore, "The plaintiffs'. . . fears of prosecution under COPA will result in the self-censorship of their online materials in an effort to avoid prosecution, and this court has concluded . . . such fears are reasonable given the breadth of the statute."[39] Reed concluded by writing that no one, including the government, has an interest in enforcing an unconstitutional law and that the public is not served by the enforcement of such a law.[40] Based on this rationale, Judge Reed formally granted a TRO against enforcement of the COPA (February 1, 1999).

The Justice Department appealed the decision. But unlike the CDA, which had a fast track to the Supreme Court, the COPA may slowly wind its way through the federal appeals process.[41]

Finally, the crux of the CDA and COPA debates is not new, but rather, one that links this debate with the new technologies. Where do you draw the line in restricting information that adults can "legally" access? When is free speech diminished? How do you ensure that children are protected from online abuses? It is a balancing act, but one in which free speech concerns must be adequately addressed while ensuring that children are afforded their rights. As of this writing, this national and international debate continues.

Other Issues

We should note that the CDA and COPA were not the only online obscenity cases. While they ran into legal roadblocks, other obscenity regulations still apply. To date, most of these online criminal obscenity cases have involved online interstate transportation (from one state to another state) of child pornography and have been decided by lower courts.

The most highly publicized case is *United States v. Robert and Carleen Thomas.* Husband and wife Robert and Carleen Thomas, were convicted of federal obscenity charges (18 U.S.C. Section 1462/1465, 1996) in the U.S. District Court for the Western District of Tennessee. They used their bulletin board system to transport obscene computer files from California to other states. The recipients included an undercover agent in Tennessee.[42]

The Thomas's appealed, and the Court of Appeals for the Sixth Circuit agreed to review the case. After careful scrutiny of the lower court decision, the Sixth Circuit Judges ruled the convictions should stand.[43]

FILTERING

A sidebar online obscenity issue also arose. Should mandatory software be installed on public library computers to block access to Internet sites with sexually explicit content? Various library boards installed or considered installing blocking software to prevent minors from accessing various categories of

speech including, "hate speech, criminal activity, sexually explicit speech, adult speech, violent speech, religious speech, and even sports and entertainment."[44]

The ACLU, the American Library Association, and others have opposed such software use. They believe that library computers provide links to diverse and global information, which is subsequently made available to all segments of the American public. This includes users who might not otherwise have access to the Internet.[45]

Recent statistics indicate that 60 percent of the nearly 9000 U.S. public libraries provide Internet access to their patrons. A Nielsen survey further reveals that many people specifically go to the library to use the Internet.[46] In many instances, users opt to go online because it affords them more privacy when accessing information on assisted suicide and other, sensitive and controversial topics.

Library blocking software opponents generally agree that children should not be able to gain access to sexually explicit information, but believe that parents, teachers, and librarians should guide their Internet use—not restrictive and oftentimes flawed filtering software.[47] Proponents include library boards, the American Family Association, parents, librarians, and politicians.[48] They indicate the Internet can provide information, images, and chat rooms that are inappropriate for minors—the online world can be a dangerous place for children. They believe that blocking software could help protect minors from these pitfalls.

A Case and Congress

The first case to examine the constitutionality of using blocking software on public library computers was heard in 1998 by the U.S. District Court for the Eastern District of Virginia. In *Mainstream Loudoun et al. v. Board of Trustees of the Loudoun County Library,* Judge Leonie M. Brinkema had to decide if the library board policy requiring mandatory blocking of child pornography, obscene material, and material harmful to juveniles violated the First Amendment.[49]

Citing *ACLU v. Reno*, the precedent-setting online obscenity case, Judge Brinkema ruled that this policy was unconstitutional for these reasons:

- The policy was not narrowly tailored.
- There were less restrictive options to prevent minors from accessing online sexually explicit material.
- The policy restricted adults from obtaining materials that were deemed inappropriate for children.[50]

The Loudoun County Library Board considered appealing the decision, but the ACLU convinced them that their policy was unconstitutional. Instead, the board developed a new policy that allows both adults and minors to use either a filtered or an unfiltered computer.[51] While the ACLU has also convinced other library boards not to implement blocking software, library communities in some states continue the practice of library computer Internet filtering.

Congress has also played a role in this controversy, and legislative sessions continue to produce blocking software bills. The McCain bill (106th Congress) is one such example. It would require public libraries and schools, which received certain federal support, to use blocking software to protect minors from online pornography and child predators.[52] The ACLU and other free speech groups oppose these bills since they generally:

- Restrict access to protected speech.
- Allow the federal government not parents and teachers to guide a child's Internet use.
- Prevent minors who do not have a home computer from full access to the Internet.[53]

Since 1995, 25 states have either enacted or are considering the passage of state Internet censorship legislation. This is in addition to the proposed federal legislation.

To date, federal courts have struck down state Internet censorship laws in New York and Georgia. But similar legislative initiatives continue in other states.[54]

INTERNATIONAL RESPONSE TO ONLINE OBSCENITY

CompuServe

Besides the United States, other countries have developed their own rules governing sexually explicit online material. For an online company, this situation has become a legal nightmare. If an operation is international in scope, how do you comply with each country's laws?

For example, in December 1995, the German government informed CompuServe that a number of its sexually oriented newsgroups contained materials that were unsuitable for children and violated German law. A German prosecutor insisted that CompuServe block this access in Germany.

CompuServe informed the German government it was not responsible for, and had no authority over, a newsgroup's content. Nevertheless, CompuServe's employees working in Germany were threatened with arrest if blocking was not initiated.[55]

CompuServe's CEO had to decide if he should comply with German law or keep the newsgroups and risk his employees being arrested. In the end, lacking the technological means to isolate German users, CompuServe was forced to block access to these newsgroups for all its subscribers. Basically, German obscenity standards were globally imposed.[56]

To some individuals, the German government's decision was an echo of its past, a time when publishers were ordered to halt distribution of certain information based on its content. Some saw the German prosecutor's action as a "digital book-burning" with global implications.[57] "What could have been the most democratic medium in the world all of a sudden becomes the lowest common denominator of what is acceptable to all governments."[58]

The Germans denied they forced CompuServe to block the newsgroups and claimed they simply informed the company that local German law would be enforced. They also said they were not trying to make their standard a global legal standard.

CompuServe was criticized for its actions. Some individuals believed the company should have anticipated this problem. It had more than a million non-U.S. users and a history of similar problems with Germany prior to the 1995 incident.[59]

Given this past history, critics claimed that CompuServe and other U.S. online services should have been prepared for this latest legal salvo. CompuServe's lack of insight was costly, and its technicians were under pressure to quickly find a solution to this problem.

The Case

While CompuServe blocked these newsrooms until suitable parental control software could be deployed, the chief German prosecutor indicted Felix Somm, the managing director of CompuServe in Germany, in 1997. The charge? Allowing the trafficking of violence, child pornography, and bestiality via various CompuServe newsgroups during 1995 and 1996.[60] These indictments mark the first time that Western authorities have filed charges against a commercial online service based on the content of material the service did not create.[61]

In May 1998, Somm was found guilty on 13 counts of assisting in the dissemination of pornographic writings and for negligent dissemination of publications morally harmful to youths (*People of Munich v. Felix Somm*).[62] The court ruled that Somm had violated German law that criminalizes making ". . . 'simple' pornography (softcore nudity) available to children and 'hardcore' (child pornography and bestiality) available to anyone, . . . and with allowing the dissemination of nazi, neo-nazi materials and violent computer games . . . in Germany."[63]

Judge Wilhelm Hubbert found Somm guilty on all 13 counts of complicity, handing down a 2-year suspended jail sentence with probation. CompuServe appealed the conviction, and in November 1999, a 3-day hearing was convened in Munich.[64]

At this time, Chief Judge Laszlo Ember reversed the lower court, overturning Somm's conviction. He ruled that, at the time of the violation, CompuServe did not have the

technology to block the publication of the cited material. The judge further held that Somm had done all he could and he was merely a slave to his parent company. Finally, Judge Ember cited Germany's newly enacted multimedia law as an additional legal rationale for reversing Somm's conviction since it would only hold a service provider legally liable for material if it had the technical capability to block it.[65]

Other Countries

Online censorship has not been restricted to any one part of the world, however. Government control over online content is a global problem.

In the Middle East, various governments are attempting to control user access to sexual, political, and religious online information. For example, consider these situations:

- Jordan has installed a costly filtering system to block sites containing sexually explicit content.
- Kuwait is developing its Internet capabilities, but has stressed that pornographic and politically subversive material will not be permitted.
- In Saudi Arabia, only universities and medical institutions have Internet access. The rationale for this action is to shield the general public from pornographic web sites and sites that are construed as harmful.[66]
- Iran has mandated the use of filters to block access to web sites with sexual, religious, and political content. In one case, medical students cannot gain access to sites that contain information concerning the human anatomy.[67]

Internet access in China has also been tightly controlled. Users are monitored, and according to Chinese law, are expected to register with government authorities. Violations are treated seriously with possible imprisonment. For instance, in 1999, a computer technician received a 2-year jail sentence for providing a dissident web site with the e-mail addresses of 30,000 Chinese online subscribers. Chinese officials have also shut down 300 cybercafés to keep users from accessing controversial information.[68]

The Malaysian government also regulates access to Internet content. Individuals who violate its policies can be imprisoned. In one case, three Malaysian online users were arrested, according to authorities, for spreading false reports concerning internal country rioting. This criminal violation carries a 2-year prison sentence.

In Turkey, an Internet user received a 10-month suspended sentence for using an online forum for criticizing the police for roughly treating a group of blind protesters. Russian authorities, for their part, are developing a plan that would monitor all information sent via the Internet within Russian borders.[69]

In sum, proposed and adopted Internet regulation is an international situation—from the United States to Russia. It is also ironic. For the first time in human history, millions of individuals can readily exchange ideas and information by tapping the Internet and other information sources. But now, government regulation may stifle this free and open forum.

ONLINE LIBEL

Besides examining online obscenity, it is important to explore First Amendment implications with regard to online libel. As described in previous chapters, the online universe is an important part of our information and communications infrastructure. Yet with the growing list of applications and information pools, there are potential legal hot spots. These include defining online libel and determining if an online service should be treated as a distributor or a publisher. But before we begin, the following terms, as used in this chapter, must be defined.

- *Libel*: Any expression that damages your reputation within the community, causes others to disassociate themselves from you, or attacks your character or professional ability.[70]

- *Publisher and distributor*: The CompuServe and Prodigy cases, covered in the next section, reflect how online libel has been handled. A critical point was determining if a service was either a publisher (originator) or an information distributor. A publisher of a libelous remark is legally responsible for the act. A distributor is not legally responsible if it was made by a third party without the distributor's knowledge or consent.[71]

The Cases

CompuServe.   In *Cubby Inc. v. CompuServe Inc.*, the U.S. District Court for the Southern District of New York (19 Med. L. Rptr. 1525. 1991), had to decide if CompuServe had acted as a publisher or distributor of libelous material. Specifically, was CompuServe legally responsible for remarks published by a third party in *Rumorville USA*, a component of the journalist forum? *Rumorville* was a daily, electronic newspaper included on the forum.

CompuServe independently contracted with a third party to "'manage, review, create, delete, edit and otherwise control the contents' of the Journalism Forum 'in accordance with editorial and technical standards and conventions of style established by CompuServe.'"[72] Furthermore,

- CompuServe had no contractual or direct tie with the forum or Rumorville.
- The third party accepted full responsibility for Rumorville's content.
- The setup made it technically unfeasible for CompuServe to monitor Rumorville prior to its uploading.
- CompuServe did not receive money for Rumorville or pay a fee for its inclusion on the service. The only fee was the standard subscriber charge.[73]

In presenting its case, CompuServe did not question whether the statements were libelous. Rather, it argued that like a news vendor, bookstore, and library, it was a distributor and not a publisher. Based on current law, CompuServe could not be legally responsible for the libel committed by its independent contractor unless it played an active part in the libel.[74]

Cubby, in turn, alleged that CompuServe was a publisher and was responsible for publishing the libel. Based on the language of *Cianci v. New Times Publishing Co.* (639 F.2d. 54), anyone who repeats or "republishes" the original libel is considered a publisher who is equally guilty and can also be sued.[75]

The court disagreed and found that CompuServe should be treated as a distributor. It compared CompuServe's services to an electronic, for-profit library that provided its subscribers with access to numerous publications. The court further held that it was impossible for CompuServe to be aware of the content of all its information services.

For instance, although CompuServe can decline to carry a publication, once it does decide to carry one, it has little or no editorial control over its content. This is especially true when the publication is part of a forum that is managed by an unrelated company.

CompuServe has no more editorial control over a publication than does a public library, bookstore, or newsstand, and it would be no more feasible for CompuServe to examine every publication it carries for potentially defamatory statements than it would be for any other distributor to do so.[76]

The court cited *Smith v. California* (361 U.S. 147) and other cases as precedents. In *Smith*, the U.S. Supreme Court struck down a law that would hold a bookstore owner legally responsible for carrying an obscene book even though the owner was unaware of the book's content. "Every bookseller would be placed under an obligation to make himself aware of the contents of every book in his shop. . . . And the bookseller's burden would become the public's burden, for by restricting him the public's access to reading matter would be restricted.'"[77] Consequently, the court found that CompuServe had acted as a distributor and not a publisher. As such, it was not guilty of libeling Cubby.

Prodigy. In *Stratton Oakmont et al. v. Prodigy Inc.* (63 USWL. 2765. 1995), the New York

Supreme Court examined whether Prodigy was the publisher or distributor of the online libel of Stratton Oakmont, Inc. Specifically, did Prodigy act as the publisher of the online libel that appeared on *Money Talk*, a financial investment and information resource?

The court was also asked to determine if *Money Talk's* board leader (BL) had acted in an official capacity as Prodigy's representative. BLs sign an agreement with Prodigy that spells out their responsibilities. These include an obligation to work with members using the service and to follow Prodigy's procedures.

Stratton Oakmont also claimed that Prodigy promoted itself as a service that exercised some degree of editorial control. For example:

Users are requested to refrain from posting notes that are 'insulting' and are advised that notes that harass other members or are deemed to be in bad taste or grossly repugnant to community standards, or are deemed harmful to maintaining a harmonious online community, will be removed when brought to PRODIGY's attention; the Guidelines all expressly state that although 'Prodigy is committed to an open debate and discussion on the bulletin boards, . . . this doesn't mean that anything goes.'[78]

Prodigy also employed a software system that prescreened bulletin board content for offensive material. BLs could also use an emergency deletion function.[79] Consequently, because Prodigy exercised this degree of editorial control, Stratton Oakmont asked the court to declare the company a publisher and not just a distributor. As a publisher, Prodigy would be legally responsible for repeating or republishing the libel originating on Money Talk. In its defense, Prodigy claimed the court ruling in *Cubby v. CompuServe* set a precedent. It should be treated as a distributor.

In reaching its decision, the court distinguished the CompuServe case from the Prodigy case on three points:

1. CompuServe had no opportunity to monitor their boards' content prior to distribution.

2.   CompuServe had not promoted itself as a service that exercised Prodigy's type of editorial control.

3.   Prodigy used an automatic screening program and designed BL guidelines.[80]

Based on these factors, the court ruled that Prodigy was a publisher of the libel and was legally responsible. According to the court, Prodigy "has virtually created an editorial staff of Board Leaders who have the ability to continually monitor incoming transmissions and in fact do spend time censoring notes."[81]

The court further held that Prodigy's own policies, technology, and personnel decisions were responsible for the service's classification as a publisher. Its choice to gain the benefits of editorial control also "opened it up to a greater liability than CompuServe and other computer networks that make no such choice."[82]

The court then turned its attention toward deciding if *Money Talk*'s BL had acted as Prodigy's agent, for the purpose of the libel. The court said "Where one party retains a sufficient degree of direction and control over another, a principal-agent relationship exists."[83] Because Prodigy required BLs to follow their guidelines and performed administrative duties, the court ruled the BL was Prodigy's legal agent.

The 1996 Telecom Act

The *Prodigy* decision had a chilling effect in the online community, even though Prodigy was differentiated from CompuServe by its operational guidelines. The 1996 Telecom Act, however, included a "Good Samaritan" blocking and screening clause concerning offensive material: "No provider or user of an interactive computer service shall be treated as the publisher or speaker of any information provided by another information content provider."[84] An information content provider is an information producer.

Similarly, a provider (for example, Prodigy) or a user would not be legally responsible for "any action voluntarily taken in good faith to restrict access to or availability of material that the provider or user considers to

be obscene, lewd, lascivious, filthy, excessively violent, harassing, or otherwise objectionable, whether or not such material is constitutionally protected . . . ."[85] This is exercising editorial control.

Libelous remarks, another key point, are said to fall within the "otherwise objectionable material" clause of this section of the act. If this holds up, legal scholars, media organizations, and public interest groups have indicated the *Prodigy* decision would be overturned.[86]

In sum, the "Good Samaritan" defense may have prevented Prodigy from being punished for trying to police its service, that is acting as a publisher. To encourage this type of behavior, Congress also refused to hold Prodigy and other such companies to a higher standard than companies that do not police their services.[87]

However, despite this defense, an online service that was aware of a libelous remark and intentionally and/or recklessly "repeated" or "republished" it, would not be protected. An online service could also be held legally responsible for any libelous statement that it originated.

The Cases

Since the passage of the 1996 Telecom Act, online libel cases have been heard by federal and state courts. Two important federal cases are *Kenneth M. Zeran v. America Online* (958 F. Supp. 1124. 1997. 129 F. 2d 327. 1997) and *Sidney Blumenthal et al. v. Matt Drudge and America Online* (992 F. Supp. 44. 1998). *Zeran v. AOL* was heard by the U.S. District Court for the Eastern District of Virginia and by the Fourth Circuit Court of Appeals. The case was appealed to the U.S. Supreme Court but was not accepted for review.[88] *Blumenthal v. Drudge and AOL* was heard by the U.S. District Court for the District of Columbia and is being appealed to the circuit court.[89]

Zeran v. AOL.   In *Zeran v. AOL*, the district court had to decide if AOL, a service provider, could be held legally responsible for libelous remarks made by a third party against Kenneth Zeran. An AOL subscriber, posing as Kenneth Zeran, advertised the sale of t-shirts and other items that contained slogans praising the murder of the 168 Oklahoma City bombing victims. This offensive online message, which included Zeran's home telephone number, was posted without his knowledge or consent. Zeran became aware of the message when he began receiving abusive and threatening telephone calls from outraged AOL users.[90]

Upon discovering the message, Zeran contacted AOL demanding its prompt removal and a retraction identifying him as the victim of the online hoax. AOL deleted the message but refused to post a retraction—this did not fall within company policy. Unfortunately for Zeran, additional notices of a similar nature continued to appear on AOL for several more days and the threatening telephone calls continued.[91]

Zeran responded by suing AOL for negligence. He claimed that AOL failed to adequately respond to the online hoax after it was contacted. He also contended that AOL was legally responsible for the libelous message posted by an unknown third party on its service.[92]

Zeran based his legal claim on Virginia state law that permits legal action against the distributor of libelous material if the information provider knew or should have known of its existence. AOL, on the other hand, argued to the court that the 1996 Telecom Act preempted Virginia state libel law and as such, no service provider could be held legally responsible for the libelous remarks of a third party.[93]

After considering each party's complaint, the U.S. district court ruled in favor of AOL on the grounds that treating the service provider as the publisher of the online libel violated section 230 of the Telecom Act. Zeran responded by appealing the case to the Fourth Circuit Court of Appeals.[94]

On appeal, Chief Judge Wilkinson affirmed the decision of the lower court, holding that, "by its plain language, section 230 creates a federal immunity to any cause of action that would make service providers liable for information originating with a third party user of the service."[95] Thus, the Fourth Circuit Court

of Appeals upheld the language of the 1996 Telecom Act, and the U.S. Supreme Court refused to accept the case for review.

Blumenthal v. Matt Drudge and AOL. In 1998, another important online libel case challenging the 1996 Telecom Act was heard by a federal court. The case involved top Clinton aide Sidney Blumenthal, political reporter Matt Drudge, and AOL, the service provider for Drudge's electronic reports.[96] This case is of particular importance since it marks the first time the courts have examined section 230 in a case where the online service provider paid the content provider for its information.[97]

Blumenthal filed suit against Matt Drudge and AOL for allegedly libelous remarks made in the Drudge Report. AOL subsequently removed all the remarks from past editions of the Report, which were stored in its electronic archives, and Drudge retracted the story.[98] Nevertheless, Blumenthal proceeded with the lawsuit.

Blumenthal contended that AOL was also responsible for these reasons:

- The company was more than a mere service provider. Drudge was paid a flat monthly royalty payment of $3000 by AOL for his online reports.[99]
- The remarks did not come from an unknown person sent over the Internet anonymously.[100]
- "AOL reserves the right to remove, or direct [Drudge] to remove, any content which, as reasonably determined by AOL . . . violates AOL's . . . terms of service."[101]

AOL responded by filing a court motion to have itself dismissed from the suit. This would leave Drudge solely responsible for the allegedly libelous remarks. Citing section 230 of the 1996 Telecom Act, AOL argued that as a service provider, it was immune from third-party remarks.[102]

U.S. District Court Judge Paul Friedman granted AOL's request for dismissal from the suit, but included an opinion that was highly critical of AOL. Friedman clearly agreed, however, that the language of section 230 protected AOL from liability. He wrote that Congress had provided, "... immunity even where the interactive service provider has an active, even aggressive role in making available content prepared by others."[103] Sidney Blumenthal is appealing Judge Friedman's decision.

State Cases

Two other online libel cases, *Kempf v. Time, Inc.* (No. BC 184799. California. 1998) and *Ben Ezra, Weinstein & Co. v. America Online* (No. 97-485. N.M. 1999), were heard in 1998 and 1999, respectively, by state courts. In *Kempf v. Time*, a California court dismissed a libel suit against the service provider, Time, Inc., on the grounds that it was not involved in creating or developing the libelous content in question and was, in fact, protected according to the language of section 230 of the 1996 Telecom Act.[104]

In 1999, a New Mexico court deciding *Ben Ezra, Weinstein & Co. v. AOL* ruled that as a service provider, AOL was not legally responsible for false and libelous online remarks concerning the company's financial situation. The court, citing section 230 of the 1996 Telecom Act, ruled in AOL's favor stating it had Internet service provider immunity.[105]

Summary

As of this writing, since the passage of the 1996 Telecom Act, courts have upheld the immunity protection afforded to Internet service providers. This is based on the language contained in section 230 of the act.

It may also be a mixed bag of benefits, depending on which side of the fence you are on. A company was now protected from carrying libelous statements made by a third party. On the flip side, some individuals believed the clause removed all responsibilities from a company even if its policies permitted the removal of content, as was the case with AOL.

CONCLUSION

First Amendment rights and the legal system are key components of the communication revolution. At times, it is almost a tug of war—

how do you preserve your constitutional rights while protecting children and someone's good name? It is also a complex issue.

The controversy surrounding filtering software, for instance, does not solely involve libraries and children. Businesses have experienced their own browsing problem—employees spending work time cruising the Internet. This raises efficiency and potentially legal issues. For the latter, if "sexually explicit material" is left on a computer and observed by a coworker, this could trigger a sexual harassment lawsuit. Consequently, filtering programs have also being slated for the workplace.[106]

Other issues continue to crop up. Online stalking (*cyberstalking*), for example, has become more prevalent. As implied, you could be the target of harassing e-mail and an individual who follows you around in cyberspace—impersonating you in a chat room to incite other members to react to typically volatile information. As the target, you could receive hostile e-mail responses. This is only the tip of the proverbial iceberg.

You should also think about how you use e-mail. It has, to a great extent, become the electronic/written equivalent of picking up a telephone. But now, you can reach tens—hundreds—even thousands of sites and people at the push of a button. Cyberstalkers have tapped this capability to harass and threaten others.

Another factor is e-mail's anonymity. As stated in a government report, "whereas a potential stalker may be unwilling or unable to confront a victim in person or on the telephone, he or she may have little hesitation sending harassing or threatening electronic communications to a victim."[107] Geographical and time elements are also removed, and you can sign-up for free e-mail without any verification on the company's part.

Legally, what can you do? As of this writing, not all states have online antistalking laws. In essence, unless there is a direct threat to your life, or other such serious communication, the harassment may continue.

Another factor is this crime's fairly recent appearance. In some cases, a local law enforcement agency may be ill equipped (training-wise) to handle the situation.

In response, cyberstalking support groups have appeared on the Internet. It is also important for states to adopt appropriate legislation. California and Pennsylvania are pioneers in this area. Other locales have or are setting up special computer crime units.

As has been stated, it is a balancing act. You must protect an individual's rights, including First Amendment rights. But the legal system should also be responsive to individuals and what may initially be novel situations that are the fallout of the communication revolution.

REFERENCES/NOTES

1. Brock Meeks, "The Obscenity of Decency," online document, 1995, 1.
2. Ibid.
3. Telecommunications Act of 1996. P.L. 104-104, 66.
4. Electronic Frontier Foundation, "Constitutional Problems with the Communications Decency Amendment: A Legislative Analysis." Downloaded from www.eff.org, June 16, 1995.
5. Ibid.
6. *Miller v. California* (413 U.S. 15. 1973).
7. *Sable Communications v. Federal Communications Commission* (492 U.S. 115. 1989).
8. Electronic Frontier Foundation, "Constitutional Problems," 4.
9. Ibid.
10. *American Civil Liberties Union v. Attorney General Janet Reno,* U.S. District Court Eastern District of Pennsylvania, June 11, 1996. No. 96-963. No. 96-1458. 14.
11. Ibid, 16.
12. Ibid, 3.
13. Ibid, 3.
14. Ibid, 2.
15. Ibid, 4.
16. Ibid, 33.
17. Ibid, 31.
18. Ibid, 33.
19. Ibid, 32. *Note:* Twenty-five percent of all new HIV cases in the United States occur in the 13- to 20-year-old age group.
20. Ibid, 32. *Note:* This section was concerned with the community standards/vagueness issue.
21. Ibid, 35.
22. Ibid, 36. *Note:* The speaker was also charged with the responsibility for screening potentially inappropriate online material. Therefore, content providers would have been required to decide what would constitute inappropriate materials for minors.
23. Ibid, 37.
24. Ibid, 37.
25. Ibid, 65.
26. Ibid, 66.
27. Ibid, 65.
28. Ibid, 65.
29. *Reno et al. v. American Civil Liberties Union et al.* 117 S.Ct. 2329. 1997. Downloaded from www.aclu.org.
30. Ibid.
31. Ibid.
32. *American Civil Liberties Union et al. v Janet Reno et al.* No. 98-5591. U.S. District Court Eastern District of Pennsylvania, February 1, 1999. Downloaded from www.aclu.org.
33. Ibid.
34. Ibid.
35. "ACLU Files Suit Challenging the Child Online Protection Act," *Tech Law Journal.* Downloaded from www.techlawjournal.com, October 23, 1998.
36. *American Civil Liberties Union v. Janet Reno.* No. 98-5591. U.S. District Court Eastern District of Pennsylvania, November 20, 1998. Downloaded from www.aclu.org.
37. Ibid.
38. *ACLU v. Reno.* No. 98-5591, February 1, 1999. Downloaded from www.aclu.org.
39. Ibid.
40. Ibid.
41. ACLU. "*ACLU v. Reno,* Round 2: Internet Censorship Battle Moves to Appeals Court," April 2, 1999. Downloaded from www.aclu.org.
42. *United States v. Robert Thomas and Carleen Thomas* (1996 WL. 30477. Sixth Circuit Tennessee. January 29, 1996.), 1.
43. Ibid.
44. ACLU. Cyber-Liberties. "Censorship in a Box: Why Blocking Software Is Wrong for Public Libraries," 1998. Downloaded from www.aclu.org.
45. Ibid.

46. Ibid.

47. Ibid.

48. Ibid.

49. *Mainstream Loudoun v. Board of Trustees of the Loudoun County Library.* No. 97-2049-A. 1998, *Tech Law Journal.* Downloaded from www.techlawjournal.com.

50. Ibid.

51. ACLU. Cyber-Liberties. "Censorship in a Box."

52. Ibid.

53. ACLU. Action Alert. "Congress Tries to Force the Use of Clumsy Filtering Software," February 1999. Downloaded from www.aclu.org.

54. ACLU. Cyber-Liberties. "Censorship in a Box."

55. Christopher Stern, "CompuServe Shuts Down Sex After German Protest," *Broadcasting & Cable* (January 8, 1996): 69.

56. Steven Vonder Haar, "Censorship Wave Spreading Globally," *Inter@ctive Week* 3 (February 12, 1996): 6.

57. Ibid.

58. Stern, "CompuServe Shuts Down Sex After German Protest," 69.

59. Ibid.

60. In the Name of the People Judgment of the Local Court Munich in the Criminal Case v. Somm, Felix Bruno. Local Court [Amtsgericht] Munich. File No. 8340 Ds 465 Js 173158/95.1998. Downloaded from www.cyber-rights.org, March 2000.

61. Stuart Biegel, "Indictment of CompuServe Official in Germany Brings Volatile Issues of Cyber-jurisdiction Into Focus," UCLA Online Institute for Cyberspace Law and Policy. 1997. Downloaded from www.gse.ucla.edu.

62. *People v. Somm.*

63. Andrew Gelman, "Sting Time for CompuServe in Germany: Online Service Head Indicted for Pornography." 1997. Downloaded from www.excitesearch.netscape.com.

64. "Update: CompuServe Ex-official's Porn-Case Conviction Reversed," Associated Press. November 17, 1999. Cyber-Rights & Cyber-Liberties (UK). Downloaded from www.cyber-rights.org.

65. Edmund L. Andrews. "German Court Overturns Pornography Ruling Against CompuServe," *New York Times* on the web, November 18, 1999. Downloaded from www.nytimes.com.

66. The Middle East, downloaded from www.cwrl.utexas.edu.

67. Letters of Protest and Press Release., Twenty Enemies of the Internet, downloaded from www.rsf.fr/uk/.

68. Ibid.

69. Special Issues and Campaigns. Freedom of Expression on the Internet, Human Rights Watch, 1999, downloaded from www.hrw.org.

70. Kent Middleton and Bill Chamberlin. *The Law of Public Communication* (White Plains, N.Y.: Longman Publishing, 1994), 70.

71. *Cubby Inc. v. CompuServe Inc.,* U.S. District Court for the Southern District of New York (19 Med. L. Rprt. 1527. 1991).

72. Ibid, 1526.

73. Ibid, 1526.

74. Ibid, 1528.

75. Ibid, 1527.

76. Ibid, 1528.

77. Ibid, 1528.

78. *Stratton Oakmont et al. v. Prodigy Inc.,* Supreme Court of New York (63 USWL. 2767. 1995).

79. Ibid, 2767.

80. Ibid, 2767.

81. Ibid, 2770.

82. Ibid, 2770.

83. Ibid, 2772.

84. Telecommunications Act of 1996, 71.

85. Ibid.

86. Kenneth Salomon and Todd Gray, "Online Content Liability," *American Council of Education* (February 14, 1996): 6.

87. Ibid.

88. (U.S. Supreme Court. Cert. Pet. 97-1488. 1997). "*Zeran v. America Online,*" Tech Law Journal, 1998. Downloaded from www.techlawjournal.com.

89. "Judge Friedman Grants Summary Judgment to AOL in Defamation Case," *Tech Law Journal,* April 24, 1998. Downloaded from www.techlawjournal.com.

90. *Kenneth M. Zeran v. AOL.* 958 F. Supp. 1124. 1997. Downloaded from www.loundy.com.

91. Ibid.

92. Ibid.

93. Ibid.

94. *Kenneth M. Zeran v. AOL.* 129 F 2d 327. 1997. Downloaded from www.law.emory.edu.

95. Ibid.

96. *Sidney Blumenthal et al. v. Matt Drudge and AOL.* 992 F. Supp. 44. Tech Law Journal, 1998. Downloaded from www.techlawjournal .com.

97. "Judge Friedman Grants Summary Judgment to AOL."

98. Ibid.

99. Ibid.

100. *Sidney Blumenthal v. Drudge and AOL.*

101. Ibid.

102. Ibid.

103. Ibid.

104. The Perkins Coie Internet Digest. "Defamation," 1999. Downloaded from www .perkinscoie.com.

105. Ibid.

106. Rosalind Retkwa, "Corporate Censors," *Internet World* 7 (September 1996): 62.

107. U.S. Department of Justice, "1999 Report on Cyberstalking: A New Challenge for Law Enforcement and Industry." Downloaded from www.usdoj.gov.

SUGGESTED READINGS

Case Law

American Civil Liberties Union v. Attorney General Janet Reno, U.S. District Court for the Eastern District of Pennsylvania. June 11, 1996. No. 96-963. No. 96-1458. 1–76.

Cubby v. CompuServe, U.S. District Court for the Southern District of New York (19 Med. L. Rprt. 1527. 1525-1533 1991).

Miller v. California (413 U.S. 15. 15-22. 1973).

Stratton Oakmont et al. v. Prodigy Inc., Supreme Court of New York (63 USWL. 2765-2773. 1995).

United States v. Robert Thomas and Carleen Thomas (1996 WL. 30477. Sixth Circuit Tennessee. January 29, 1996.), 1-68.

Legislation

Computer Security Act of 1987. P.L. 100-235. 101 Stat. 1724. 1724-1730.

Telecommunications Act of 1996. P.L. 104-104. 1-76.

Journals, Magazines, and Online Documents

ACLU. "*ACLU v. Reno,* Round 2: Broad Coalition Files Challenge to New Federal Net Censorship Law." October 22, 1998, downloaded from www.aclu.org.

ACLU. "*ACLU v. Reno,* Round 2: Internet Censorship Battle Moves to the Appeals Court." April 2, 1999, downloaded from www.aclu.org.

ACLU. "*ACLU v. Reno,* Round 2: Rejecting Cyber-Censorship, Court Defends Online 'Marketplace of Ideas.'" February 1, 1999: 1–2, downloaded from www.aclu.org.

ACLU. Civil-Liberties. "Online Censorship in the States." 1998: 1–4, downloaded from www .aclu.org.

"ACLU Files Suit Challenging the Online Protection Act." *Tech Law Journal.* October 23, 1998, downloaded from www.techlawjournal.com.

Amalfe, Christine A., and Kerrie R. Heslin. "Court Starts to Rule on Online Harassment." *The National Law Journal* (January 24, 2000). Downloaded from Lexis; Hatcher, Michael and Jay McDannel, et al. "Computer Crimes." *American Criminal Law Review* (Summer 1999) (36 Am. Crim. L. Rev. 397). Downloaded from Lexis; U.S. Department of Justice, "1999 Report on Cyberstalking: A New Challenge for Law Enforcement and Industry." Downloaded from www.usdoj.gov. Information on cyberstalking and hints for victims. *Note:* As stated in the chapter, numerous web sites offer support and information. The Department of Justice report is a good source of sites (for example, cyberangels at www.cyberangels.org).

Biegel, Stuart. UCLA. Online Institute for Cyberspace Law and Policy. "Indictment of CompuServe Official in Germany Brings Volatile Issues of Cyber-Jurisdiction into Focus." April 27, 1997, downloaded from www.gse.ucla .edu.

Boles, Garry. "Online Obscenity: Decency by Law." *Inter@ctive Week* 2 (April 10, 1995): 15–16.

Crocker, Steven. "The Political and Social Implications of the Net." Online document, July 5, 1994, 1–14.

"Defamation." Perkins Coie Internet Case Digest. 1999, downloaded from www.perkinscoie .com.

DiLeleo, Edward. "Functional Equivalency and Its Applications for Freedom of Speech on Computer Bulletin Boards." *Columbia Journal of Law and Social Problems* (Winter 1993): 199–247.

Electronic Frontier Foundation. "Constitutional Problems with the Communications Decency Amendment: A Legislative Analysis." June 16, 1995, 1–10.

Elmer-Dewitt, Philip. "On a Screen Near You: Cyberporn." Online document, July 3, 1995, 1–3.

Faucher, John. "Let the Chips Fall Where They May: Choice of Law in Computer Bulletin Board Defamation Cases." *Davis Law Review* (Summer 1993): 1045–1078.

Freedom Forum Online. "Online Libel Bibliography," downloaded from www.freedomforum .org.

Freedom Forum Online. "Justice Department Appeals Ruling That Blocked Enforcement of COPA." April 5, 1999, 1–2, downloaded from www.freedomforum.org.

"German Net Porn Conviction." *Wired News* (May 28, 1998): 1–2, downloaded from www.wired .com.

Godwin, Mike. "Internet Libel: Is the Provider Responsible?" *Internet World* (November/December 1993): 1–3.

Godwin, Mike. "Libel, Public Figures, and the Net." *Internet World.* Online document, June 1994, 1–4.

"Judge Friedman Grants Summary Judgment to AOL in Defamation Case." *Tech Law Journal* (April 24, 1998), downloaded from www.tech-lawjournal.com.

Lewis, Anthony. "Staving Off the Silencers." *New York Times Magazine* (December 1, 1991): 72–75.

"Loudoun Library Board Decides Not to Appeal Filtering Decision." *Tech Law Journal* (April 21, 1999): 1–3, downloaded from www.tech-lawjournal.com.

Steiner-Threlkeld, Tom, and Brock Meeks. "Free Speech Defenders Rally in Cyberspace." *Inter@ctive Week* 2 (March 13, 1995): 36–37, 59–60.

Stern, Christopher. "CompuServe Shuts Down Sex After German Protest." *Broadcasting & Cable* (January 8, 1996): 69.

GLOSSARY

Chilling Effect: Any regulation or threat of regulation that makes the speaker think twice about providing information that might be considered inappropriate by some. In such a situation, the speech is halted before it ever has a chance to be heard.

Communications Decency Act (CDA): Part of the Telecom Act of 1996. It was designed to regulate the use of obscene as well as indecent online material.

Filtering: Software used to block access to specific types of web sites (e.g., pornographic). Filtering has raised various legal questions.

Indecent Material: Nonobscene language that is sexually explicit or includes profanity.

Libel: Any expression that damages your reputation within the community, causes others to disassociate themselves from you, or attacks your character or professional ability.

Obscene Speech: Language that appeals to prurient interest, is patently offensive, and lacks serious literary, artistic, political, or scientific value.

Spectrum Scarcity: Theory based on the belief that since the electromagnetic spectrum is finite, only a limited number of individuals could receive a license.

Future Visions

This chapter serves three purposes. First, it introduces holography and virtual reality (VR), one older and one younger technology, both of which are still maturing. Second, it explores the growth of a paperless society, a growth that has been fueled by many of the technologies covered in the book. Third, it serves as a footnote, a way to wrap-up some of the book's major points.

The first three topics have been gathered under the future visions banner because they are a fitting end to our exploration of the communication revolution. They will help propel us toward a future with an array of new tools and services. From a nontechnical standpoint, they have also captured the popular imagination.

We finish with a summary that raises questions that will continue to challenge us as we head into the twenty-first century.

HOLOGRAPHY

Holographic techniques were discovered in the 1940s by Dr. Dennis Gabor. A research engineer, he won a Nobel Prize in 1971 for his pioneering work as the father and creator of holography. But the field did not fully blossom until the 1960s. The development of a suitable light source, the laser, helped provide the key.

Even though it is beyond our scope to explore holography's physical properties, we can still present a working definition. A *hologram*, a product of holography, can be considered a record of the optical information that composes a scene. It can store information about a three-dimensional object, and unlike a standard photograph, which

records light intensity . . . the hologram has the added information of phase . . . to show depth.

When a person looks at a tree . . . he is using his eyes to capture light bouncing off the object and then processing the information to give it meaning. A hologram is just a convenient way to recreate the same light waves that would come from an object if it were actually there.[1]

This capability can be startling. Objects are quite lifelike, and they may appear to "jump out" of a scene.

A standard photograph also records a scene from a single perspective. A hologram breaks these boundaries.

Finally, many holograms are created through a special film and are chemically processed. But newer techniques have enhanced this task.

Applications

A statue can be recorded as a cylindrically shaped hologram.[2] You can see the recorded

Figure 22.1

Corrosion monitored: an example of holographic interferometry. Holography can, in effect, let you visualize the interior of a pipe from the outside. The circular fringe pattern, due to a small pressure change, indicates weakened regions of the pipe wall. (Courtesy of the Newport Corp.)

statue as you would the original object—from various views. This process could prove valuable for preserving a 3-D object's characteristics and for distributing copies for research and pleasure (for example, viewing).

The advertising and security fields are also supported. Attention-grabbing ads can be produced, and small, inexpensive embossed holograms have been placed on credit cards for security purposes.[3]

Another application area is holographic interferometry. Unlike artistic endeavors, a conventional hologram is not created. Rather, in one variation, a double exposure of the same object is made. During one exposure, the object is stationary and at rest. In the second, the object is subject to stress through the introduction of the physical forces it may experience in actual use.[4]

When processed, the hologram will depict a series of fringes or lines across the object. They resemble the contour lines on a map. The fringes reveal the small differences between the exposures—the differences between the object while at rest and under stress.[5] Consequently, the hologram serves as a visual map of the areas that may be deformed or affected by the operation.

Holographic interferometry has been used to inspect products for manufacturing defects. Metals and other materials have also been subjected to stress to test for possible flaws.[6]

Holography has also provided us with the holographic optical element (HOE), a hologram that functions, for example, as a lens or mirror.[7] HOEs are especially useful when a standard optical component may not fit or may be too heavy.

In one application, a head-up display, an image of an instrument panel could be displayed before a pilot's eyes. The pilot would not have to move his or her head to view the instruments.

The holographic field may also bring massive data storage capabilities to the desktop. Its unique properties can make a holographic-based system a highly effective storage system.[8]

The future may also witness 3-D television. Experimental work has already been conducted, and if combined with CD-quality sound, the system would support realistic computer games as well as entertainment and educational programs.

Figure 22.2
A VR outfit. (Courtesy of NASA, Ames Research Center.)

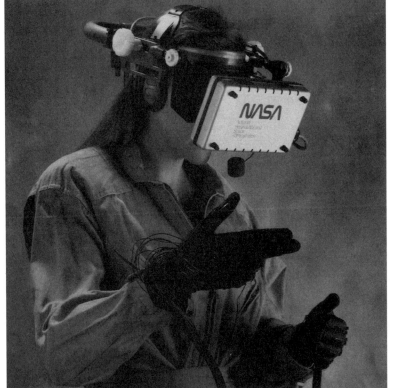

VIRTUAL REALITY

On the Threshold of a Dream

Virtual reality (VR) is a footnote to many of the book's topics. These range from computers to teleconferencing to information processing. The implications also cut across technological, social, and ethical lines:

In brief, VR can be defined as a display and control technology that can surround a person in an interactive computer-generated or computer-mediated virtual environment. Using head-tracked head-mounted displays, gesture trackers, and 3-D sound, it creates an artificial world of visual . . . and auditory experience. With a digital model of an environment, it creates an artificial place to be explored with virtual objects to be manipulated.[9]

As just described by Michael W. McGreevy, a VR pioneer, you can use a system to enter into and interact with a computer-generated environment. The passage to this realm can be through different VR platforms.

In the configuration we are probably the most familiar with, you wear special "goggles" to view the computer-generated images and a glove to navigate through and interact with this environment.[10] Realistic sound may also be a key element, and head motions are tracked so the scene shifts as you move your head.

Outfitted with this equipment, the real world can be blocked out as you are fed images of a room. You can move through this space with a gesture of your hand. You can also grab and move objects and even turn wall switches on and off. The objects can act as they would in real life. But in this reality, the action is taking place in a virtual world, the world generated by the computer.

Applications

The early 1990s saw the full emergence of VR on the public stage in books, magazines, and television reports. Some sample applications include the following:

1. *Telepresence.* Telepresence is the "ability to interact in a distant environment through robotic technology."[11] In one example, NASA is experimenting with remote robotic systems, controlled by distant operators, for space-based construction and repairs. Instead of simply pressing buttons to carry out commands, the robotic device becomes a natural extension of its human operator.

2. *Teleconferencing.* With conventional teleconferencing, you may only see another individual on a monitor. In a VR setup, your virtual image could shake hands with your counterpart's virtual image. You could also "meet" in New York City or on a tropical beach.

3. *Architecture.* You could walk through a building before it is constructed, so you could judge the design's effectiveness, and if

necessary, make modifications. This may be one of the closest sensations to living in a building and trying it out before it is built. A VR system could also be used to help design buildings that better accommodate the needs of individuals with disabilities (for example, someone who uses a wheelchair).[12]

4. *Medicine.* The medical applications are numerous. A surgeon could practice a procedure with a virtual patient. A medical student could take a tour through the human body.

In a more specialized application, VR has been used to study a disease's molecular structure and "to design drug molecules which could shut down its defense systems."[13] The setup allowed the researchers to see, and in a sense, to experience the data in new ways. The data were not simply viewed on a flat, conventional monitor.

5. *Education.* Applications include the medical tour, meeting historical figures, and investigating an ancient city or even the Solar System.

6. *Personal freedom.* This is a potent VR application. You could assume a new identity or enter environments you may never be able to experience in real life. NASA, for instance, is already using outer space probe data to explore other worlds. Although scenery generator programs can create and display a realistic scene on a monitor, a VR system pushes this application to a higher level. For example, as you turn your head, the scene realistically shifts.

7. *Other applications.* Other application areas include military war games, enhanced scientific visualization, art, and virtual sex. A VR system could also serve as a more natural human–computer interface, especially if it incorporates speech recognition capabilities.

If you like video games, you could be in for the ride of your life. You could be immersed in the action. Early video game users had a taste of this potential through Mattel's Power Glove for the Nintendo system. VR arcades have also been launched.[14]

While not strictly VR, *virtual sets* have been adopted by the broadcast industry. Instead of building elaborate and expensive sets, you can use a computer-generated environment. You

Figure 22.3

The STAR*TRAK RF wireless motion capture system. It can capture the motion of two performers in three-dimensional space. Interfaces of this type enhance our human–machine capabilities. In one application, you could capture a golf swing or dance steps for various applications. (Courtesy of Polhemus, Inc.)

can also tap effects and capabilities you may not be able to match with a "real" set. CBS News, in fact, touted this feature in its 1996 presidential election coverage.

Like the book's other topics, VR raises ethical and social questions. In one case, a VR configuration could be used for torture. In another example, some individuals have claimed that VR may have a powerful addictive and "intelligence-dulling" power, much like contemporary television.[15]

Other concerns have touched on its potentially insulating quality. If we use realistic simulations, could we become desensitized to a war in the real world where real people die?

In essence, could war turn into a VR game? As a precedent, many of the bombing runs in Operation Desert Storm, as portrayed on television, took on the appearance of a video game.[16]

In another situation, will people be able to strike a balance between the virtual world and the real world? Other questions and concerns exist as well.

Summary

Like other technology-driven fields, VR's tools will continue to mature. The equipment should become lighter, more mobile, and less expensive. The graphics quality should also improve, in response to a current complaint, while feedback developments should keep pace. You may experience a physical response similar to the one you get when you pick up a real ball or other object.

Building on contemporary systems, you may also enter a room, much like the holodeck in the *Star Trek* television series. In this setting, you would be unencumbered by equipment. As portrayed in different episodes, a fine line may also emerge between what is reality and what is a virtual reality.[17]

Finally, while VR's potential is enormous, the time frame for its general acceptance is still unknown. Adopted for specialized applications, manufacturers have yet to create the ultimate consumer product. This could change, however, through the Internet. As described, VR reached the Internet through VRML. It could introduce a broad user base to this field and should help accelerate its overall development.

At this point, we may all, as stated in the Moody Blue's album title that opened this section, be on the "threshold of a dream."

PAPERLESS SOCIETY

Much like VR, a paperless society is a product of different technologies and applications. They run the gamut from optical disks to the Internet to multimedia productions.

In brief, in a paperless society, information is increasingly created, exchanged, and

stored in an electronic form. It is also a society that is no longer dependent on paper for writing and storing memos, checks, and other documents.

Benefits

A switch to a paperless society could have some benefits:

1.   We would witness the creation of an enhanced communications system. Information could be read, stored, and retrieved with the accuracy and control afforded by a computer.

2.   New software tools are helping to make this goal a reality. For example, documents can be transported across multiple computer platforms and used without an arcane conversion process. This makes the technology somewhat transparent to the user.

3.   Physical space would be conserved because information would no longer be saved on bulky paper. Environmentally, fewer trees would be cut down for paper, and less waste material would be generated to further clog our landfills.

4.   Electronic publications could be quickly revised. This is an important factor for industries with rapidly shifting technology bases that necessitate the continual updating of manuals.

The advent of cost-effective optical recording systems also opens up electronic publishing to more people. If you have a mass storage demand, but not a mass distribution requirement, a CD-R system may, for instance, prove useful. You could create the disk and use it with conventional CD-ROM players.[18] The same concept can be extended to recordable DVD systems.

5.   A paperless society can save money. Checks could potentially be eliminated, doing away with bank handling and other associated charges. The Internet has accelerated this process through e-commerce and other functions.

In fact, this development was given impetus by a legislative initiative in mid-2000, the *Electronic Signatures . . .* Act. Basically, the use of digital or electronic signatures will now be

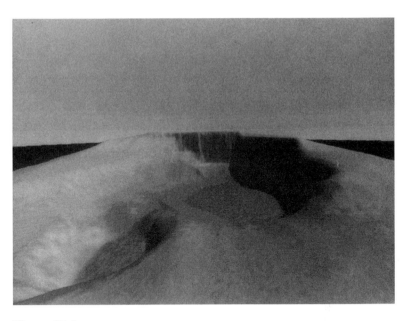

Figure 22.4

An example of a scene you can now generate, and more importantly, can explore with your PC: Olympus Mons on Mars. (Software courtesy of Virtual Reality Laboratories; Vista Pro.)

Figure 22.5

New software has made it possible to exchange documents across different platforms. Acrobat helped pioneer this application, particularly at the PC level. The full document (right) and individual pages, represented by icons (left), provide easy access to the information. (Software courtesy of Adobe Systems, Inc.; Acrobat.)

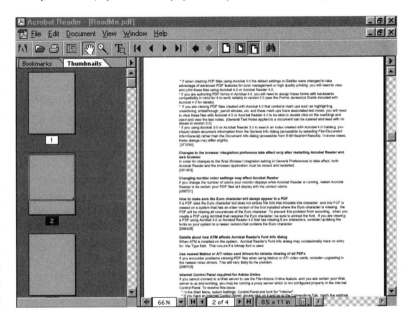

permitted "in many types of contracts and business transactions done on the Internet."[19] On a positive note, this will speed up many Internet-based transactions since you potentially "sign" legal documents via your computer instead of waiting for the paper copy. On the flip side, while safeguards were taken to help ensure your electronic signature's integrity, there was some concern about what can be termed electronic signature pirating.

6. CD-ROM and DVD producers may benefit. The electronic distribution of information can be cost effective. A CD-ROM, for instance, becomes an inexpensive, mass distribution and multimedia tool. By the mid-1990s, after a master was produced, the cost for CD-ROM copies generally hovered around the $1 range per disk.[20]

We can also establish our own web sites. It is now possible to distribute electronic documents and media types (for example, graphics).

7. Information in an electronic form is elastic. It can be stored diversely and communicated through different retrieval mechanisms. Digital clips and other information can also work together, as in a multimedia presentation.

8. A PC's processing power can be tapped to conduct data searches. Instead of flipping through hundreds of pages to find a Sherlock Holmes' quote, use a PC. Load in the appropriate CD-ROM, type in a keyword, and the PC can find the line. Depending on the interface, you may also discover complementary information.

Or use the PC to find a term. Click on the word and bring up the definition in a window. You could also search through an on-screen table of contents, bring up a passage in one window, and retrieve and then view a graphic or even a digital clip in another.[21]

This type of interface, which may include an electronic note writing option, enhances the search process. You can simultaneously retrieve a variety of information, and you aren't limited to static images.

9. A paperless society affords you access to an almost unlimited information resource.

By tapping the Internet, you can explore this ever-expanding data pool and learn about King Arthur, view the latest space telescope images, and communicate with a friend who lives half a world away.

10. A paperless society can speed up your work. You may be conducting research about privacy. In this electronic environment, you could use a search mechanism to locate relevant information. If available online, you could have it in a matter of minutes.

Pitfalls

Despite the benefits, a paperless society does have potential pitfalls. Paper is a better choice for some uses. Newspapers, books, and other paper documents are cost effective, easily mass produced, and can be read almost anywhere. People are also used to working with paper.

Paper can also speed up certain jobs, such as quickly scanning through a stack of books and magazines. While you can somewhat duplicate this process with a computer, you may face hardware and software constraints, especially if you have a small monitor. You may only be able to read small chunks of information rather than the broad sweeps you can take in by quickly scanning a page(s) in a book(s).

Electronic information systems, for their part, call for machine-dependent operations to store, retrieve, and exchange information. If standards are not adopted, technical roadblocks may impede this information flow.[22]

Paperless or electronic resources can also be relatively expensive. Like a thirsty swimmer in an ocean, you may be surrounded by a sea of information, but may not be able to use it. You may lack the prerequisite knowledge, skills, or money.

This scenario brings up two questions. If information is reduced to an electronic form, will it be readily accessible? What is a realistic timetable for the technology to trickle down to all levels of society?

A paperless society also thrives on fast communications channels and complementary information services. Yet when will they be widely available and affordable?

For the latter, a partial solution may be offered by a Teletel-type system. It helped extend access to a large user base, even though it is a paid and not a free service.

Analogous schemes have been floated for the United States. In one example, the Electronic Frontier Foundation's president Mitchell Kapor, articulated a proposal for a high speed-network, the National Public Network (NPN). Besides serving information needs, it could help preserve First Amendment rights and would be easy to use. The last point is critical. If a system is available but complicated, it may not be used by the person it was designed to serve.[23]

As outlined in the preceding chapter, the 1996 Telecom Act also addressed the accessibility issue. It opened the door to enhanced telecommunications services across the United States and to provide support for schools and other institutions.[24] The Internet is also a cornerstone in this plan.

As described, though, concerns were expressed about availability as well as the increased subsidies that may be required to build an maintain such an operation.[25]

A paperless society also has privacy implications. E-mail may only be the tip of the proverbial iceberg.

GENERAL TRENDS

As indicated in the chapter's opening, this section is a general summary. It also looks at some potential trends.

On the political front, the Internet could nurture an electronic democracy while providing access to vast information resources. CD-ROMs and other optical media will serve as information and electronic publishing sources in their own right.

PC developments will blossom, and new systems may sport GUIs and other interfaces that will make the human–computer link even more transparent. Speech recognition may also play a prominent role in this field, and the application could be

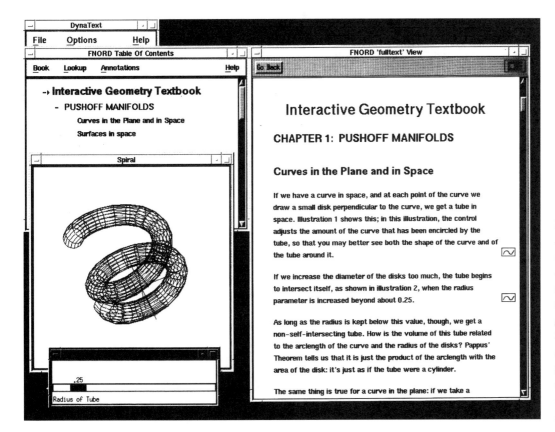

Figure 22.6

A sophisticated electronic book. Multiple views and hypertext links may be supported, including in this case, a link to a program that illustrates the theorem mentioned in the text. It may also be possible to create your own electronic notes. (Courtesy of Electronic Book Technologies, Inc.)

extended to television sets and other everyday appliances.

The convergence between technologies will continue, and the resulting synergy will be fertile ground for new products and applications. One such area is the convergence between computers and audio-video equipment, as is the case with the still shifting HDTV/digital television market.

Entire communications systems may be similarly influenced. As introduced in Chapter 9, we may witness the development of integrated information and entertainment highways. This may be matched by a growing number of intra- and interindustry mergers.

Digital technology will continue to play a key role in our future information and communications systems. But as described at different points, the analog world is still very much alive.

Personal media options will grow. By using desktop publishing and video systems, we become producers and not just consumers. This facet of the communication revolution cannot be overstated. We now possess a remarkable set of personal communications tools. You can let your imagination soar as you create a 3-D graphic, and you can gain access to tailored information and entertainment pools.

By extension, our communication options will deepen, especially in the area of wireless systems. We may also hold a videoconference from either the office or home. For more people the home may, in fact, become their offices.

Information will remain an important resource. As our information society matures, the production of and the demand for information will increase. We'll also develop new ways to manipulate and use this information. The tools include existing and newer configurations (for example, VR).

Despite the growth of the information sector, however, it is also important to remember that heavy industries are still vital—someone has to make the steel and other products we use.

The integration of and the correct balance between an information and industrial base could help propel a country toward the future. If the balance is lost, a country's economy could suffer.

Other Issues

Some people believe the new technologies may be too much of a good thing. Will electronic clutter become so pervasive that these tools become more of a distraction than an aid? How do you strike a balance between the need for quiet, reflective periods and your communications and information requirements?

With the possible proliferation of VR systems, and the Internet's growth, will people become more divorced from reality? What of potential electronic addictions and obsessions? Will more people spend more time online than in the real world? Are we catching a glimpse of Asimov's future world in our present society?[26]

While this may appear to be a remote possibility, think about the potential. With the Internet, for instance, you have access to an almost unlimited and ever-changing data pool. Some people have likened potential addiction to smoking. But instead of one more cigarette, there may be one more site to visit. And there will always be one more site.

Although current and projected applications are exciting, they must also be examined with a dose of reality. One of the problems with predictions is that they often remain predictions even after a number of years. In one classic example, artists and authors in the 1950s and 1960s painted a bright picture for the space program. Moon bases would be sprawled over the lunar surface, and huge space stations, much like the one portrayed in *2001: A Space Odyssey*, would orbit the Earth. But many of these predictions have not materialized. Part of the problem was the lack of money. Another may have been the technical and/or political inability to transform these visions into a reality.

The same sequence of events could affect the new communications technologies. Although PCs can serve as educational tools, will schools be equipped with enough systems to make a difference? Even though we will have greater technical support to maintain the free flow of information, will government regulations intercede?

In addition, ethical and legal questions will loom ever larger. Pertinent topics include intellectual property, digital image manipulation, and e-mail/privacy.

The First Amendment and its attendant rights will also be challenged and further defined for the electronic media. In one case, it has been suggested that CDA-type acts could be forestalled by taking reasonable means to protect children from potential, electronic online abuses. This could include, when properly implemented, filters that would prevent children from gaining access to certain sites.[27] But as covered in Chapter 21, constitutional questions are raised when mandatory blocking is adopted. Consequently, how do you reconcile the two viewpoints? Based on court cases, it appears that more narrowly tailored regulations, in tandem with appropriate safeguards, may be the answer. In one library dispute, computers were made available with and without filtering software. Another factor is parents and educators assuming greater responsibility in helping their children to navigate the Internet.

In essence, technological changes will continue to clash with the legal system. In some cases, technological developments will outstrip the law's ability to react to these changes.

Some people also believe the legal system may overreact. As covered in the satellite chapter, do you introduce tighter controls on imaging, which some have categorized as vague and a First Amendment violation? Or do you work with the media in regard to national security issues?

As an employer, you may have to make your own, hard decisions. Do you adopt a published e-mail policy? Or do you use the legal system, as of this writing, to monitor your employees at your discretion?

The issues are complex and important. They may affect our workplaces and our personal lives.

CONCLUSION

The new communications technology universe is rapidly expanding. New applications are introduced each day, and they influence almost every facet of our lives. The connotations can also be positive or negative, depending on the impact, and potentially, your viewpoint.

Nevertheless, it is a fascinating time to be alive. Our communications tools are extending our vision from the ocean's floor to the very edge of the universe. We also have the opportunity to express ourselves in new ways and to tap information pools that were inconceivable only a few years ago.

REFERENCES/NOTES

1. Brad Sharpe, "Hologram Views," *Advanced Imaging* 2 (August 1987): 28.

2. Thomas Cathey, *Optical Information Processing and Holography* (New York: John Wiley & Sons, 1974), 324.

3. These holograms would be difficult to duplicate and may have high expertise and physical plant requirements.

4. Edward Bush, "Industrial Holography Applications," *ITR&D* (n.d.): 30.

5. Cathey, *Optical Information Processing and Holography,* 324.

Figure 22.7
The disk mounted on the Voyager *spacecraft that may eventually carry information about Earth to a distant star; it's potentially one of our first calling cards to another world. (Courtesy of NSSDC.)*

6. James D. Trolinger, "Outlook for Holography Strong as Applications Achieve Success," *Laser Focus/Electro-Optics* 22 (July 1986): 84.

7. Jose R. Magarinos and Daniel J. Coleman, "Holographic Mirrors," *Optical Engineering* 24 (September/October 1985): 769.

8. Tom Parish, "Crystal Clear Storage," *Byte* 15 (November 1990): 283.

9. Michael W. McGreevy, "Virtual Reality and Planetary Exploration," paper presented at 29th AAS Goddard Memorial Symposium, March 1991, Washington, D.C., 1.

10. "The Future Is Now," *Autodesk* (1992): 32.

11. Linda Jacobson, "Virtual Reality; A Status Report," *AI Expert* 6 (August 1991): 32.

12. Ben Delaney, "Where Virtual Rubber Meets the Road," *AI Expert,* Virtual Reality 93: Special Report (1993): 18. *Note:* This special edition of *AI Expert* focuses on the VR field/applications.

13. Elliot King, "The Frontiers of Virtual Reality and Visualization in Biochemical Research," *Scientific Computing & Automation* (July 1996): 20.

14. Howard Eglowstein, "Reach Out and Touch Your Data," *Byte* 15 (July 1990): 283–290; other articles describe how to interface the Power Glove to a PC.

15. Howard Rheingold, *Virtual Reality* (New York: Summit Books, 1991), 355.

16. See Rheingold, *Virtual Reality,* 357–362, for some of the implications.

17. Tom Reveaux, "Virtual Reality Gets Real," *NewMedia* 3 (January 1993): 34.

18. Jon Udell, "Start the Presses," *Byte* 18 (February 1993): 132. *Note:* The article provides a detailed look at CD-ROM publishing and the CD-R. DVD systems will also affect this market.

19. Todd R. Weiss, "Senate Joins House in Passing Digital Signatures Bill," downloaded from www.computerworld.com, June 14, 2000.

20. Disc Manufacturing, Inc., Wilmington, Delaware, CD-ROM price quotation list. *Note:* Although this did not include a nominal fee for a Tyvek sleeve or other protective package, it still made the CD-ROM a cost-effective distribution medium, especially in light of its storage capabilities. As the number of disks per order increases, the per-disk price decreases.

21. Louis R. Reynolds and Steven J. Derose, "Electronic Books," *Byte* 17 (June 1992): 268.

22. There is also an archival issue. If you use a proprietary system, and the company goes out of business, what do you do if your system breaks?

23. Testimony of Mitchell Kapor, president, the Electronic Frontier Foundation, before the House Subcommittee on Telecommunications and Finance, October 24, 1991. Excerpts.

24. This included reduced rates.

25. Charles S. Clark, "Regulating the Internet," *CQ Researcher* 5 (June 30, 1995): 573.

26. There are also related health issues (for example, carpal tunnel syndrome).

27. There are other provisions.

SUGGESTED READINGS

Holography

Amadesi, S., A. D'Altorio, and D. Paoletti. "Real-Time Holography for Microcrack Detection in Ancient Gold Paintings." *Optical Engineering* 22 (September/October 1983): 660–962; Carelli, P., D. Paoletti, and G. Schirripa Spagnolo. "Holographic Contouring Method: Application to Automatic Measurements of Surface Defects in Artwork." *Optical Engineering* 30 (September 1991): 1294–1298. Holography and the art restoration world.

Hand, Aaron, J. "Is Holographic Storage a Viable Alternative for Space?" *Photonics Spectra* 32 (June 1998): 120–124. Holographic storage systems and technology.

Higgins, Thomas V. "Holography Takes Optics Beyond the Looking Glass." *Laser Focus World* 31 (May 1995): 131–142. Excellent tutorial about holography.

Ih, Charles S. "A Holographic Process for Color Motion-Picture Preservation." *SMPTE Journal* 87 (December 1978): 832–834. Using holography as a film preservation system.

Jeong, Tung H., and Harry C. Knowles. *Holography Using a Helium-Neon Laser.* Bellimawr, N.J.: Metrologic Instrument Corp., 1987. A manual that explores holographic principles and production techniques. It is a companion to a Metrologic holography kit.

N.A. "Holography Drives Ford Car Design." *Laser Focus World* 36 (February 1999): 14–16; Parker, R. J., and D. G. Jones. "Holography in an Industrial Environment." *Optical Engineering* 27 (January 1988): 55–66; Rosenthal, David, and Rudy Garza. "Holographic NDT and the Real

World." *Photonics Spectra* 21 (December 1987): 105–106; Holography and industrial applications.

Virtual Reality

The following cover VR systems, operations, applications, and implications.

Aukstakalnis, Steve, and David Blatner. *Silicon Mirage*. Berkeley, Calif.: Peachpit Press, 1992.

Benedikt, Michael, ed. *Cyberspace: First Steps.* Cambridge, Mass.: The MIT Press, 1991.

Bove, Tony. "On Location." *NewMedia* 7 (September 1997): 40–45; Hodges, Mark. "Virtual Reality in Training." *Computer Graphics World* (August 1998): 45–52; Mahoney, Diana Philips. "Better than Real." *Computer Graphics World* (February 1999): 32–40; Wheeler, David L. "Computer-Created World of 'Virtual Reality' Opening New Vistas to Scientists." *Chronicle of Higher Education* 37 (March 13, 1991): A6, A12, A13. Various VR applications and issues.

Heid, Jim. "VR: QuickTime's New Edge," *Macworld* (July 1995): 98–103. QuickTime and VR.

King, Elliot. "The Frontiers of Virtual Reality and Visualization in Bioechemical Research," *Scientific Computing & Automation* (July 1996): 20–22; Taylor II, M. Russell, and Vernon L. Chi. "Take a Walk on the Image with Virtual-Reality Microscope Display." *Laser Focus World* 30 (May 1994): 145–149. VR medical applications.

Reveaux, Tony. "Virtual Set Technology Expands." *TV Technology* 17 (July 28, 1999): 14; Tubbs, David A. "Virtual Sets: Studios in a Box." *Broadcast Engineering* 40 (November 1998): 90–92. Overviews of virtual set systems.

Rheingold, Howard. *Virtual Reality.* New York: Summit Books, 1991. Excellent introduction and discussion about the topic.

Wenzel, Elizabeth M. *Three-Dimensional Virtual Acoustic Displays.* NASA Technical Memorandum 103835, July 1991.

Wylie, Philip. *The End of the Dream.* New York: DAW Books, Inc., 1973. See pp. 162–165 for a precursor look at a virtual sex-type setup. The book, a science-fiction novel, primarily focuses on environmental issues and the Earth's potential future.

Paperless Society

Murray, Philip C. "Documentation Goes Digital." *Byte* 18 (September 1993): 121–128. Digital delivery of reference materials.

Retkwa, Rosalind. "Corporate Censors." *Internet World* 7 (September 1996): 60–64; Venditto, Gus. "Safe Computing," *Internet World* 7 (September 1996): 48–58. Highly recommended for information about Internet filtering software for children and the workplace.

Also see references in the desktop publishing, Internet, and hypertext chapters.

GLOSSARY

Hologram: A hologram is a record of the optical information that composes a scene. It can be used for applications ranging from advertising to security.

Holographic Interferometry: A holographic industrial application (for example, for material testing).

Holographic Optical Element (HOE): An application where a hologram functions as a lens or other optical element.

Paperless Society: A society in which information is increasingly created, stored, and exchanged in an electronic form.

Virtual Reality (VR): A system where you can enter into and interact with a computer-generated environment.

23 Afterword: The V-Chip and the First Amendment

Besides providing us with new applications, the new communications technologies have raised social and political questions. One such product that raised a storm of controversy is the V-chip.

Designed to give parents control over their children's television viewing options, critics claimed the V-chip had potential censorship implications. More importantly, it was a re-hash of a scenario that took place some 30 years ago. But this time, according to some critics, the television industry caved-in to political pressures—accepting the V-chip in trade for the spectrum required for digital television.

The V-chip is placed in the Afterword because it serves as an example of how the new technologies are clashing with the legal system and the Constitution. This arena will become the focal point of other conflicts as new products both enhance our lives and, potentially, raise constitutional questions. Consequently, this general topic area merits special attention. As such, it is an appropriate end to the book.

BACKGROUND

There have been public outcries about violence in the media for years. For example, it is estimated that American children witness thousands of televised murders and acts of violence by time they're 14.[1] Studies have also pointed to a significant correlation between viewing violent programming and committing violent acts.[2] In one case, "children exposed to violent video programming at a young age have a higher tendency for violent and aggressive behavior later in life than children not so exposed. . . ."[3]

Congress addressed this issue as part of the 1996 Telecom Act. The television and cable industries were expected to develop what has been dubbed "V-chip" (violence chip) technology. As envisioned, it would enable consumers to electronically block out any program, channel, or time period.[4]

V-chip proponents applauded the new technology and claimed it empowered the viewer, particularly parents of young children. In the context of our present discussion, they also claimed it did not violate the First Amendment. It was analogous to the use of a remote control or to parental decision making.[5]

But opponents were troubled by potential First Amendment abridgements. Programming would have to be rated for the V-chip technology to work. Ratings categories would also have to be created, and programs would have to be appropriately slotted.

The television and cable industries had to develop a "voluntary" ratings system that would identify violent, sexually oriented, and indecent programming inappropriate for children. Programming would not be rated, however, based on political or religious content.[6] If they failed, the FCC would develop its own guidelines, based on the recommendations from an advisory committee composed of parents, industry representatives, and others. The committee would be politically balanced and would represent diverse viewpoints.[7]

FCC intervention was not necessary, however, since the television and cable industries created guidelines prior to the deadline. In

March 1998, the "TV Parental Guidelines" were adopted by the FCC.[8] In addition to developing a multicategory content rating system, the industry also agreed to provide on-screen rating icons and establish an oversight monitoring board.[9]

Broadcasters were supposedly caught off guard about the V-chip. Industry representatives claimed the amount of violent and sexually oriented programming aired by the commercial networks had actually declined. They supported this contention with a UCLA study that was conducted on the industry's behalf.[10] The industry had also earmarked monies to develop acceptable V-chip alternatives without requiring the use of a program ratings system.[11]

In operation, a program would be rated, and this information would be placed in the television signal's VBI.[12] When the program is received by a viewer's television set, the chip would determine if that particular program content had been blocked.[13]

Finally, the actual cost for the plan's implementation was disputed. According to Representative Edward Markey (D-Mass.), who was instrumental in the act's final design, a television set could be equipped with a V-chip device for $1. He also indicated that like the closed-caption chip, V-chip technology would eventually get cheaper.[14]

Markey's contentions were disputed by the Electronic Industries Association (EIA), a trade organization for television set manufacturers. The EIA indicated that "the chip could add as much as $72 to some low-end TV sets ($20 to $50 for additional memory capability and another $10 or more for channel-blocking capabilities)."[15]

MONITORING

Because the V-chip plan called for a ratings system, programming would have to be monitored and assigned to certain categories.[16] Opponents indicated that this system violated the First Amendment. There was also a practical concern.[17] Unlike the film industry, which only has to rate some 700 to 800 films a year, thousands of hours of television programming would have to be evaluated.[18] If parents do not have the time to evaluate their own children's programming, does the industry? Senator Markey indicated the networks had the capability:

The networks show three hours of programming at night. Each of them has thousands of people working for them. Each of these programs has been inspected and digested by dozens, if not hundreds, of people . . . before it goes on the air. Judgments have been made . . . over an extended period of time, from every possible angle . . . it would defy explanation that they would be unable to give it some rating in regard to whether or not it is appropriate for a six-, seven-, or eight-year-old boy or girl.[19]

He also suggested the television and cable industries should examine the Motion Picture Association of America's (MPAA) film ratings system.[20] The television and cable industries had initially worked with the MPAA to develop a constitutional and usable V-chip alternate. But this venture was unsuccessful.

THE V-CHIP AND OTHER PROGRAMMING

The television and cable industries grew alarmed at the "snowball" effect associated with the V-chip rating plan. Initial legislation called for regulating "violent" content only. However, the act's actual language expanded this scope and also encompassed the regulation of "violent, sexual and indecent material."[21]

This type of regulation "creep" concerned First Amendment advocates. But during a meeting between the television industry and President Clinton, news and sports programming were exempted from the ratings process.[22] For journalists, this would protect their rights to make story content decisions. It also made legal sense. Any move to expand V-chip jurisdiction to include news program content would not withstand constitutional scrutiny.[23]

Representative Markey agreed with this analysis and indicated that controlling newsroom decisions was not the act's intent.[24] He also promised to protect First Amendment rights, but insisted that a viewer's rights, parents and children in particular, must be given equal protection.[25]

More ominously, though, at least in the eyes of V-chip opponents, was his opinion that news programming was too violent:

I don't think anyone will deny that the rule of thumb in too many newsrooms in the U.S. is 'if it bleeds, it leads.' Too many newsrooms use that as their guide because they think it might draw more viewers, in the same way that other drivers slow down when they see a car crash on the highway.[26]

Based on this viewpoint, could regulation be extended to news in the future? This statement also has the potential to "chill speech." A broadcaster may think twice before airing an important but violent story.

CONCLUSION

In 1998, the FCC adopted the rating system developed by the television and cable industries. The TVG General Audience category, for instance, indicated that "Most parents would find this program suitable for all ages . . . most parents may let younger children watch this program unattended. . . ."[27] The rating would appear on the television screen and would trigger the V-chip (when available).

Some proponents hailed this development as an empowering tool that would place television control back in the hands of consumers and parents. There were, however, two categories of opponents. The first group thought the ratings were too diluted—they did not provide adults with enough detailed information.

The second group viewed this entire scenario as a First Amendment violation. As one writer put it, "the thought of putting a device in every television set in the country with an umbilical that can lead back to the government, scares the hell out of us."[28]

Finally, other issues are still unresolved. Will future legislation be more restrictive? Although news is currently exempt, will this hold true down the road? As was the case with the 1996 ratings, which didn't satisfy numerous critics, would the ratings system be a moving target that could change every few years?

For example, in 1998, Thomson Consumer Electronics, the largest single supplier of television sets in the United States, announced possible plans to make sets that would enable the viewer to block news, sports, and commercial advertisements via the V-chip. Thomson manufactures 20 to 25 percent of all U.S. sets and if they actually make good on this proposed plan, the industry could be faced with the possibility of voluntarily rating such material.[29]

Other concerns were expressed by Don West, the editor of *Broadcasting & Cable* magazine. He was involved in a similar controversy in the late 1960s. His insights provide an historical framework for exploring the First Amendment connection. They also provide an argument for the preservation of First Amendment rights. His discussion, titled "Eyewitness to History—Broadcasting's Finest First Amendment Hour," *Broadcasting & Cable* (February 26, 1996) is reprinted below with permission from *Broadcasting & Cable*.[30]

An editor's personal perspective on the Fifth Estate's fight for First Amendment freedom, and a modest proposal to help keep it.

Twenty-seven years ago, the government's first attempt to police television programs for sex and violence was stopped in its tracks by one man of principle and courage. The year was 1969, the Congress was represented by Senator John Pastore of Rhode Island, and the industry executive who wouldn't say yes was Frank Stanton, then president of CBS. The behind-the-scenes story of that fateful encounter has never been reported until now.

This is a first-person account. The writer, now editor of *Broadcasting & Cable*, was then Stanton's assistant at CBS and participated in the events described. I am going public now because the history of 1969 so closely parallels what is happening between the industry and the government in 1996. If there are lessons to be learned from the past, they should not be withheld from the present.

It began on March 12, 1969. Pastore, then technically chairman of the Senate Communications Subcommittee . . . was having one of his periodic hearings on sex and violence in television. . . .

Stanton was testifying. . . .

But this time there was a difference. At the end of Stanton's testimony, Pastore leaned forward and asked: "Dr. Stanton, will you allow CBS programs to be prescreened by the National Association of Broadcasters Code Authority?" It was a request the senator had been making since 1962, and CBS had always led the resistance. This time, Stanton committed a rare misstep. He responded: "Mr. Chairman, I'll take it under advisement. . . ."

From that moment, and for the next two weeks, the fat was in the fire. . . . CBS's historic defense of the First Amendment had begun to unfold by the time we flew back to New York.

Stanton called the first of many meetings on the subject. . . .[31] My position was already certain: We had to say no. "The five people in this room do not have the right to give away 50 years of broadcasting freedom," I remember declaring. They did, of course, have the power. . . .

There was a quid pro quo in 1969, too. Pastore had introduced a bill that would protect licensees from challenge at license renewal, and the popular view was that his bill's passage depended on the networks' prescreening for the NAB authority. NBC and ABC had already indicated compliance; only CBS was in doubt. Telexes and phone calls from affiliates were bombarding Bill Lodge, then the head of station relations, imploring Stanton to yield. It was the same kind of pressure that would be brought to bear that same year on the Smothers Brothers, forcing them off the air.

By the second week I could sense my position eroding, and I began to search for an alternative response—something that would serve the senator's purpose and not damage our First Amendment rights. At least part of the issue then, as now, was in advising parents about what was coming up on TV, so they could control their children's viewing. Why not, I thought, prescreen CBS programs for TV critics, and let them advise the audience whether there was too much violence or sex, or whether a show was worth viewing at all? It had never been done; a

critic's reviews were read the morning after, and a viewer could only compare points of view. . . .

Armed with an alternative and my proxies, I returned to the attack. Stanton readily agreed to implement the critics' prescreening idea and the tide began to turn. After two weeks of listening, he declared his decision: We would say no to Pastore. . . .

The turndown was given to Pastore shortly before he was to address the NAB's annual convention in Washington. . . .

To give Pastore his due, he did not seek to codify the NAB prescreening into law. Because of the First Amendment, he said at the time, there was little Congress could do to prescribe standards of programming, but there was much that broadcasters could do with effective self-regulation. . . . Responding, Stanton doubted that there would be much change in programming standards, and said: "In a year we'd be back on the same subject because you would not be satisfied."

I thought the television industry was home free until the V-chip came along a quarter century later. This time the sponsor was Democratic Congressman Edward J. Markey of Massachusetts. . . . From the moment the prospect of government program control raised its head . . . this magazine went after it. . . . The difference between then and now: no Stanton to take the heat. . . .

The quid pro quo is even greater in 1996. The industry finally got its protection from license challenge without having to give up the First Amendment. This time, the stakes are higher. The deal is: Give us a V-chip-enforced rating system and we will give you the digital spectrum. As for me, I'd let them keep the spectrum. It is pottage without the First Amendment.

At this writing, the networks, cable and the programming community are far down the road of conceding the development of a ratings system. It would be strictly voluntary, they say, despite the law demanding its creation. Then, if the government should try to substitute its own ratings system, the industries would go to court. . . .

Arguably, it is too late to say no. In its rush to capitulation, the industry may well have announced a ratings system before this story goes to press. Nevertheless, as in 1969, let me offer these discussion points regarding a far more First Amendment-friendly alternative.

1.   It is a given that there should never be a government-imposed ratings system. . . .

2. Any proposal to empower parents should be informational rather than instructional. It should provide parents with the information they need to exert control over content coming into their households, without preaching to them about which values are acceptable.

3. With those considerations in mind, any informational system should focus on labeling, rather than rating programs. In other words, the industry could be encouraged to list the various attributes of a program (adult language, nudity, brief nudity, cartoon violence, mild violence, graphic violence, political content, comedy, adult drama, etc.) without having to make a judgment about whether a program may be "good" or "bad" for children. That should be left for parents to decide. Such an approach would also insulate the system from political influence. . . . (President Buchanan's idea of a proper V-chip ratings system versus President Clinton's, for example.)

4. The key consideration: Broadcasters would never "rate" programs; they would only provide the descriptions built in by the program suppliers. . . . It would not be necessary to have a bureau or committee to evaluate programming for quality or content, as does the current Motion Picture Association of America system. Instead, the industry could be encouraged to devise an all-inclusive list of categories, and program producers themselves would be able to voluntarily provide program information based on the list.

They say it is harder, now, for one person to make a difference. CBS in 1969 was the most powerful of but three broadcast television networks. Now it is down a rung or two in a field of six broadcast TV networks, and an entire cable industry has sprung up to share the audience's attention. But if Frank Stanton alone could hold the government at bay in 1969, the 15 to 20 broadcasting, cable and programming executives who will meet next week with President Clinton can do the same. Whether the number is five or 20, the men who enter that room have no right to give away more than 65 years of broadcasting and cable freedom.

They do, of course, have the power.

REFERENCES/NOTES

1. U.S. Congress, Telecommunications Act of 1996, 104th Cong., 2nd session, section 551 (January 3, 1996).

2. Barri Lazar, "The V-Chip: Censorship or Technological Miracle," unpublished paper, April 23, 1996, 14.

3. Telecom Act, Section 551.

4. Ibid.

5. "Why the Markey Chip Won't Hurt You," *Broadcasting & Cable* (August 14, 1995): 11.

6. Telecom Act, Section 551.

7. Ibid.

8. FCC. "Commission Finds Industry Video Programming Rating System Acceptable: Adopts Technical Requirements to Enable Blocking of Video Programming [the V-Chip]," Report No. GN 98-3. March 12, 1998.

9. Ibid.

10. Christopher Stern, "Face-Off on the V-Chip," *Broadcasting & Cable* (October 2, 1995): 19.

11. Ibid.

12. See footnote 1, Chapter 17, for a VBI definition.

13. The set would be equipped with this function.

14. Art Brodsky, "The V-Chip Moves Toward Reality," *Broadcast Engineering* 37 (September 1995): 81.

15. Ibid.

16. "Why the Markey Chip Won't Hurt You," 11.

17. Ibid.

18. Sarah Haag, "The V-Chip: Censorship or Technological Miracle?," unpublished paper, April 23, 1996, 8.

19. "Why the Markey Chip Won't Hurt You," 11.

20. Kent R. Middleton and Bill F. Chamberlin, *The Law of Public Communication* (White Plains, N.Y.: Longman Publishing, 1994), 354. *Note:* The MPAA "rates films as, for example, 'G' for general audiences and 'R' for restricted to those under 17 unless accompanied by a parent/guardian."

21. Telecom Act, section 551.

22. Christopher Stern, "Valenti Pledges Government-Free Ratings," *Broadcasting & Cable* (April 8, 1996): 14. *Note:* The meeting took place shortly after the act's passage.

23. "Why the Markey Chip Won't Hurt You," 12.

24. Ibid, 15.

25. Ibid, 15.

26. Ibid, 12.

27. NAB web site, downloaded December 1996.

28. "Why the Markey Chip Won't Hurt You," 15.

29. "Ratings on a Roll," *Broadcasting* (August 17, 1998): 98.

30. Don West, "Eyewitness to History—Broadcasting's Finest First Amendment Hour," *Broadcasting & Cable* (February 26, 1996).

31. Other key participants in this process were John A. Schneider, Richard W. Jencks, E. Kidder Meade, Dick Salant, Bob Wood, Frank Smith, Mike Dann, Jack Cowden, Charles Steinberg, and Bill Tankersley.

SUGGESTED READINGS

ACLU. "Violence Chip: Why Does the ACLU Oppose the V-Chip Legislation Currently Pending in Congress." ACLU Freedom Network. 1996. Downloaded from www.aclu.org.

Bitol, Solange. "The V-Chip: Point-and-Click Parenting." Freedom Forum Online. March 16, 1998. Downloaded from www.freedomforum.org.

Canton, William F. FCC. "RE: CS Docket No. 97-55." September 10, 1997. Downloaded from www.FCC.gov.

"Content Regulation: The ACLU Backfires Self-Rating Technology." LAMBDA Bulletin. 3.05. August 27, 1997. Downloaded from www.freenix.fr.

FCC. "TV Program Ratings and the V-chip." V-Chip home page. March 17, 1998. Downloaded from www.FCC.gov.

FCC. "Video Programming Ratings." Report and Order.CS Docket No. 97-55. March 13, 1998: 1–12.

Macavinta, Courtney. "FCC Plans PC V-chip." C/NET. The NET. Downloaded from www.news.com:80.

Powell, Adam Clayton III. "FCC Clarifies: V-Chip Mandate Doesn't Apply to Internet Video." Freedom Forum Online. March 16, 1998. Downloaded from www.freedomforum.org.

Index